The Physics
of Sound

Richard E. Berg

David G. Stork

University of Maryland

The Physics of Sound

PRENTICE-HALL, INC., Englewood Cliffs, New Jersey 07632

Library of Congress Cataloging in Publication Data

BERG, RICHARD E.
    Physics of Sound.

    Includes bibliographies and index.
      1. Sound.   2. Music—Acoustics and physics.
I. Stork, David G. II. Title.
QC225.15.B47   781′.22   81–13840
ISBN 0-13-674283-1   AACR2

Printed in the United States of America

10 9 8 7 6 5 4 3 2 1

Editorial/production supervision
and interior design by Ellen W. Caughey
Cover photograph by G. Frederick Stork
Cover design by Maureen Olsen
Manufacturing buyer: John Hall

ISBN 0-13-674283-1

Prentice-Hall International, Inc., *London*
Prentice-Hall of Australia Pty. Limited, *Sydney*
Prentice-Hall of Canada, Ltd., *Toronto*
Prentice-Hall of India Private Limited, *New Delhi*
Prentice-Hall of Japan, Inc., *Tokyo*
Prentice-Hall of Southeast Asia Pte. Ltd., *Singapore*
Whitehall Books Limited, *Wellington, New Zealand*

To Erik and Dwight
and
to Nancy

Contents

2
Waves and Sound 20

3
Standing Waves and the Overtone Series 63

4
Analysis and Synthesis of Complex Waves 89

5
Electronic Music and Synthesizers · 118

6
The Human Ear and Voice 137

7
Sound Recording and Reproduction 168

8

Room and Auditorium Acoustics 203

9

Musical Temperament and Pitch 223

10
Woodwind Instruments 245

11
Brass Instruments 283

12
String Instruments 301

Preface

The Physics of Sound was written for introductory courses in acoustics and music for nonscientists. Neither a background in physics nor mathematics above high-school algebra is required. Traditionally, such courses have been tailored to music majors; nonmusicians either do not enroll, or do not fully appreciate the physical principles, because they are applied almost exclusively to musical topics. We have tried to avoid this problem by dividing the text into three main sections.

Chapters 1 through 4 present the basic physics essential for virtually all topics throughout the text: simple harmonic motion, resonance, the overtone series, wave properties, standing waves, and Fourier synthesis and analysis. No previous musical knowledge is required to appreciate these chapters, and applications and illustrations come from a variety of nonmusical and musical areas.

Chapters 5 through 8 illustrate these principles and are of general interest to the nonmusician and the musician alike. The use of musical notation and concepts has been minimized so as to retain the broadest base of appeal to the largest number of students, but still includes the more important musical aspects of each topic. The chapters on stereos and room acoustics, for example, illustrate well the principles presented in the earlier chapters and are of interest because of the part they play in our daily lives. Chapter 5 discusses synthesizers in a way that is understandable to the nonmusician. Chapter 6 discusses the physical principles of the ear and voice at an elementary level and has been used with great success for beginning hearing and speech students. The first eight chapters, in short, contain the core material for a one-semester course in the physics of sound and music.

Chapters 9 through 14 are more specialized. Each of these chapters independently treats a different aspect of *musical* acoustics, and is best (though not exclusively) understood by those with some musical experience. Any of the final six chapters could be studied in class or could naturally be assigned to students on an individual basis.

Unlike most authors of elementary acoustics texts, we have treated the historical development of instruments, paying particular attention to acoustical changes. We have also tried to relate the physical principles of contemporary instruments to performance technique; the knowledge of *how* and *why* an instrument works, and its limitations and problems, should improve one's performance.

Most chapters conclude with a set of problems and a list of resources. Those for the last six chapters sometimes require greater sophistication than the others and could form the basis for student projects.

A text is but one of the resources required for the student to learn the material. Our course as taught at the University of Maryland uses a large number of demonstrations, tapes, and films, including the set of videotapes, "Demonstrations in Acoustics," designed as a companion to the text.

When fully integrated, we hope these resources will bring to the nonscience student a deeper appreciation of acoustical phenomena and science in general.

Acknowledgments

We are grateful to the reviewers for their detailed comments and suggestions: J. Gerard Anderson, University of Wisconsin—Eau Claire; Stanley H. Christensen, Kent State University; and Bruce Daniel, Pittsburg State University.

We would like to thank Professors John Guillory and Angelo Bardasis, who taught using our manuscript in a preliminary form. Several of their insightful comments and ideas have been incorporated into the final version of the text. Similarly, Professor Willard Larkin of the University of Maryland Psychology Department reviewed much of a late version and gave many useful suggestions. We appreciate the typing assistance of Anne Green Hellman and Betty Alexander, as well as the graphic work of Joan Wright and Lionel Watkins. Deep thanks go to our students, many of whom have given thoughtful comments on both the course and our manuscript.

RICHARD E. BERG
DAVID G. STORK

1

Simple Harmonic Motion and Applications

In this chapter we shall first define the physical terms that will be used throughout the book and then discuss *the* single concept most basic to the study of sound—simple harmonic motion—and its direct application to sound and the concept of resonance. Finally, to help establish and review our use of these elementary concepts, we shall study Lissajous figures, a type of construction found in some artwork, which is based on simple harmonic motion.

1.1 *Fundamental Definitions*

To understand physics, we must first understand its language. Its definitions are exact and are formulated in the language of mathematics. For our purposes, however, more qualitative and operational definitions will suffice. The most fundamental quantities in physics are discussed next; they will be used regularly in our study of acoustics. Other important quantities will be defined throughout the book as the need arises.

Position, length, or distance (symbols x and y) This quantity is simply a measure, in the normal everyday sense, of how far one point is from another. For our purposes we shall primarily study motion in one dimension, as shown in Fig. 1-1. Position relative to the origin of coordinates ($x = 0$) can be positive (to the right) or negative (to the left); dots have been placed on the line at $x = 6$ cen-

1

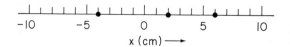

Figure 1-1 Graph showing points positioned along a line.

x (cm) ⟶

timeters, (cm), $x = 2$ cm, and $x = -4$ cm. Units of length can be miles, feet, inches, meters, and so on. We shall regularly use the metric system; in this system the prefix deci- means 1/10, centi- means 1/100, milli- means 1/1000, micro- means 1/1,000,000, kilo- means 1000, and mega- means 1,000,000. Some metric length units are meters (m), kilometers (km), centimeters (cm), and millimeters (mm), where 1 cm = 0.01 m, 1 mm = 0.001 m (so 1 m = 100 cm = 1000 mm), and 1 km = 1000 m. One meter is approximately 39.37 inches; 1 inch (in.) is approximately 2.54 centimeters.

Time (symbol t) For our purposes, we simply observe that time, as we know it, passes and is divided into seconds, hours, days, and so on. One unit of time that will be used regularly is the millisecond (ms); 1 ms = 0.001 sec, so 1 sec = 1000 ms.

Velocity or speed (symbol v) Velocity or speed (which will be used synonymously) is defined as distance traveled per unit of time. For example, if the dot in Fig. 1–1 moves uniformly from $x = 2$ cm to $x = 6$ cm in 1 sec, its speed is 4 centimeters per second (4 cm/sec). If it moves from $x = 6$ cm to $x = -4$ cm in 2 sec, $v = -10$ cm/2 sec or $v = -5$ cm/sec. Convince yourself that 1 m/sec = 1 mm/ms. The speed of sound is 345 m/sec or 1100 ft/sec; the speed of light is 300,000,000 m/sec or 186,000 miles/sec.

Acceleration (symbol a) Acceleration is defined as change in velocity per unit of time. For example, suppose a ball rolls down a ramp starting with zero velocity (that is, it starts at rest), and after 1 sec its velocity is 1 cm/sec, after 2 sec its velocity is 2 cm/sec, after 3 sec its velocity is 3 cm/sec. The increase in velocity is 1 cm/sec for each second it is rolling down the ramp, which can be written $a = 1$ (cm/sec)/sec or $a = 1$ cm/sec^2. Deceleration, or slowing down, is acceleration in the direction opposite to the motion.

Mass (symbol m) Mass is a measure of the amount of matter in an object. A 1-gram (g) mass of iron, for example, contains a certain number of iron atoms, which remains constant on the earth, the moon, or in outer space. The most often used units of mass are the gram (g) and the kilogram (1 kg = 1000 g). One kilogram weighs about 2.2 pounds (lb) on the surface of the earth. A 5-cent coin has a mass of about 5 g.

Force (symbol F) Force can be thought of as a push or a pull. If you push or pull on a mass at rest with some net force, it will begin to move in the direction of the force.

Weight is the force of gravity pulling a mass toward the center of the earth or other celestial body and will be different on the surface of the moon or some other planet; mass will not. In outer space, where there is almost no force of gravity, all objects are weightless, but they still have mass.

The metric unit of force is the newton (N), named after the physicist Isaac Newton (1642–1727). A 100-lb boy standing on the earth is held to the ground by a downward gravitational force, his weight, of about 445 N. The surface of the earth pushes up on the boy's feet with an equal force. These two forces balance and the boy does not move up or down. On the moon, his weight would be only about 74 N because the force of gravity is less on the surface of the moon than on the earth.

Pressure (symbol p**)**   Pressure is defined as force per unit area and can be measured in units of pounds per square inch or pascals (newtons per square meter). An example of pressure can be seen by considering an elephant walking on a sandy beach. Although the elephant weighs a great deal, its feet are very large, and the pressure, or force per unit area of its feet, exerted on the sand is relatively small. The weight is spread out over the large area. The elephant therefore leaves very shallow footprints in sand. On the other hand, a woman (whose weight is considerably less than that of an elephant) wearing high-heeled shoes exerts more pressure on the sand because her smaller weight is exerted on the very small area of the heels. The prints of the heels in the sand could be deeper than the footprints of the elephant.

Air pressure is exerted on everything in contact with air; rapid changes in air pressure cause vibrations of the eardrum, which we hear as a sound wave. Usually, air pressure on the outside of the eardrum is balanced by pressure in the ear, since air can flow into the mouth and then through the eustachian tube to the middle ear, behind the eardrum. Congestion of the eustachian tube restricts the air flow from the atmosphere to the space behind the eardrum. Slow changes of air pressure in the atmosphere can then create pressure differences between the atmosphere and the space behind the eardrum, causing large forces on the eardrum, which can be painful. Such an effect is often observed at the beginning and end of airplane flights, when the air pressure changes with changes in altitude.

Again, consider the 100-lb boy. Suppose the area on the bottom of each foot is 0.01 m². His weight, 445 N, is distributed over both feet, 0.02 m², and the pressure on the bottom of each foot is 22,250 pascals (Pa). When the boy stands on one foot, the area over which his weight is distributed is halved, so the pressure on his foot doubles to 44,500 Pa.

Atmospheric pressure on the surface of the earth is about 14.7 lb/in² or about 100,000 Pa.

1.2   *Simple Harmonic Motion*

The single most important concept in the study of waves and sound is that of simple harmonic motion (SHM). We shall now study SHM and some of its properties and in Chap. 2 demonstrate how SHM is basic to waves and sound.

Periodic motion is any type of motion that exactly repeats itself after successive equal time intervals. Some examples of periodic events are twirling a rock on a string around your head, the rotation of the earth on its axis, the revolution of the moon about the earth, a mass bouncing up and down on the end of a spring, a swinging pendulum, a blinking warning light, the vibration of a tuning fork, or the vibration of the reed of a clarinet or a singer's vocal folds, producing a constant musical tone.

Simple harmonic motion is a specific type of periodic motion (for instance, a mass bouncing up and down on the end of a spring) that arises from the conditions described in the next paragraph and whose graph looks like the curve shown in Fig. 1–2. This graph shows the position of one dot in Fig. 1–1 as time goes on. Notice that the motion repeats itself after 2 sec in this example, and, as with all SHM, it has a characteristic smooth shape. Other names for SHM or waves of this character are simple, pure, sine, cosine, or sinusoidal motion or waves.

Two conditions must be met to produce SHM in a mechanical system. First, an equilibrium position must exist; that is, there must be one position at which the object executing the motion would remain at rest if placed there and to which it returns if displaced and released. Second, the force tending to pull the object back to its equilibrium position must be a *linear restoring force:* That is, the force must be linearly proportional to the distance of the object from the equilibrium position, as explained next.

As an example of these two conditions, let us consider a particular mass hanging on the end of a flexible spring attached to a fixed point, as shown in Fig. 1–3. The equilibrium position is the position of the mass when it is hanging at rest directly below its suspension point. We consider now only vertical motion of the mass along the direction of the extended spring, above and below the equilibrium position. If the mass is lifted above its equilibrium position, the force of gravity and perhaps the compression of the spring force it back down toward the equilibrium position. If the mass is pulled down, the spring tends to pull it back toward the equilibrium position. If it requires 1 N of force to lift the mass up 1 cm, it will require 2 N for a 2-cm displacement, 3 N for a 3-cm displacement, and so forth, since for this system the restoring force is linear. Likewise, to pull the mass down requires 1 N of force for each cm of displacement from the equilibrium position. This is an example of a linear restoring force with a force constant (or force per unit of stretch) of 1 N/cm. Such a linear system is said to obey Hooke's law, named after the English physicist Robert Hooke (1635–1702).

When the mass on the spring in the preceding example is lifted 3 cm above its equilibrium position and released, it begins to accelerate downward, passing

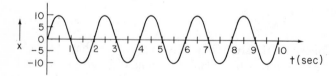

Figure 1–2   Graph of simple harmonic motion (SHM).

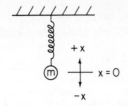

Figure 1-3 Mass hanging from a spring; equilibrium position is labeled $x = 0$.

through its equilibrium position, and executes SHM as shown in Fig. 1-4. At time $t = 0$ the mass is released from rest 3 cm above ($x = +3$ cm) the equilibrium position ($x = 0$). The mass starts moving, builds up speed as it passes through the equilibrium point, then slows down to zero velocity by the time it reaches the point $x = -3$ cm below the equilibrium point. The mass then starts upward, moving faster until it goes through the equilibrium point, then slows down to zero velocity by the time it reaches its original position, $x = 3$ cm. The motion then repeats itself every 2 sec, as shown on the graph.

The maximum displacement of the mass (in either direction) is called the *amplitude A*, and the repetition time T is called the *period* of the SHM. For this example, $A = 3$ cm and $T = 2$ sec. Note that the amplitude of any oscillation, including SHM, can be expressed in units appropriate to the system under consideration: centimeters for a mass oscillating on a spring, pressure units like pascals for a sound wave traveling through air, volts for an electrical signal in a stereo set, and so on.

The period T is related to the frequency f, which is the number of periods (or cycles) per second:

$$f = \frac{1}{T} \text{ or } T = \frac{1}{f}$$

A high-frequency oscillation has many oscillations per second; to fit many oscillations into a 1-sec time interval, the period must be short. Thus, a high-frequency oscillation has a short period; an oscillation with a long period has a low frequency. If we know that the period is 2 sec, the frequency f is

$$f = \frac{1}{T} = \frac{1 \text{ cycle}}{2 \text{ sec}} = \frac{\frac{1}{2} \text{ cycle}}{\text{sec}} = \frac{1}{2} \text{ hertz.}$$

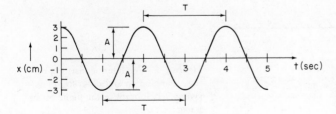

Figure 1-4 SHM with amplitude $A = 3$ cm and period $T = 2$ sec.

The unit of frequency is the hertz (Hz), which is 1 cycle/sec, named after Heinrich Hertz (1857–1894). For a 500-Hz audio-frequency oscillation, the period is

$$T = \frac{1}{f} = \frac{1 \text{ cycle}}{500 \text{ cycles/sec}} = \frac{1}{500/\text{sec}} = 0.002 \text{ sec} = 2 \text{ ms}.$$

Thus, the period of a 500-Hz oscillation is 2 ms. Check to see that if the period of an audio-frequency tone is 0.5 ms its frequency is 2000 Hz or 2 kHz.

Another example of SHM is a pendulum consisting of a mass hanging by a string from a fixed point that is free to oscillate back and forth in one plane, as shown in Fig. 1–5. The *equilibrium position* is the point directly below the suspension point at which the pendulum will hang motionless. For small displacements of the bob in either direction, the force required to pull the pendulum away from its equilibrium position is linearly proportional to the distance the bob is moved along the arc shown in Fig. 1–5. When displaced through some small angle and released from rest, the bob executes SHM in the plane of the paper. Its motion can again be described by a graph similar to the one shown in Fig. 1–4.

Some important details of SHM can be illustrated by comparison of SHM with uniform circular motion, which is the motion of an object around a circle with constant speed. Uniform circular motion, viewed from a distant point in the plane of the motion, is SHM, as can be seen using Fig. 1–6; SHM is the projection of uniform circular motion on a line in the plane of the motion. Points are shown on the circle corresponding to the position of the object at 16 equal time intervals for one period of the motion (one complete circle). The projections of these points on the line are shown and give the position of an object such as a mass on the end of a spring, executing SHM along the line; the dots represent equal time intervals. The radius of the circle is 3 units; thus the amplitude of the projected motion is 3 units.

Notice that the projection of the distance traveled in successive equal time intervals is not the same, and that therefore the velocity along the vertical line is continually changing. The velocity is least when the distance between two successive points is smallest, and, in fact, the velocity is zero when the projection is at either extreme of displacement. That is, for a brief instant of time, the point is neither moving up nor down, but is instantaneously at rest. If the x position of the point in Fig. 1–6 is plotted after each time interval, beginning with $x = 0$, the graph, shown in Fig. 1–7, is a sinusoidal or SHM curve. Starting at $t = 0$, the point moves a large distance in the positive direction in the first time interval and less in each succeeding time interval until it reaches its maximum position. Its velocity is thus large and

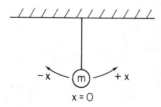

Figure 1-5 Pendulum consisting of mass hanging on end of string; mass swings left to right in the plane of the paper.

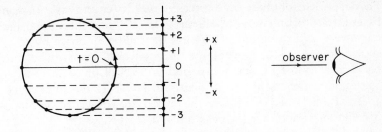

Figure 1-6 Identification of SHM as the projection of uniform circular motion along a line in the plane of the paper.

positive at $t = 0$ and decreases to zero at $t = 0.5$ sec, when the largest positive position is reached, as shown in Fig. 1-8. The velocity of the point at any given time can be determined by examining the slope of the graph in Fig. 1-7. When the slope of the graph is large, the velocity is high; a large distance is traversed in a given time. When the slope of the graph is 0, as at $t = 0.5$ sec, the velocity is 0. (The slope is graphed in Fig. 1-8.) The direction of the motion then changes, that is, the velocity becomes negative, and the point moves down through zero toward its extreme negative position. Looking at the spacing of the points, one can see that at $t = 1$ sec, as the point passes through $x = 0$, its velocity has the largest negative value, as shown in Fig. 1-8. As time progresses, the position and velocity repeat once per period, as shown. Make sure you understand how position and velocity of SHM are correlated, as shown in the figures. Note specifically that velocity as a function of time for SHM also has the unique SHM shape.

Because a circle is divided into 360°, one can analogously divide one period of

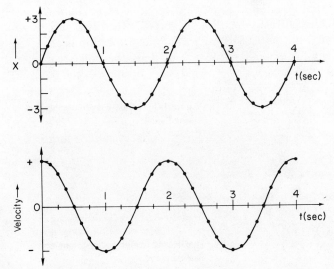

Figure 1-7 SHM position versus time at equal time intervals.

Figure 1-8 SHM velocity versus time at equal time intervals.

SHM into 360° of phase φ (lowercase Greek letter phi), as shown in Fig. 1–9. If one period is 360° of phase, the half-period is 180°, and the quarter-period is 90°. Two curves are said to be *out of phase* if they differ in phase by 180° at all times, as shown in Fig. 1–10. One of the curves would have to be moved forward or backward in time by one half-period or 180° of phase in order to make the two curves appear the same, or *in phase*. In Fig. 1–11 the solid curve is 90° in phase ahead of the dashed curve; thus the peak of the solid curve occurs earlier in time by one quarter-period than the peak of the dashed curve. In Fig. 1–12 the solid curve is 45° behind the dashed curve. Relative phase between two curves is only well defined when the two periods (and thus frequencies) are the same.

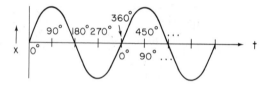

Figure 1-9 Phase of SHM curve.

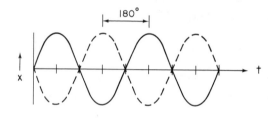

Figure 1-10 Two SHM curves differing in phase by 180° (out of phase).

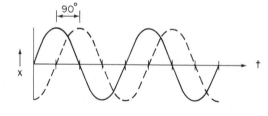

Figure 1-11 Solid curve 90° ahead of dashed curve.

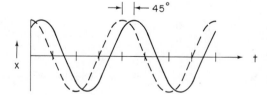

Figure 1-12 Solid curve 45° behind dashed curve.

1.3   Application to Sound

Our knowledge of simple harmonic motion can be applied directly to sound. In our study of acoustics we shall generally deal with frequencies in the audible range, approximately 20 Hz to 20 kHz. Sound waves are changes in air pressure occurring at frequencies in the audible range. The normal variation in air pressure associated with a musical instrument played quietly is about 0.002 Pa. The smallest pressure variation that can be heard is about 0.00002 Pa, whereas the pressure variation that produces pain in the ear is about 20 Pa. Normal atmospheric pressure is about 100,000 or 10^5 Pa. The maximum change in atmospheric pressure due to changes in weather is a few percent of this average value.

We can correlate the physical properties of sound waves with our perception of pitch, loudness, and tone quality. This can be readily demonstrated by using an electronic wave generator and a loudspeaker. Changing the frequency of the oscillator varies the perceived pitch; changing the amplitude of the oscillator's signal varies the loudness.

A higher frequency (or shorter period) wave sounds higher in pitch. For two waves of the same frequency, the one with the greater amplitude sounds louder. We shall discuss in Chap. 6 the variation in loudness with frequency for a constant-intensity wave and the slight variation in pitch with intensity. Shown in Fig. 1–13 are waves that (a) remain at the same pitch but get softer, (b) get higher in pitch but remain approximately at the same volume, and (c) become simultaneously softer and lower in pitch.

Wave forms other than the sine wave will play an important part in our study of sound. Shown in Fig. 1–14 are a triangular wave, a square wave, a sawtooth (or

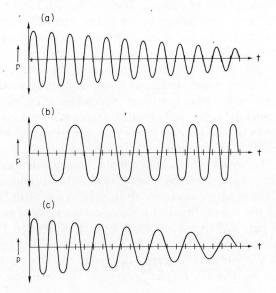

Figure 1–13 Audible waves which (a) remain constant in pitch but become softer; (b) become higher in pitch but remain at nearly the same volume; and (c) become simultaneously softer and lower in pitch.

9

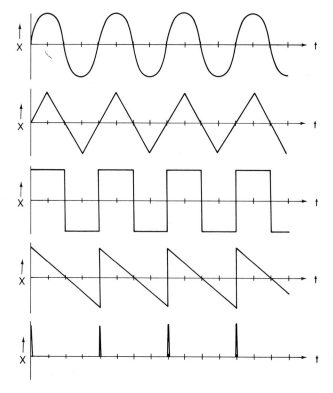

Figure 1-14   Sine wave, triangular wave, square wave, sawtooth (ramp) wave, and pulse train (series of pulses) of the same frequency.

ramp) wave, and a pulse train (series of pulses). These waves all have the same periodicity as the sinusoidal wave also shown in the figure; thus all these waves have the same frequency or pitch. The important difference between them is that they sound different; the tone quality or timbre of each is different. The variation in tone quality for different wave shapes can be demonstrated using an oscillator, which can produce different wave shapes. In Chap. 4, we shall study what makes these tones sound different and under what conditions two waves of different shape can sound very similar. In general, waves with different shapes will sound different.

Wave shapes will often be displayed using an oscilloscope. An *oscilloscope*, or "scope," is a device for displaying an electronic signal, that is, for translating it into visible form on a screen. The signal from a microphone or any other electronic signal is traced onto the oscilloscope screen by a beam of electrons produced in the scope. If no signal is fed in, the beam moves from left to right in a straight line, taking a specified time to cross the screen. Input of a signal causes vertical motion of the beam as it moves left to right and thus traces out the graph of the signal voltage as a function of time. Both axes can be calibrated, so we can observe the amplitude of the signal and how rapidly it varies. The graphs of sounds and other waves obtained using an oscilloscope are like the graphs previously shown in the various figures.

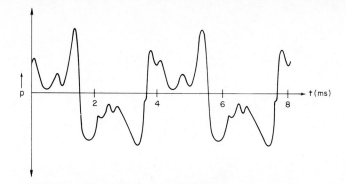

Figure 1-15 Wave form of a clarinet playing the lowest C on the instrument, which sounds at the pitch of B♭ on the piano.

Figure 1-16 A possible wave form of noise.

Just as each wave has a particular shape and tone, the sound wave of a note played on a musical instrument has a shape, called its *wave form*. A musical tone, which consists of periodic oscillations of air pressure, is converted into an electrical signal by a microphone. This electrical signal, or voltage, which is proportional to the pressure of the air in the sound wave, can be displayed, using an oscilloscope, as voltage versus time. The wave form for a clarinet playing its lowest C, obtained in the manner just described, is shown in Fig. 1-15. The period of this wave is about 4 ms; its frequency is therefore approximately 250 Hz.

All the preceding wave forms have one very important common feature: They are *periodic*. Waves do not have to be periodic; Fig. 1-16 shows a wave form of "noise," which contains no observable periodicity. This is the difference between musical sounds, which are periodic, and nonmusical sounds, or noise, which are nonperiodic. In Chaps. 4 and 5 we shall discuss the nature of several types of noise and the uses of noise in musical instruments and electronic music synthesizers.

1.4   *Damped and Driven Oscillations*

Consider a pendulum as in the example of Sec. 1.2. If it is started into motion by moving the bob to one side and releasing it from rest, it will not continue to oscillate forever, but will slowly decrease in amplitude, due to air resistance and friction in the suspension, until it stops. This process, called *damping*, is graphed in Fig. 1-17. In this case, the damping is very slow (try a pendulum), but under certain circumstances it can be very rapid, as in the damping of a pendulum under water or

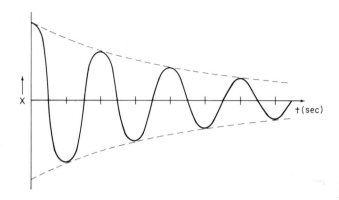

Figure 1-17 Damped harmonic motion.

damping of the sound of a plucked guitar string. For the guitar, while the amplitude decreases, the period remains almost constant until the motion stops; that is, the pitch remains almost constant while the note becomes continuously softer. In damped harmonic motion the period remains constant while the amplitude decreases.

Now consider the case of driven, or forced, oscillations of a pendulum, as illustrated in Fig. 1-18. If we pull the pendulum through a small displacement in the $-x$ direction and release it from rest at $t = 0$, it will undergo SHM, if there is no friction. Each time the pendulum reaches its extreme $-x$ position, we give it an additional sharp force, or push, in the $+x$ direction, causing the amplitude of the motion to increase slightly each cycle, as shown; this is similar to pushing a child on a swing. The times at which the force is applied are indicated by arrows. This growth in amplitude due to application of a periodic force is called *driven* or *forced harmonic motion.* While the amplitude increases with time, the period remains constant.

It is important to recognize that this continual increase in amplitude can occur only if the applied force continually comes at the right time, or at the proper phase, with respect to the motion of the pendulum bob; the force and the motion of the bob are then said to be *resonant* or *in resonance* with each other. In general, resonance can occur whenever the frequency of the driving force is the same as the natural, or normal, frequency of the oscillating system. If started at the proper phase, this phase relationship will continue.

Suppose now that the frequency of the driving force is slightly different from the natural frequency of oscillation of the pendulum. If the sharp force starts in

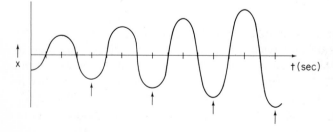

Figure 1-18 Driven harmonic motion with the frequency of the applied force the same as the natural frequency of the oscillator, so that the force remains in the same phase relationship to the motion. Arrows denote times when the driving force is applied.

phase with the swing of the pendulum, after some time the force and oscillation will become out of phase. The force, which originally was causing an increase in the amplitude of the motion, will now be opposing, or tending to decrease, the motion. This situation is shown in Fig. 1–19; again, the force is indicated by arrows. The continual phase change between the force and the motion can be easily observed.

A different type of resonance, called a *coupled* or *coupling resonance*, occurs when two mechanical oscillatory systems with the same (or simply related) natural frequency are connected or "coupled" mechanically, so that vibrational energy can be transferred from one oscillator to the other.

As an example of coupled resonance, consider a set of pendulums attached to a rod that can rotate slightly in loosely fitting holes at each end of the frame in which it is mounted, as shown in Fig. 1–20. If the pendulum at the right is started in motion (by lifting the bob out of the plane of the paper and releasing it from rest), its swinging will start the rod rotating back and forth at the same frequency. The rocking motion of the rod will then act as the driving force for the other five pendulums. The middle pendulum of the five on the left has the same length and thus the same natural frequency as the moving pendulum at right; it will therefore be driven into oscillation by the pendulum on the right. The other four, having different lengths, and therefore different frequencies, are not driven to large amplitudes. In fact, in this demonstration the motion transfers almost completely back and forth between the two identical pendulums, while the other four pendulums, not being in resonance, execute smaller oscillations similar to that shown in Fig. 1–19. Their motion goes in and out of phase with the motion of the driving pendulum at the right, so the amplitudes of their oscillations will alternately increase to a small level and then decrease to zero.

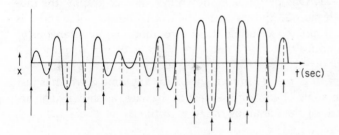

Figure 1-19 Driven harmonic motion with the frequency of the applied force different from the natural frequency of the oscillator, resulting in continuous phase change between the oscillation and the force.

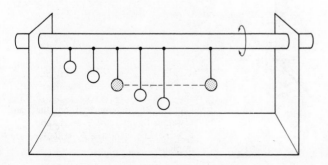

Figure 1-20 Coupled pendulum system. Suspension rod is free to rotate as indicated in holes in end support plates.

One of the most dramatic recorded forced oscillation resonances arose when, owing to the action of the wind, certain oscillations built up in the Tacoma Narrows Bridge. Eventually, the amplitude of the oscillations exceeded the elastic limit (breaking point) of the material in the bridge and caused its collapse on November 7, 1940.

Another example of resonance can be illustrated by using two identical tuning bars, say at 440 Hz. One is struck, then held near the other. A resonance condition exists between the two tuning bars with air providing the coupling, and *sympathetic vibrations* of the second tuning bar result. If either of the tuning bars is tuned slightly higher or lower, at 441 Hz for instance, the two do not have the same natural frequencies; no resonance condition exists and therefore no sympathetic vibrations would result.

1.5   Lissajous Figures

We have now studied SHM in one dimension and some of its applications to sound. A useful technique for studying the relationship between two waves involves Lissajous figures, named after the French mathematician Jules Antoine Lissajous (1822–1880). These figures record the position of a point that is executing SHM simultaneously in both the right-left (x) and up-down (y) directions. The geometrical figures obtained by plotting this combined motion are called *Lissajous figures*. In general, the amplitudes, frequencies, and phases of the two oscillations can be different, but we shall start with the simpler examples.

Consider first the combination of x and y motion with equal amplitudes, in phase, as shown in Fig. 1–21. One period T is shown in reference graphs; the motion repeats itself after each period. Shown at the right of the graphs of x and y is the curve traced out by a point executing these two motions simultaneously. Following the motion, we notice that $x = 1$ and $y = 1$ at $t = 0$; $x = 0$ and $y = 0$ at $t = T/4$; $x = -1$ and $y = -1$ at $t = T/2$; $x = 0$ and $y = 0$ at $t = 3T/4$; $x = 1$ and $y = 1$ (the same as $t = 0$) at $t = T$. In fact, $y = x$ for any time t, which is just the equation of a line at an angle of 45° with respect to the axes, as shown. The arrows show the development of the Lissajous figure with time.

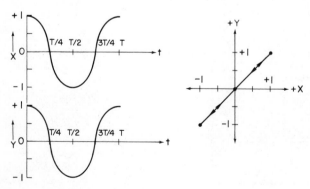

Figure 1–21 Reference graphs (left) and Lissajous figure with $A_x = A_y$ and $f_x = f_y$. The x and the y oscillations are in phase.

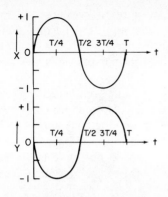

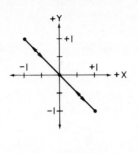

Figure 1-22 Reference graphs and Lissajous figure with $A_x = A_y$ and $f_x = f_y$. The x and the y oscillations are out of phase.

Now suppose that the relative phase of the oscillations is changed by 180° so that they are out of phase. The oscillations along with the resultant Lissajous figure are shown in Fig. 1-22. In this case, the x and y values are equal, but opposite in sign, giving the line $y = -x$: $x = 0$ and $y = 0$ at $t = 0$; $x = +1$ and $y = -1$ at $t = T/4$; $x = 0$ and $y = 0$ at $t = T/2$; $x = -1$ and $y = +1$ at $t = 3T/4$; $x = 0$ and $y = 0$ (the same as at $t = 0$) at $t = T$.

Two other phase relationships are important: the x motion 90° ahead of the y motion, and the x motion 90° behind the y motion. These two cases are shown in Figs. 1-23 and 1-24, respectively. When the phase of the x oscillation leads that

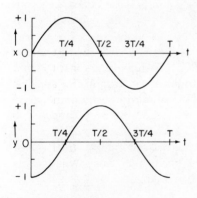

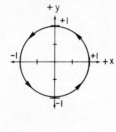

Figure 1-23 Reference graphs and Lissajous figure with $A_x = A_y$ and $f_x = f_y$. The x oscillation is 90° in phase ahead of the y oscillation.

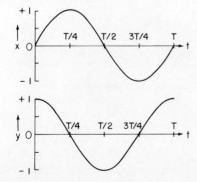

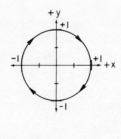

Figure 1-24 Reference graphs and Lissajous figure with $A_x = A_y$ and $f_x = f_y$. The x oscillation is 90° in phase behind the y oscillation.

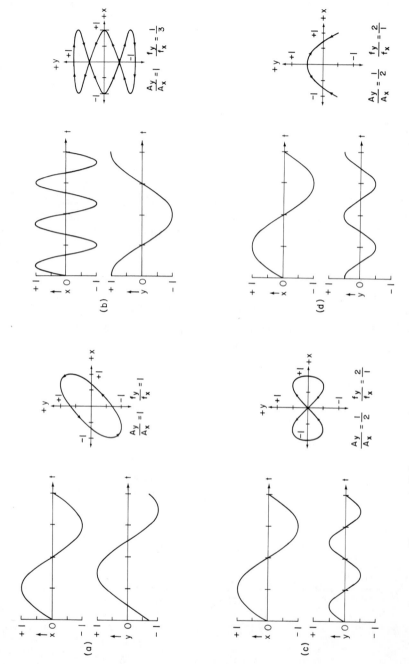

Figure 1-25 Reference graphs and associated Lissajous figures with amplitude and frequency relationships shown. In (a) the x oscillation is $45°$ in phase ahead of the y oscillation. Figures (c) and (d) differ only in the relative phase between the x and the y oscillations.

of the y oscillation (Fig. 1–23), the motion can be analyzed as follows: $x = 0$ and $y = -1$ at $t = 0$; $x = +1$ and $y = 0$ at $t = T/4$; $x = 0$ and $y = +1$ at $t = T/2$; $x = -1$ and $y = 0$ at $t = 3T/4$; $x = 0$ and $y = -1$ (the same as $t = 0$) at $t = T$. The motion then repeats. Thus, if the x motion leads the y motion by 90°, the point moves counterclockwise in a circle. A similar analysis shows that the case where y leads x by 90° results in clockwise motion in a circle. If the amplitudes of the x and y oscillations are not equal for the preceding cases of 90° phase differences, the circles become ellipses. If the phase difference is some value other than those mentioned, the Lissajous figure becomes a tilted ellipse, as in Fig. 1–25(a).

Of course, an infinite number of possible figures exist: various amplitude ratios, various frequency ratios, and various phase relationships. Some of the possibilities are shown in Fig. 1–25.

Any Lissajous figure can be analyzed to determine the ratio of the amplitudes A_x and A_y of the x and y oscillations and the ratio of the frequencies f_x and f_y of the x and y oscillations. The amplitude is the maximum displacement in the x or y direction and can be read directly off the graph. In the example of Fig. 1–26, $A_x = 2$

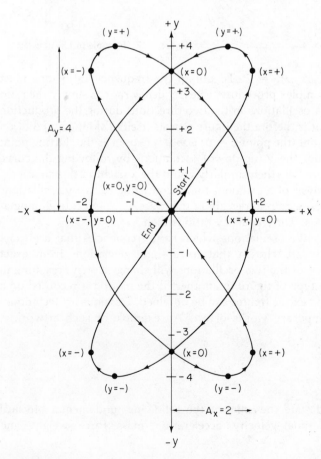

Figure 1-26 Lissajous figure with $A_y/A_x = 2$ and $f_y/f_x = 2/3$. Amplitude ratio is determined by noting maximum displacement in each direction. Frequency ratio is determined by counting the number of complete oscillations (0 to + to 0 to − to 0) in each direction during the completion of one entire figure (start to end).

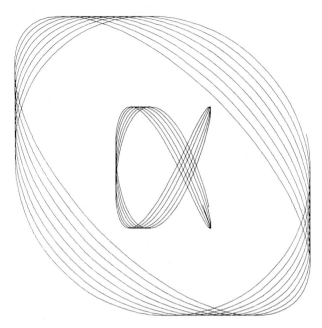

Figure 1-27 Example of Lissajous art.

and $A_y = 4$, so the ratio $A_y/A_x = 2$. To obtain the frequency ratio, one must follow a direct but more complex procedure. The frequency ratio is simply the ratio of the number of complete oscillations in the two directions during the production of the complete figure, that is, before the figure repeats itself. Taking any point on the figure as the starting point (the point $x = 0$, $y = 0$ is chosen as the starting point in Fig. 1-26), we first count the y (up-down) oscillations by following the curve through one complete figure. We then similarly count the x (right-left) oscillations. Figure 1-26 shows the analysis of a Lissajous figure consisting of two y oscillations and three x oscillations whose amplitudes are in the ratio $A_y/A_x = 2$, beginning at the origin of the coordinates. The frequency ratio is then $f_y/f_x = \frac{2}{3}$.

If, rather than using two oscillations which have frequencies that are exact multiples, two frequencies are chosen that have slight deviations from exact multiples, the relative phase of the two oscillations will change slowly, resulting in gradual changes from one type of figure to another. If the motion is recorded on a graph plotter, beautiful and exotic figures can be obtained. Lissajous art techniques have been used in contemporary works of art. An example of such artwork is shown in Fig. 1-27.

EXERCISES

1. Define, discuss, and state the units of the following fundamental physical quantities: position, time, velocity, acceleration, mass, force, weight, and pressure.

2. What is a linear restoring force? How is simple harmonic motion related to linear restoring forces?

3. Give examples of motion that are periodic but not simple harmonic. Give examples of simple harmonic motion. Is the brightness of a flashing automobile signal light simple harmonic? Is it periodic?

4. The frequency limits of human hearing are about 20 Hz to 20 kHz. What are the periods of these oscillations? What is the frequency of a tone that has a period of 0.01 sec?

5. Draw a triangular wave of period $T = 5$ ms and amplitude $A = 2$ V. What is the frequency of this wave?

6. Draw a 100-Hz sine wave of 3-V amplitude. For each of the following parts, draw a wave that differs from this sine wave in the characteristic specified: (a) greater in frequency, (b) lower in amplitude, (c) different in wave shape, (d) greater in period, (e) different in phase, (f) greater in intensity, (g) different in tone quality, (h) lower in pitch, (i) louder.

7. (a) Draw a curve of damped harmonic motion.
   (b) Draw a graph of motion that decreases in *both* period and amplitude as time progresses. Is this damped harmonic motion? Why or why not?

8. In general, which would have a higher frequency, a violin note or a cello note? A trumpet note or a trombone note? Which tone would have a longer period, one from a bassoon or one from a piccolo?

9. Draw a graph of two sinusoidal waves of the same frequency and amplitude that differ in phase by 180°. What term do we apply to two waves that have this phase relationship? Draw a graph of two sinusoidal waves of the same frequency but (a) different in amplitude, and (b) with a phase difference of 90°. Which wave is ahead in phase? Do the same for two waves that differ in phase by 45°.

10. Describe physically the relationship between the motions of two pendulums whose oscillations are (a) in phase, (b) out of phase, and (c) differ in phase by 90°.

11. Define "resonance." Give some examples of resonance from music and other fields.

12. Draw the reference graphs and Lissajous figures for two sinusoidal oscillations of the same frequency with $A_y = 2A_x$ when they are (a) in phase, (b) out of phase, (c) the x oscillation is 90° ahead of the y oscillation, and (d) the x oscillation is 45° behind the y oscillation.

13. Draw the reference graphs and Lissajous figures for two sine waves starting out of phase with $A_y = \frac{1}{2}A_x$ and $f_y = 2f_x$.

2

Waves
and Sound

In this chapter we shall investigate the important features of waves and wave motion and apply the concepts introduced in the first chapter to sound and ultrasound. Each musical and nonmusical topic we consider will be related to the fundamental physical law or laws that govern it. Examples and applications will be taken from familiar areas, areas in which the nonscientist may not yet understand the phenomena. Finally, we shall study the addition of waves and some of its important applications. Most of the concepts studied in this chapter will be relevant to the understanding of the advanced topics in later chapters.

2.1 Transverse and Longitudinal Waves

We define a *wave* as possessing the following properties:

1. A wave is a disturbance within some medium; the disturbance propagates (or travels) with some velocity, which depends on the medium. Most of the waves that we shall study will be periodic.

2. A wave transfers energy; by virtue of the energy it carries from its source to its destination, a wave can perform work. For example, a rope wave could perhaps move a pump handle up and down to pump small amounts of water, a sound wave can make an eardrum vibrate to create the sensation of sound, waves from the sun can heat and illuminate the earth, and radio

waves can make electrons in a radio antenna oscillate to enable reception of the signal.

Two basic types of waves, transverse and longitudinal, are distinguished by the direction of the wave oscillation (that is, the direction of the motion of matter in the medium) relative to the direction of propagation of the wave.

In a *transverse wave* the oscillations forming the wave are perpendicular, or transverse, to the direction of propagation. Examples of transverse waves are rope waves, waves in slinky springs where the wave is created by motion perpendicular to the slinky, and the pulse traveling down a cowboy's whip as he cracks it. Each point of the whip (the medium) moves up and down while the wave pattern propagates, say, left to right.

A very important class of transverse waves is electromagnetic (EM) waves, which are unique in that they require no medium in which to propagate. They are exemplified by light rays from the sun, which pass through the vacuum of outer space on the way to the earth. The various types of electromagnetic waves are similar in all their properties except frequency (or equivalently, as we shall soon see, wavelength). In order of increasing frequency, the types of EM waves are radio, television, radar, infrared, visible, ultraviolet, X-rays, and gamma rays.

In a *longitudinal wave* the motion of the points of the medium (forming the wave) is in the same direction as the direction of propagation of the wave pattern. Examples of longitudinal waves are slinky spring waves created by expansion and/or compression of the spring coils, sound waves, and ultrasonic waves (sound waves of frequencies above the audible range). Shock waves, such as sonic booms, are also longitudinal in nature. The propagation of sound waves, and longitudinal waves in general, always requires a medium for reasons that will become obvious later.

A demonstration that illustrates a fundamental difference between sound and light waves is the "bell in vacuum," illustrated in Fig. 2–1. A ringing bell is suspended by a string inside a glass jar from which the air is evacuated by means of a mechanical vacuum pump. The string suspension minimizes direct transmission of the sound to the glass jar. As the air is pumped out, the sound of the bell becomes softer; when the jar is almost entirely evacuated, the ringing of the bell becomes inaudible. We are not surprised that the bell remains visible in the jar even when the air is entirely evacuated. Thus, we verify that light waves (and EM waves in general) do not require a medium in which to propagate, whereas sound waves (and longitudinal waves in general) do require a medium, air in particular for this experiment, in which to propagate. All transverse waves other than EM waves and all longitudinal waves require a medium.

One of our goals will be to understand the low-amplitude transverse waves in a guitar or violin string. To help visualize these transverse oscillations, we shall first consider the larger amplitude, slower moving transverse waves in a rope.

Let us consider how a transverse rope wave is produced by rapid sinusoidal transverse oscillation of one of its ends. Consider the rope shown in Fig. 2–2, where

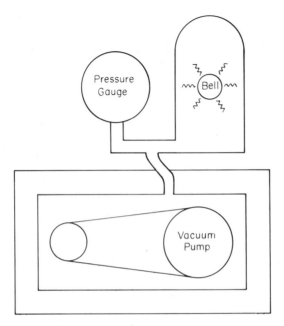

Figure 2-1 "Bell in vacuum" demonstration. Sound of bell diminishes as air is pumped from jar.

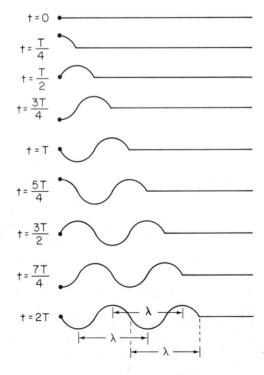

Figure 2-2 Transverse wave in a rope, created by sinusoidal oscillation of the end of the rope.

at time $t = 0$ the left end of the rope is suddenly started in simple harmonic oscillation in the vertical direction with a period T (frequency $f = 1/T$).

In the first time interval of $T/4$, the end of the rope will move up to its maximum positive value, causing a displacement of the rope. The wave thus produced moves from left to right, the direction of propagation. After one half-period, $T/2$, the oscillating end of the rope returns to the equilibrium point, and the corresponding sinusoidal displacement of the rope moves along the rope as shown. The resultant wave is then shown after successive time intervals of one quarter-period, $T/4$, until two periods have passed.

Notice that the sinusoidal motion of the end of the rope produces a periodic wave that is also a sine curve; each point in the medium oscillates vertically in SHM. In fact, the time variation in the position of the left end of the rope has been converted into a variation in shape along the length of the rope. The distance along the rope after which the spatial variation repeats itself is called the *wavelength*, symbol λ (lowercase Greek letter lambda), and is shown on Fig. 2–2.

It can also be observed that in a time interval of one period T, the wave has traveled a distance of one wavelength λ along the rope. The speed of propagation v of the wave along the rope is then equal to the distance traveled, λ, divided by the time interval T:

$$v = \frac{\lambda}{T}$$

Since $T = 1/f$ or $f = 1/T$, it follows that

$$v = \frac{\lambda}{T} = \lambda f$$

The mathematical formula $v = \lambda f$, where v is the wave speed, λ the wavelength, and f the frequency, is true for any wave of any type and frequency. For instance, the preceding discussion also applies to both longitudinal and transverse waves in long slinky springs. The speed of the wave depends upon the type of wave (rope, sound, EM, and so on) and the medium through which it is propagating.

The representation of a transverse wave in a rope is relatively easy to visualize and to draw: It is like a photograph of the rope at any given time. The succession of drawings in Fig. 2–2 can be thought of as photographs of the rope at the times indicated.

Now let us think about longitudinal waves. Consider a slinky spring, at rest, stretched and held so that it has some (equilibrium) length, as shown in Fig. 2–3(a); shown are successive turns of the spring. If a *compression* is formed at the center of the spring by squeezing several of the turns together, they will expand when released to return to equilibrium turn spacing. Similarly, a *rarefaction*, an extended region of coil layers, will compress toward its equilibrium spacing when released. These two conditions are shown in Fig. 2–3(b) and (c).

A longitudinal wave in a slinky spring is formed by a succession of such com-

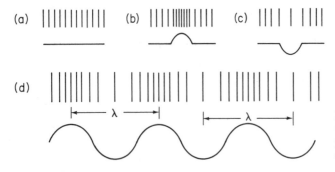

Figure 2-3 Transverse representation of (a) a spring at equilibrium, (b) a compression, (c) a rarefaction, and (d) a series of compressions and rarefactions.

pressions and rarefactions moving along the direction of the spring. Because such a longitudinal wave is difficult and time consuming to draw, it is usually represented by a transverse wave graph, as shown in Fig. 2–3(d). The upward arc of the transverse wave graph represents a compression in the longitudinal wave (spring spacing smaller then equilibrium spacing), while the downward arc of the transverse wave graph represents a rarefaction in the longitudinal wave (spring spacing greater than equilibrium spacing). Again, as in the case of the transverse wave, we can identify the wavelength λ, the distance after which the wave shape repeats itself. When a sinusoidal wave propagates along the spring, each turn of the spring moves in SHM.

As in the case of the transverse wave of Fig. 2–2, the production of a longitudinal wave in a spring by longitudinal vibrations of the end of the spring is illustrated in Fig. 2–4. The wave, consisting of sinusoidal variations in the spring coil spacing, possesses a certain wavelength λ and moves to the right with a wave speed v characteristic of the spring. Also shown is the transverse representation of the longitudinal spring wave.

The longitudinal waves in a slinky spring can serve as a model of a sound wave in air. In this simplified analogy, the slinky spring turns represent successive "layers" of air molecules. As in the case of the slinky, if a compression of air layers is created on the left (say, by a loudspeaker) it will tend to expand, pushing the next air layers to the right, and so on, thereby causing a compressional pulse to propagate to the right. A succession of compressions and rarefactions then forms the wave, as in the case of the slinky. Again the wave possesses the period T and frequency f of the source, but now moves with the speed of sound in air, for which we shall use the symbol S. The value of the wavelength can be calculated using $S = f\lambda$ or, equivalently, $\lambda = S/f$.

One method for determining the speed of sound uses an electronic signal generator, a loudspeaker, a microphone, and a dual-trace oscilloscope, that is, an oscilloscope that can simultaneously display two signals. The experimental setup shown in Fig. 2–5 will now be described; the exact numerical values of the signal frequencies are not crucial, but are presented as used in one particular experiment.

The complex signal produced by the signal generator is a 10,000 Hz sine wave that is on for 0.5 ms and then off for 4.5 ms, as shown in Fig. 2–6. The effect is that

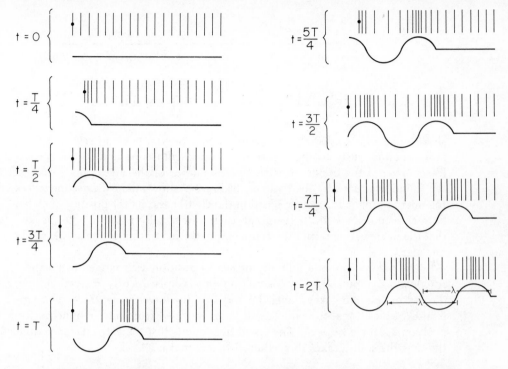

Figure 2-4 Production of a longitudinal wave in a slinky spring by longitudinal oscillation of one end of the spring, shown at time intervals of one-quarter period. Also shown is the transverse representation of the longitudinal wave.

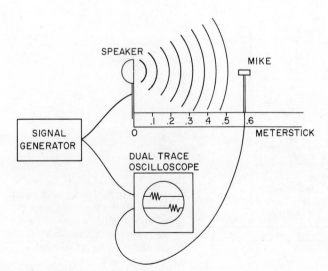

Figure 2-5 Experiment for determination of the speed of sound in air. Position of microphone along meter stick is adjustable.

25

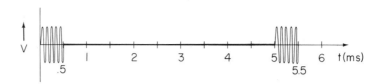

Figure 2-6 Wave pattern input to speaker of Fig. 2-5.

short bursts of a 10,000-Hz sine wave are generated at 5-ms intervals. This signal is sent simultaneously to the top trace of the oscilloscope and to the speaker. The lower trace of the oscilloscope shows the output of the microphone that picks up the sound produced by the speaker. Figure 2-7 shows the oscilloscope traces when the microphone is in its near position (labeled 1) and its far position (labeled 2). In both cases the scope starts the graph each time a pulse is sent to the speaker. Note that the microphone response occurs before the next pulse to the speaker is begun 5 ms later.

As the microphone, initially placed in position 1, is moved to position 2, the response signal picked up by the microphone is consequently delayed in time, as indicated by the two curves labeled 1 and 2 on the lower scope trace. The time $t_2 - t_1$ is just the additional time the sound takes to travel from microphone position 1 to position 2, labeled $d_2 - d_1$. The speed of sound is just the extra distance divided by the time the sound takes to go that distance; that is,

(a)

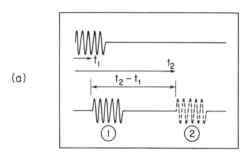

(b)

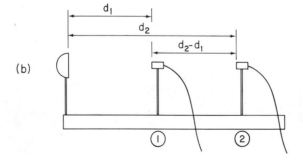

Figure 2-7 Part (a) shows the face of the oscilloscope for the speed of sound experiment. When the microphone is placed back a distance $(d_2 - d_1)$ as shown in part (b), the sound arrives at the microphone later by an amount $(t_2 - t_1)$.

$$S = \frac{d_2 - d_1}{t_2 - t_1}$$

If we measure $d_1 = 10$ cm at $t_1 = 0.30$ ms and $d_2 = 60$ cm at $t_2 = 1.80$ ms, then the speed of sound is calculated to be

$$S = \frac{(60 - 10) \text{ cm}}{(1.80 - 0.30) \text{ ms}} = \frac{50 \text{ cm}}{1.50 \text{ ms}} = 33{,}333 \text{ cm/sec} = 333 \text{ m/sec}.$$

Our experimental determination gives fair agreement with the accepted value of 345 m/sec for the speed of sound at room temperature.

2.2 *Basic Phenomena and Properties of Waves*

We begin our study of waves by surveying the general laws and phenomena that govern wave behavior:

(1) Huygens's principle

(2) Superposition

(3) Inverse square law

(4) Reflection

(5) Refraction

(6) Interference

(7) Diffraction

(8) Polarization

These phenomena (with the exception of polarization) apply to all three-dimensional waves. We shall first take familiar examples of various types of waves to illustrate wave behavior; then in each case we shall apply these principles to the domain of sound and ultrasonic waves. We shall see that the first three principles governing waves are more fundamental than the next four, which largely arise by application of Huygens's principle and the law of superposition.

Throughout our study of wave behavior, illustrations will be made by the use of a *ripple tank*, a shallow-water tank with a small ball or straight bar that is dipped slightly beneath the surface of the water and vibrated; the ball creates outgoing circular waves, and the bar creates plane waves. The water waves in a ripple tank are two-dimensional waves, that is, they propagate on the two-dimensional (plane) surface of the water. The rope waves discussed in the last section are one-dimensional as they propagate in the one-dimensional (line) medium, the rope. Figure 2–8 shows ripple-tank representations of a circular wave and a plane wave; successive wave

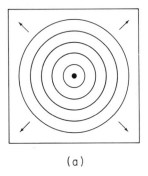
(a)

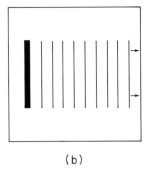
(b)

Figure 2-8 Circular waves (a) produced by a small vibrating ball, and plane waves (b) produced by a vibrating bar in a ripple tank. Periodic vibration of the source perpendicular to the surface of the water produces uniformly spaced waves. The lines in both (a) and (b) represent the crests of successive waves.

crests are represented by the lines; the spaces between the lines represent troughs. Any of the circular waves is similar to the wave obtained by dropping a pebble onto a calm water surface; the continuous succession of circular waves could be obtained by dropping additional pebbles at equal time intervals. Later it will also be necessary to insert various barriers into the ripple tank to reflect or stop the waves or to allow only certain parts of the wave to pass.

Huygens's principle   Let us examine in some detail the motion of waves in the presence of the barriers represented by the cross-hatched horizontal bars in Fig. 2-9. Here the barrier is a breakwater formed by a rectangle of material resting on the bottom of the tank and extending above the surface of the water so that the wave is only able to pass through the small opening, or slit, between the barriers. We observe that the wave emanating from the slit does not continue unchanged but emerges from the barrier in a circular wave with the slit at its center. This general result, the same for all waves, constitutes the experimental basis for Huygens's principle, named for the Dutch physicist Christian Huygens (1629–1695). This principle can be stated as follows: All points on a wave act as sources of circular waves (or spherical waves in three dimensions); the total wave pattern at some later time will be the sum of all the individual Huygens's wavelets.

The intensity of the circular wave emanating from any point (such as the point at the slit) is greatest in the direction of the original wave propagation and gradually decreases with angle, becoming zero in the opposite direction. This too can be observed in the ripple tank, but is not shown in Fig. 2-9.

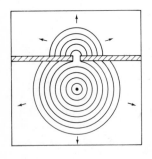

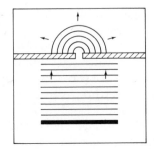

Figure 2-9 Circular and plane waves in a ripple tank. The waves are interrupted by a barrier, allowing only a small part of the wave to pass. This opening acts as a source of Huygens's wavelets.

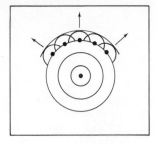 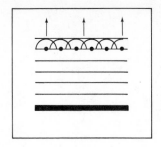

Figure 2-10 Huygens's wavelets originating from a succession of points along circular and plane wave fronts. The leading edge of all the individual wavelets forms the next circular or plane wave front.

In many circumstances, analysis using Huygens's wavelets is a convenient theoretical technique for understanding the propagation of waves. We can now see how Huygens's principle applies to the propagation of a circular or a plane wave, illustrated in Fig. 2-10. If the circles represent the position of the wave crests at time intervals of one period, then one can break up any one of the circular waves into a set of points, each of which acts as a source of circular Huygens's wavelets whose sum intensity is greatest in the original direction of propagation of the wave. This set of Huygens's wavelets emanating from points along the original wave front is shown in part in Fig. 2-10. The crests of these wavelets from successive points on the original waves are lined up, forming a shape identical to the original wave after one period. Thus, the overall effect of the addition of all these wavelets after one period is propagation of the original wave. The wavelets propagating in directions different from the original direction of propagation consist of a wide range of phases and will cancel each other.

We can see, however, that this argument cannot hold at the edges of the plane wave of Fig. 2-10. As a result of the wavelets at the edges of the plane wave, any narrow plane wave such as the one drawn will spread sideways as it propagates. This phenomenon will be treated again in the discussion of diffraction.

Another important result of this wave behavior is that the direction of propagation of a wave always remains perpendicular to the wave front. The wave front is the region defined by the farthest extension of the wave propagation, and is a straight or curved line in two dimensions and a plane or curved surface in three dimensions. For instance, in the previous example of a stone dropped into water, the wave front is an expanding circle. For sound in air or a flash of light in three-dimensional space, the wave front is an expanding sphere. This will be important to remember in our study of refraction and will be dealt with in more detail in that section.

Superposition The law of superposition can be stated as follows: The existence of one wave does not affect the existence or properties of another wave, even if they are in the same place at the same time. This is equivalent to the statement that waves add algebraically; that is, the displacement of the sum wave, $A + B$, is equal to the displacement due to wave A added to the displacement due to wave B at the same point and time. This is valid for small-amplitude waves,

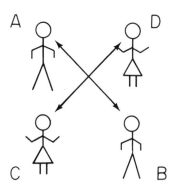

A   D

C   B

Figure 2-11   Sound waves from the conversation between persons *A* and *B* pass unaffected through the same point in space as the waves from the conversation between persons *C* and *D*. This illustrates the principle of superposition.

which create limited stress on the medium. This clearly distinguishes waves from material things, no two of which can occupy the same place at the same time. Waves can pass through each other without affecting each other.

One can also define superposition in terms of the addition of waves, a topic we shall cover shortly. In that approach, waves are said to obey the superposition principle if the resulting displacement at a point is algebraically equal to the sum of the displacements of the individual waves at that point.

An illustration of this is shown in Fig. 2-11. Persons *A* and *B* can talk with each other at the same time persons *C* and *D* talk with each other. The sounds or words from either conversation are completely unaffected by those of the other; the voice of *A* will not change when *C* produces waves that pass through the same crossing point. Not only are sound waves present in this example, but light waves of all colors (the people see each other), radio waves (a portable radio at the center of the foursome will pick up stations), radiating heat waves, and others. In fact, at any point in space there is in general an infinity of waves; all space is filled with a limitless number of waves of all kinds and wavelengths, none of which is affected by the existence of the others. If the persons in our example have trouble hearing each other owing to external sound waves, it is due to physical or psychological processes in the auditory system and not to any exception to the rule of superposition of waves. The superposition principle will be used to help explain several other properties of waves, such as the addition of waves and interference.

Inverse square law   Everyone is familiar with the fact that the farther one is from a source of sound the softer it sounds, or, equivalently, the farther one goes from a sound source, the more intense the source must be made in order for you to perceive the same loudness. The inverse square law tells exactly how the intensity of a sound wave (or any other type of wave in three dimensions) decreases as the distance between the source and the observer increases.

Consider the two-dimensional wave created by dropping a stone into a pool of water; as time goes on, the wave expands in a circle of increasing radius, as shown in Fig. 2-12(a). The energy transferred to the water by the stone when it hits the surface is contained in the wave. The wave front, the circumference of the cir-

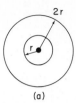

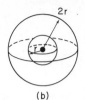

(a)   (b)

Figure 2-12 (a) Expansion of a two-dimensional water wave, illustrating the intensity law $I \propto 1/r$ for a two-dimensional wave. (b) Expansion of a three-dimensional wave, such as a sound wave, illustrating the intensity law $I \propto 1/r^2$ for a three-dimensional wave.

cle, gets longer as the circle gets larger. The same energy is "spread out" over a larger circumference. In fact, if the circle doubles in radius, then the circumference, $C = 2\pi r$, is twice as large, the energy is spread out over a wave front twice as long, and the intensity, the power per unit length, of the wave is then half as much. Stated mathematically, $I \propto 1/r$, or the intensity is inversely proportional to the radius of the circle. Thus for two-dimensional waves the intensity is inversely proportional to the distance of the observer from the source.

Now consider the extension of this concept to three dimensions, the usual case for sound waves, light waves, and others, which expand in all directions. Rather than expanding in circles, the wave will expand in spherical surfaces of increasing radius, so the energy of the wave is spread out over the spherical surface as shown in Fig. 2-12(b). The area A of a sphere of radius r is given by the formula $A = 4\pi r^2$, so when the radius of the sphere is increased to twice its original radius, the area is *four* times the original area. The same amount of energy is therefore spread out over four times the original area, so the intensity, in this case, the power per unit area, is only one-fourth of the original intensity. Similarly, an increase of a factor of 3 in radius increases the area of the sphere by a factor of 9, resulting in one-ninth of the original intensity. Stated mathematically, $I \propto 1/r^2$, or the intensity decreases in proportion to the square of the distance from the source—the inverse square law.

All types of waves obey this law when they are emitted and allowed to travel freely away from the source, with no focusing or other means of confinement: an ordinary light bulb emitting light, a church bell ringing, a radio transmitter broadcasting, and so on. One interesting light source that emits light that does not decrease in intensity according to the inverse square law is the laser. Because it is *coherent* light and is confined to a very narrow beam rather than spreading out in all directions, the intensity does not decrease as rapidly as would be expected according to the inverse square law. These features allow a powerful laser beam to retain sufficient intensity to be observable after being reflected back to earth by a mirror on the moon.

Reflection Everyone is familiar with an echo, a reflection of sound off some surface, the audio analog to a light wave reflecting off a mirror. The simplest type of reflection is that occurring when you stand facing the flat wall of a building and create a sound, which reflects back to you as an echo. Similar reflectors are useful if placed in back, on the sides, and on top of an orchestra in concert, since they reflect the sound waves out toward the audience.

32   *Waves and Sound*

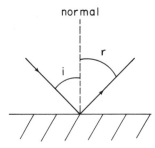

Figure 2-13 A wave incident in the direction shown reflects as indicated from a flat surface. The angle of incidence *i* and the angle of reflection *r* are equal. The normal is perpendicular to the reflecting surface at the point of reflection.

A mathematical feature of reflection is that the angle at which an incident wave approaches the reflecting surface is the same as the angle at which the reflected wave leaves the surface. This angle is generally given with respect to the *normal,* a line perpendicular to the surface at the point of reflection, as shown in Fig. 2-13. We can understand this more thoroughly by drawing a sequence of Huygens's wavelets, as in the example given in Fig. 2-10; such a construction is shown in Fig. 2-14. In the case of the plane waves, the angles defined are between the normal and the directions of the incident and reflected waves. Shown in Fig. 2-13 are incident and reflected rays. A *ray* marks the propagation of a single point on the wave. A ray is always perpendicular to the wave front, and for plane waves the ray is a straight line showing the direction of propagation of the wave.

If waves are reflected from irregularly shaped surfaces, the reflection must be examined in detail to determine the appropriate normals and angles. At each point along a curved surface there is a normal line. The law holds for a wave striking the different parts of such a surface. The normal is different at each point, whereas for

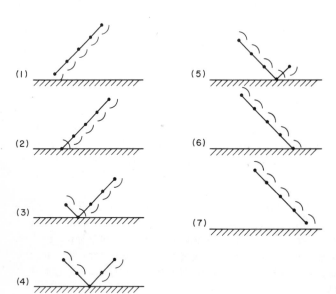

Figure 2-14 Huygens's wavelet construction showing how a wave is reflected from a smooth, flat surface. The drawings show the advancing wave front and the Huygens's wavelets at a succession of times.

the case illustrated by Fig. 2–14 the normal at any point along the surface points in the same direction.

A more complex example of reflection is the focusing of plane waves to a point by a parabolic reflecting boundary, that is, a surface with the shape of a parabola. Figure 2–15 shows the propagation of a plane wave created by dipping the straight element at the left quickly in and out of the water in a ripple tank. The point where the entire wave comes together is called the *focus* or *focal point* of the parabola.

Figure 2–16 shows the same phenomenon, except now plane waves are being continuously created at the left and focused. If they are not stopped at the focus, they will continue on through that point and expand as a sequence of circular wave segments. It is important to point out here that to achieve a good focus the reflector must have a *parabolic* shape, not a circular shape. In fact, in three dimensions, the reflector must be a paraboloid of revolution, obtained by rotating a parabola about a line through the focus and the center of the reflector. Such a mirror (in the case of light) or sound reflector is thus called a *parabolic mirror* or *parabolic reflector*.

The parabolic reflector has many important applications in the fields of both light and sound. A high-quality reflecting astronomical telescope requires a parabolic reflector to obtain the good focus necessary to resolve closely spaced celestial objects. Parabolic microphones are formed by placing a microphone at the

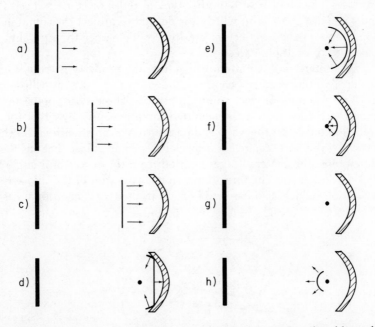

Figure 2–15 A plane wave pulse is created at the left of the ripple tank and focused to a point by a parabolic reflector. The wave is shown at a succession of unequal time intervals.

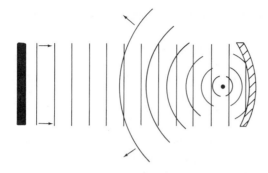

Figure 2-16 A plane wave in a ripple tank is focused by a parabolic reflector and continues to expand as a circular wave segment.

focus of a large parabolic sound reflector, which is often made of clear plastic. The microphone is placed at the focus facing the reflector (*not* the source of the sound) such that all the reflected rays are collected. The signal from the microphone can be fed into an amplifier and loudspeaker to make it audible. Such microphones are regularly used at football games to capture the signals from the quarterback and are used to record the calls of distant birds and other soft sounds.

A parabolic reflector can also be used to obtain a beam of light from a small source, as in the case of a flashlight or large searchlight, by the inverse procedure from focusing a beam of light. The source is placed at the focus and the reflector produces a beam of light. Another type of reflector is used in certain loudspeakers and some megaphones in which it is desired to project the sound in a certain direction. Some of these reflectors are hyperbolic rather than parabolic in shape and make use of multiple reflections.

An interesting variation of this focusing property is the *whispering chamber*. In the "two-dimensional" chamber shown in Fig. 2-17, an ellipse has been drawn and the positions of two points indicated. These two points are called the *foci* (plural of focus), since waves emitted simultaneously by either focus in any direction will bounce off the walls and pass through the other focus at the same time, as shown by the rays. A circle is a special case of an ellipse that has the two foci at the same point, the center. Waves emitted from the center of a circle will be focused back to the center. The three-dimensional extension of the ellipse is the ellipsoid of revolution, which also has two foci, as in the two-dimensional case. Two people,

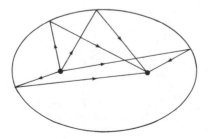

Figure 2-17 Rays emitted at the same time from one focus of an ellipse will converge simultaneously at the other focus.

one at each focus, can whisper intelligibly to each other, while people elsewhere in the room either cannot hear or cannot understand the words because of the low sound level at their location.

Thus far we have examined cases where the sound wave incident on a barrier was reflected. In reality this does not happen. Rather, when a sound wave is incident on a barrier between two media in which the velocity of sound is different, part of the wave will reflect off the boundary and part of the wave will be transmitted into the second medium. For example, when a sound wave reflects off a wall, part of the wave travels into the wall, possibly allowing someone on the other side to hear the sound. How much of the wave is reflected and how much passes into the second medium depends on the nature of the two media and the direction from which the wave was incident. More details on waves passing between two media will be discussed in the section on refraction.

Several interesting illustrations of reflection arise with respect to sound waves and ultrasonic waves, which in some navigational applications are referred to as sonar (sound navigation and ranging). Bats emit short bursts of ultrasonic waves (sound waves with frequencies above human sensitivity), listen for the reflections, and are thus able to locate objects or barriers. This procedure allows bats to fly safely in caves or dark attics.

With the exception of very low frequency radio waves, radio or radar waves are not useful for underwater applications, as they are rapidly absorbed by the water. Sound and ultrasonic waves, on the other hand, can be used to determine the locations of obstacles in water and, in a more sophisticated way, to determine the velocity of moving objects. Use of sonar to determine the velocity of a moving object will be discussed in the section on the Doppler effect. Other applications of sonar include location of large schools of fish by fishing trawlers and determination of the depth of the ocean by measuring the time for the reflected wave to return to its source on a surface ship. Dolphins possess an ultrasonic sonar system to aid them in navigation and to identify danger.

A more sophisticated application of sonar involves the location of various layers in the ground. When pressure waves from a small explosion pass downward into the earth, they will reflect off any boundary between different underground layers. By examining the amplitudes and arrival times of the returning pulses, various geological layers and their depths can be identified. Use of this technique, along with others, aids in the discovery of oil, natural gas, coal, and other minerals. A similar phenomenon occurs in the atmosphere, where marked differences in the speed of propagation of sound waves exist between the various atmospheric layers.

Refraction   It was indicated in the previous section that, when a wave is incident upon a boundary between two different media, some of the wave is reflected and some passes into the new medium. We consider now the portion of the wave that passes into the second medium. The passing of part of the wave from one medium into another is accompanied by a change of direction of the wave front, except when the wave approaches the boundary perpendicularly or along the normal

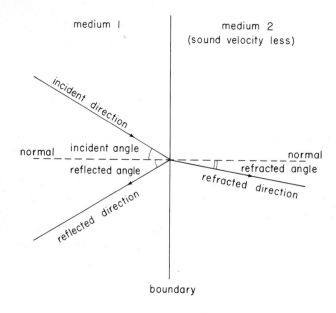

Figure 2-18 Relationship between various wave directions and the normal when a wave travels from one medium into another with smaller wave velocity.

to the boundary. The bending of a wave as it passes between two different media, called *refraction*, is illustrated in Fig. 2-18.

The nature of this bending can be understood by considering the following example. Suppose that a band is marching on a dry field, approaching the boundary of a muddy region at some angle of incidence (defined with respect to the normal to the boundary) as shown in Fig. 2-19. The positions of any rank after successive steps are indicated by the ×'s in the figure. If the same tempo (time interval for each step) is maintained, the steps must get smaller when the marcher steps into the mud, owing to the increased difficulty of marching. After each successive step in the mud, that column will fall farther behind those which enter the mud later. The band must always move perpendicular to the front rank; therefore, the band will change direction at the boundary.

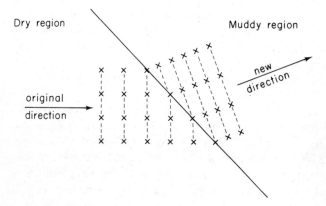

Figure 2-19 A marching band crosses the boundary from a dry region into a muddy region, maintaining the same tempo. A change in direction is necessary to keep the bank marching perpendicular to the front rank.

If the reason for the changing direction of the marching band seems confusing, we can clarify this by performing an analogous experiment using waves in a ripple tank. The concept of importance here is that the wave does in fact move in a direction perpendicular to its wave front. This situation is illustrated by a sequence of Huygens's wavelets in Fig. 2-20. After the wave hits the boundary, its velocity becomes smaller, and it goes a shorter distance in the same time interval, so the circular wavelets are smaller. As before, the line joining the extreme parts of the circular waves emitted at the same time becomes the new wave front. This process repeats itself to yield the resultant bending of the plane wave. The angle of bend is observed here to be toward the normal for the case where the wave moves into a medium with slower wave velocity. Conversely, the direction of propagation bends away from the normal when a wave enters a medium with faster wave velocity. You can understand this by considering the example worked backward in time. In that case, the wave begins in the slow medium, and, upon reaching the fast medium, one side of the wave front speeds up before the other. Thus the wave bends away from the normal.

In a ripple tank, different wave velocities are obtained by varying the depth of the water. The shallower the water, the slower the wave velocity because of friction between the water and the tank bottom. The same situation applies to water waves approaching the ocean shore, only in this case the depth changes continuously, creating a slow continuous change in the velocity of waves as they come in from the sea. Such a situation is illustrated in Fig. 2-21. As the wave approaches the shore, moving at some angle with respect to the shore, the part of the wave in the shallower water closer to the shore will move at a lower velocity. As a result, the section of the wave front more distant from the shore travels faster and will advance relative to the part of the wave closer to shore. The wave therefore bends toward the shore, as shown.

This change in wave speed with depth is also responsible for the breaking of ocean waves. As a large swell travels directly toward the shore, the front part is in water slightly shallower than the water that the rest of the wave is in. Therefore the

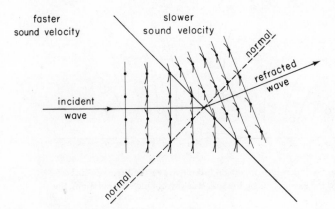

Figure 2-20 Refraction of waves, illustrated using Huygens's wavelets, analogous to the marching band example. The arcs show the leading edges of the Huygens's wavelets.

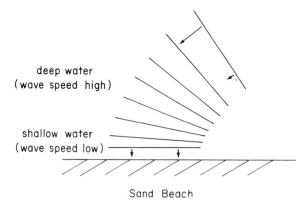

deep water
(wave speed high)

shallow water
(wave speed low)

Sand Beach

Figure 2-21 Waves approaching a beach will become parallel to the shore, because the wave velocity of the part of the wave closest to the shore is smaller than that of the part of the wave farthest from the shore.

back "catches up" to the front, until the wave shape is very steep. At a critical point, the wave breaks.

A similar effect occurs in the refraction of sound waves by the atmosphere each night and under certain weather conditions. Under normal daytime conditions, the sun heats the ground, which then radiates heat into the atmosphere; the temperature of the air therefore decreases as elevation above the surface of the earth increases. It is warmer near the ground. At nighttime, and during the day whenever a thick cloud cover exists, the sun does not heat the ground directly, and the opposite situation occurs. Hot air rises and cold air moves to the ground, resulting in a temperature that rises with elevation above the surface of the earth. This condition is called a *temperature inversion*, because the temperature gradient in the atmosphere is reversed from the normal daytime condition.

The velocity of sound is greater in warmer air, which is higher above the ground during an atmospheric temperature inversion. A condition therefore arises which is similar to that of water waves at a beach. Waves emitted by a sound source at even large upward angles are bent back to earth, owing to the change in wave speed, and thus increase significantly the distance at which sounds can be heard, as illustrated in Fig. 2-22. The change of medium here again is slow and con-

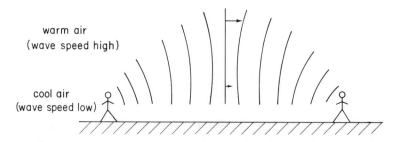

warm air
(wave speed high)

cool air
(wave speed low)

Figure 2-22 During a temperature inversion, sound waves emitted upward are refracted back toward the earth, causing the sound to travel much further along the ground.

tinuous. At night, voices carry a great deal farther than during the day, particularly over a placid lake, which enhances the temperature inversion; sound from highways or outdoor concerts can often be heard for miles. This is not a psychological effect or a result of lower background noise during the night hours, but a physical phenomenon due to the refraction of sound in the atmosphere.

Another refraction effect occurs when you try to speak either into or against the wind. If a steady wind is blowing, the air will drag against the ground; this air friction causes the wind velocity to be slightly lower near the ground than higher above the ground, as illustrated in Fig. 2–23. The wind changes the velocity of sound with respect to the ground; the sound wave moves through the air at the speed of sound in air, and the air itself is moving. As a result, the velocity of sound increases with elevation above the ground when the sound wave propagates in the direction in which the wind is blowing. A situation similar to that during a temperature inversion therefore occurs when you speak with the wind; sounds are refracted back to the earth and your voice "carries" with the wind. On the other hand, when you speak into the wind, the speed of sound decreases with elevation; sounds are therefore refracted upward and your voice is "lost." This effect is illustrated in Fig. 2–24.

Refraction of sound in the atmosphere is also responsible for there being a maximum distance over which thunder can be heard. During conditions that obtain near a thunderstorm, the usual temperature gradient in the atmosphere is found. The speed of sound is therefore greater near the surface of the earth where the warmer air is found, and waves will bend upward, as shown in Fig. 2–25. Most thunderheads are about 4 km high, and under the usual atmospheric conditions a

WIND →

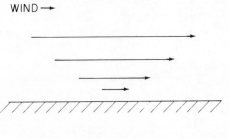

Figure 2-23 During a steady wind the air velocity is faster for points higher above the ground, due to friction of the air with the ground.

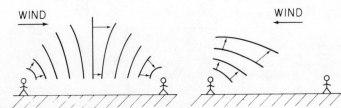

Figure 2-24 Refraction of a sound wave due to motion of the air. Speaking "with the wind" your voice carries, while speaking "into the wind" the sound of your voice is lost.

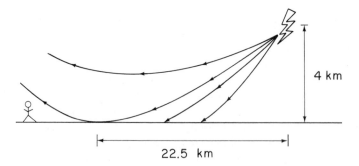

Figure 2-25 A person more than 22.5 km from a thunderhead cannot hear the thunder, because the sound is refracted back up into the atmosphere.

maximum range of about 22.5 km is attained before the sound is refracted back up-ward into the atmosphere. Thunder will not be heard beyond this range, in general.

Interference Interference has a special meaning in the physics of waves; it refers to the combining or addition of two similar waves. Interference can be either destructive, resulting in the effective disappearance of the waves, when they are out

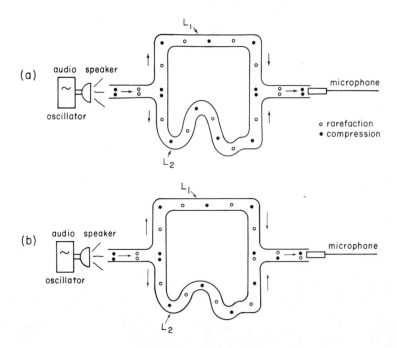

Figure 2-26 Quincke's interference tube. Part (a) shows the case where $L_2 - L_1 = \lambda$, so the waves are in phase at the microphone. Part (b) shows the case where $L_2 - L_1 = \lambda/2$, so the waves are out of phase at the microphone.

of phase, or it can be constructive, resulting in the enhancement, when they are in phase. Interference does not refer to one wave affecting the properties of another, which would be a violation of the principle of superposition. Instead, it refers to the addition of two waves at some point or points in space.

A demonstration of the interference of sound waves involves a Quincke's tube, which is shown in Fig. 2–26. An audio-frequency sine wave is fed into the end of a tube, which branches into two parts and then recombines, with the resultant wave picked up by a microphone. If the path length difference $L_2 - L_1$ is one wavelength, as shown in Fig. 2–27(a), the two waves will be in phase when they arrive at the microphone and the resulting sound will be loud. This also occurs if the path difference is zero or any integral number of full wavelengths. If, on the other hand, the path-length difference is one half-wavelength, the two signals will arrive at the microphone exactly out of phase, as shown in Fig. 2–27(b), canceling each other, and *nothing* will be heard. This will also occur whenever the path-length difference is any odd number of half-wavelengths. The two waves are out of phase if, whenever a compression from one path arrives at the recombination point, a rarefaction from the other path arrives simultaneously, or vice versa. Two identical waves arriving at the same point at the same time combine to produce the sum of their effects; when a compression and a rarefaction of equal amplitude arrive simultaneously, their effects cancel.

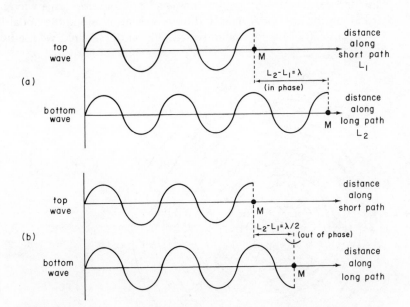

Figure 2-27 If $L_2 - L_1$ in the Quincke's tube is an even number of wavelengths, the signals interfere constructively at the microphone to produce a sound almost as intense as the original sound. If the path length difference is an odd number of half wavelengths, the two waves interfere destructively to produce almost no sound at the microphone.

A more sophisticated example is interference in two dimensions between two identical point sources; an example is shown in Fig. 2–28 for a ripple tank. In the sound analog, the lines represent compressions and the centers of the spaces between represent rarefactions. Consider the ripple-tank waves to be photographed at an instant of time in the positions shown in the drawing. Observe that there are lines called *antinodal lines*, labeled *A*, along which first peaks from both sources coincide, then troughs from both sources coincide, and so on, so that the motion is greatest along these lines. Between each pair of antinodal lines is a *nodal line*, where a peak from one source always coincides with a trough from the other, and there is no motion of the surface of the water. These lines are nearly straight at large distances from the sources. The two cases are illustrated in Fig. 2–29; the bottom curves show the sum of the waves along the antinodal and nodal lines at the particular instant of time chosen, beginning at the start of the antinodal and nodal lines, respectively, near the center.

As these waves are moving outward, they will "interfere" with each other in such a way that the nodal and antinodal lines will remain fixed. One period later the "picture" is identical to that drawn. The waves from the two sources propagate outward so that at any point along an antinodal line they are always in phase, producing a sum wave greater in amplitude than either individual wave. Conversely, at any point along a nodal line, the waves from the two sources will always be out of phase and cancel each other. In summary, you observe large waves moving outward along the antinodal lines and no waves at all along the nodal lines.

If, instead of the two point sources in the ripple tank, we use two speakers as

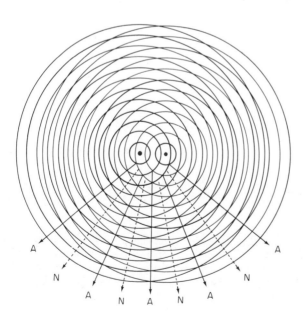

Figure 2-28 Interference between two identical point sources. *A* labels each antinodal line; *N* labels each nodal line.

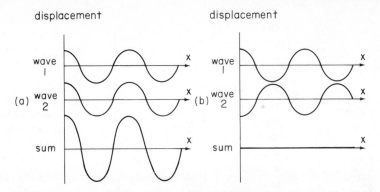

Figure 2-29 Sum of the two waves along (a) antinodal lines, and (b) nodal lines, for the interference of two point sources as shown in Fig. 2-28. The coordinate *x* denotes the distance from the point midway between the two sources.

sources of sinusoidal sound waves, the nodal and antinodal lines can be found by listening at various places around the speakers. The sound will be loud along antinodal lines and soft along nodal lines. This experiment works best out-of-doors, where there are no reflections from walls and ceiling.

The source separation and wavelength of the sound affect the pattern of nodal and antinodal lines. This can be demonstrated visually with a moiré pattern produced by superposing two sets of concentric circles with equal radius spacing that are printed on clear plastic. As the centers are separated, the nodal and antinodal lines get closer together; as the centers are brought together, fewer nodal lines appear, and they are separated by larger angles. This is analogous to changing the distance between the two sources; the closer the sources, the greater is the angular separation between nodal and antinodal lines, and vice versa.

Keeping the source separation constant, we can investigate the effect of changing wavelength. Using concentric circles with greater radial spacing, the nodal and antinodal lines spread apart. When concentric circles with small radial spacing are used, the angles between the nodal and antinodal lines decrease. Analogously, increasing the wavelength of the sound (decreasing the frequency) spreads out the pattern of nodal and antinodal lines, while decreasing the wavelength decreases the angles between the lines.

If the two sources are out of phase, rather than in phase, the nodal lines and the antinodal lines are reversed from the pattern just described. This is of considerable practical consequence in the wiring of stereo loudspeakers. With the speakers properly wired, for bass notes (long wavelength) there should be a very wide antinodal region covering the entire area in front of the speakers. If the two speakers are wired out of phase, the area in front of the speakers will become a nodal region where the low-frequency waves from the two speakers will interfere destructively. Figure 2-30 shows an experimental method for determining if you

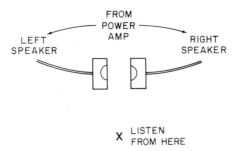

Figure 2-30 Setup for determining the proper phase relation between stereo loudspeakers. System should be set to MONO on preamp controls.

have wired your speakers in phase or out of phase. Setting the preamplifier to the monaural mode, music with a loud bass sound should be played with the speakers facing each other a few inches apart, a distance much shorter than the wavelength associated with bass frequencies. First use the normal wiring; then reverse the wires on one (only one) of the speakers. If the speakers are in phase, the bass signals from both speakers add to give a good bass sound. If they are out of phase, the bass signals will tend to cancel each other, resulting in inadequate bass intensity. The proper phase of the two is that which gives the louder bass sound (speakers in phase).

Diffraction   We have studied refraction, the bending of waves due to a change of the wave velocity. Waves do not necessarily need a change of medium to cause them to bend; bending in the absence of a change of medium is called *diffraction*. Sounds can be heard around corners or behind barriers that cut off direct view of the source of sound. This bending of waves around corners in the same medium is an example of diffraction, and it too arises from Huygens's principle.

In reality, the drawing of the plane wave in Fig. 2-8(b) is an oversimplification. While it is accurate for the middle of the wave, near the edges the Huygens's wavelets spread out in all directions and cause the wave to develop curved edges and spread out, as shown in Fig. 2-31. This, too, is an example of diffraction. The amount of diffraction will depend upon the nature of the waves and their wavelengths. For example, you can hear around corners, but you cannot see around corners, because sound waves diffract more than light waves. Low-

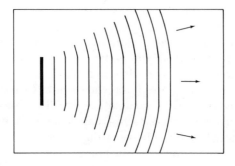

Figure 2-31 Diffraction at the edges of a plane wave.

frequency sound waves will spread out over a larger angle than high-frequency waves. This is one reason why voices sound muffled, or reduced in high-frequency content, when heard around corners or in adjacent rooms. Also, we hear the drums and tubas of a marching band as louder than the other instruments if the band is around the corner of a building. This is not because the drums and tubas are playing louder than the other instruments; it is because their low-frequency waves diffract around the corner more easily than the higher-frequency waves from the flutes, for instance. Likewise, sounds from directional speakers take on a different quality when heard from other than their preferred direction.

An interesting experiment demonstrating the diffraction and interference of sound waves involves a small unmounted loudspeaker and a thin square board about 60 cm on each side, serving as a baffle. A circular hole the size of the speaker is cut in the center of the board. If music is played with the speaker alone, the sound lacks volume and has very little bass. If the speaker is placed behind the board, facing out through the hole, the sound heard by a listener in front of the board is now much louder than before the board was present, particularly in the low frequencies. Why is this so?

As the diaphragm of the speaker moves rapidly forward and backward, it creates alternating compressions and rarefactions moving in the forward direction. The signal emitted from the rear of the speaker is exactly out of phase with the signal emitted from the front, because when the speaker cone moves forward it will create a compression in front and a rarefaction behind, and vice versa. If the baffle is not present, the waves from the rear will diffract around the edge of the speaker and travel away from the front of the speaker. The total signal from the speaker is the sum of the two waves, those from the front and those from the rear. Because they are out of phase, they interfere destructively and tend to cancel; the sound heard is thus reduced in amplitude. With the board present, the cancellation process is largely eliminated for two reasons. First, owing to the geometry, it is harder for the waves emitted by the back of the speaker to diffract all the way around to the front, and, second, owing to the extra distance traveled by the wave emitted from the back, it will not be exactly out of phase with the wave emitted from the front (in general) and total destructive interference will not occur. Therefore, the sound is louder when the baffle is in place. Because the long wavelengths (low frequencies) diffract more readily than the short wavelengths, the greatest effect is in the bass range.

The board in this experiment is a type of loudspeaker baffle. We shall study more about the effects of baffles and enclosures on speakers in Chap. 7.

Polarization Whereas all the properties discussed previously apply to both longitudinal and transverse waves, polarization is a physical property unique to transverse waves. It is therefore not directly applicable to the study of sound waves in air, but we include it for completeness and because it is applicable to the study of the coupling of sound waves from the string to the sounding board of a piano. A simple transverse wave always has a single plane in which the wave is vibrating; for example, the wave of Fig. 2–2 is oscillating in the plane of the paper. This is known as its *plane of polarization*. Another possible plane of polarization for a rope wave like that shown in Fig. 2–2 would be into and out of the paper. In either case, the plane of polarization would be determined by the direction of the oscillations forming the wave. If we cause all the waves in a beam of light, or some other collection of waves, to vibrate in a single plane or direction, the waves are said to be *polarized*. Polaroid sunglasses filter out light polarized in the horizontal direction. Because glare off horizontal surfaces, such as water and streets, is mostly horizontally polarized, Polaroid sunglasses remove most of this glare.

2.3   *Addition of Waves*

In our study of waves and wave motion, we shall regularly find it necessary to add two waves together graphically to obtain a sum wave. If the graphs represent the waves themselves (say displacement or velocity versus position along the wave), we say that we are adding the waves together when we add their graphs. If the graphs represent motion at a fixed position in the medium (say displacement versus time), we say that we are adding the motions that the graphs represent. Sometimes these two types of wave addition will look the same except for the labels on the two axes of the graph. We shall refer to either of these two types of graphical addition as the addition of waves.

Before discussing the addition of wave trains and waves of different periodicity, we shall examine how individual pulses add in the case of transverse rope waves. Consider two equal pulses, each consisting of one half-wavelength, that come together in a single rope as shown in Fig. 2–32. The succession of five pictures is like snapshots taken at equal time intervals of one half-period showing how the rope would appear as the pulses approach and pass each other. Here, of course, we are graphing displacement versus position. For each instant of time, the two individual pulses are shown on the first and second lines, with the sum of the two shown on the third line. The total displacement of the rope, its overall shape, is the sum of the displacements of the component waves at that point. Notice that the superposition of two identical pulses *in phase* with each other (both positive or both negative) produces a single pulse of twice the amplitude, whereas the superposition of two pulses *out of phase* with each other (one positive and one negative) produces a net effect of zero; their effects cancel.

Now let us consider two full wavelengths of identical sine waves passing each other as shown in Fig. 2–33. The time interval between successive pictures is one quarter-period, so each wave moves one quarter-wavelength between pictures. When two positive or two negative pulses are superimposed, the resultant is a pulse with amplitude equal to the sum of the two individual pulses; when equal but opposite pulses are superimposed, the resultant is zero.

The preceding discussion is consistent with the principle of superposition as

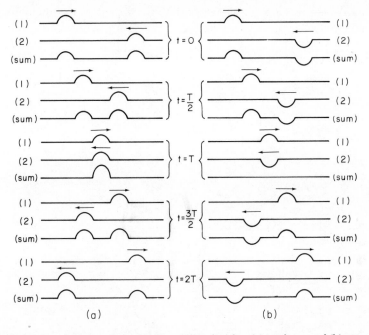

Figure 2–32 Addition of equal half-wavelength pulses (a) in phase, and (b) out of phase, as they pass each other on a rope, shown at equal time intervals of one half-period. At each point in time the individual pulses are shown, along with the overall shape of the rope due to the addition of the two pulses.

discussed in Sec. 2.2, and the presence of one pulse does not affect the other. If the two pulses affected each other, the shape of each pulse would not be the same before and after their interaction, or they would not add in the manner described here.

Suppose that we want to add the two waves shown in Fig. 2–34(a). The value of the sum wave at any given time is just the algebraic sum (positive above the horizontal axis and negative below the horizontal axis) of the two individual waves at that time. For instance, in Fig. 2–34(a) at $t = 0$ ms, both waves 1 and 2 have a value of zero. The sum wave will therefore have a value of zero. Similarly, the sum wave has a value of zero at times $t = 2$ ms and $t = 4$ ms. Now consider the situation at time $t = 1$ ms. Each component wave has a value of 1 volt (V), so the sum wave will have a value of 2 V. Similarly, the value of the sum wave at time $t = 3$ ms is $- 2$ V. In fact, since the two waves have exactly the same value at all points in time (they are identical waves), the value of the sum wave is exactly twice the value of either of the component waves, as shown in Fig. 2–34(a). If the two waves are out of phase, as shown in Fig. 2–34(b), the sum of their two values at any point in time is zero, and the sum wave is zero at all times.

A more complex situation obtains when the two component waves have dif-

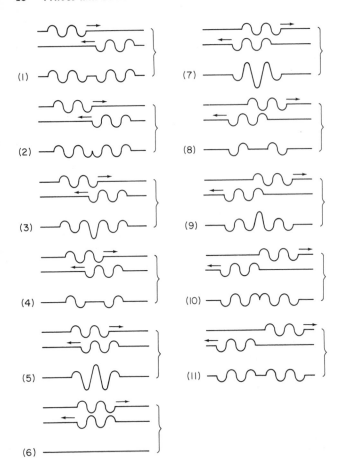

Figure 2-33 Two wavelengths of identical waves passing each other in a rope, shown at one quarter-period intervals. Each wave is shown individually, along with the resultant sum wave, which represents the overall shape of the rope.

ferent periodicity, as illustrated in Fig. 2–35, although the principle and procedure by which the waves are added remain the same. As before, the procedure to be used in the addition of these waves is to add the two component waves, point for point. Since the two waves are smooth and continuous, it is only necessary to add the waves at a limited number of points. A good rule of thumb is to choose points spaced in time at one quarter of one period of the wave with the shortest period, as shown in Fig. 2–35; in this case that is 1 ms. This spacing ensures that all the crests and troughs of the resulting wave will be seen. After adding the two component waves at each of the chosen points, the sum wave points are plotted, and a continuous smooth curve is drawn through the points. Table 2–1 shows the addition of the two component waves of Fig. 2–35. For example, at time $t = 1$ ms, the value of wave 1 is $+1.4$ and the value of wave 2 is $+1.0$; therefore, the sum wave has a value of $+2.4$. You should check the addition and look carefully at how the sum wave curve has been drawn through the points.

The addition of waves is extremely important and will be used in our study of

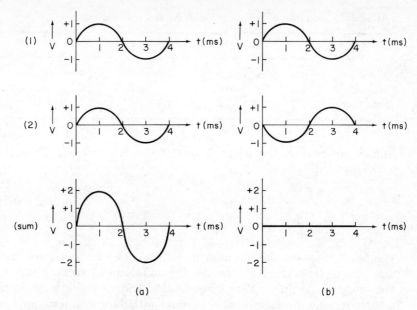

Figure 2-34 Addition of two waves of the same amplitude and frequency (a) in phase and (b) out of phase.

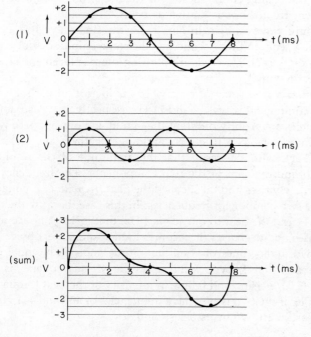

Figure 2-35 Addition of two waves of different amplitude and frequency.

TABLE 2-1   **Addition of waves of Fig. 2–35 at 1-ms intervals**

t(ms)	0	1	2	3	4	5	6	7	8
Wave 1	0.0	1.4	2.0	1.4	0.0	−1.4	−2.0	−1.4	0.0
Wave 2	0.0	1.0	0.0	−1.0	0.0	1.0	0.0	−1.0	0.0
Sum wave	0.0	2.4	2.0	0.4	0.0	−0.4	−2.0	−2.4	0.0

beats in the next section and the study of standing waves in Chap. 3 and complex waves in Chap. 4.

2.4   Beats

We have discussed the addition of two waves with the same frequencies and amplitudes. Now we shall consider the addition of two pure waves of the same amplitude, but slightly different frequencies. Those who have played instruments in a band or orchestra are probably familiar with the sound resulting when two identical instruments play the same note slightly out of tune with each other, a sum note that wobbles in loudness, where the frequency of the wobbling decreases as the two notes become better in tune. This wobbling can be used by musicians to tune their instruments; when the wobbling disappears, the instruments are in tune. The physical phenomenon of the change in amplitude, or loudness, of the sum of two tones with slightly different frequencies is known as *beats*.

Consider the two pure waves drawn in Fig. 2–36; the first wave has a frequency of 9 Hz (period of $\frac{1}{9}$ sec), and the second has the same amplitude but a frequency of 11 Hz (period of $\frac{1}{11}$ sec). Here we are plotting the displacement of one physical point with respect to time.

The resultant wave has a frequency of exactly 10 Hz (period of 0.1 sec), the average value of the two component waves (9 and 11 Hz, respectively). Also notice that because their frequencies are slightly different they go in and out of phase with each other, alternately interfering constructively (adding together) and destructively (canceling each other), with the result that the amplitude of the resultant wave varies between zero and the sum of the two components. The frequency at which the amplitude alternately rises and falls, called the *beat frequency*, is equal to the difference between the two component frequencies. In this case, because the two components are at 9 and 11 Hz, the beat frequency is 2 Hz; that is, the amplitude of the resultant wave increases and decreases (it becomes louder and softer) twice per second. In general, for two audio tones of equal amplitude and nearly equal frequencies of f_1 and f_2 (f_2 greater than f_1), you will hear a tone whose frequency is the average value $F_{\text{tone}} = (f_1 + f_2)/2$ rising and falling in intensity at the beat frequency $f_{\text{beats}} = f_2 - f_1$. For example, if the two waves have frequencies of 439 and 441 Hz, a tone of 440 Hz will be heard with two beats per second.

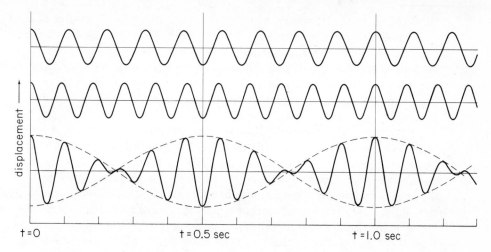

Figure 2-36 The addition of two waves of equal amplitude but slightly different frequencies results in a wave whose frequency is the average value of the two original frequencies and whose amplitude varies between zero and the sum of the amplitudes of the original waves. This phenomenon is called beats. The envelope is indicated by the dashed lines.

The *envelope* of the resultant wave, the curve showing the resultant amplitude as time passes, is indicated by the dashed lines in Fig. 2-36. The envelope is sinusoidal in shape whenever the two original waves are sinusoidal. The full envelope consists of two sine waves of frequency $\frac{1}{2} f_{\text{beats}}$, out of phase with each other.

2.5 Doppler Effect

Virtually everyone has experienced the rising and falling of the pitch of a train whistle as the train passes by a railroad crossing. This change of pitch, occurring whenever there is relative motion between the source and observer, is an example of the Doppler effect, named after the Austrian physicist Christian Johann Doppler (1803–1853). The effect becomes clearly audible when the relative velocity between source and observer is at least a few percent of the velocity of sound (1100 ft/sec or 750 miles/hour). We therefore observe the Doppler effect for sound whenever there is relative motion with speeds above about 15 to 20 miles/hour. We do not normally observe the Doppler effect for light, which would cause everything to change color, since the velocity with which we travel never approaches a few percent of the velocity of light (186,000 miles/sec). Relative motion between our galaxy and other galaxies is of this magnitude, and the Doppler effect for light plays an important role in the study of the relative motions of galaxies.

As a boat moves through water, it creates waves; one can observe these waves from shore, from a standing boat, or from a moving boat. In any case, it seems reasonable that once the wave is created it moves over the water at a speed that is independent of any motion of the source or of the observer. That is, the wave velocity is dependent only upon the medium through which it is traveling, not upon the motion of either source or observer. Likewise, all sound waves travel through the air at the speed of sound in air, independent of the speed of the source or observer.

We can study the propagation of waves and the Doppler effect using a ripple tank. We have seen how a stationary point source produces circular waves in the ripple tank as in Fig. 2–8(a). Now we shall study the waves emitted from such a source moving at constant velocity. Because the source is moving at constant velocity, it moves the same distance between the times at which it emits waves; the dots in Fig. 2–37 mark the points at which the source emits each successive circular wave. Figure 2–37 also shows the circular wave fronts. The only difference between Figs. 2–37 and 2–8(a) is that, as time has passed, the center of each successive (smaller) wave has moved to the right by the amount shown in Fig. 2–37, causing the circles to squeeze together on the right and spread apart on the left. The velocity of all the waves is the same with respect to the surface of the water and independent of the source velocity. A stationary observer counting passing waves in a specified time interval would count a larger number if he or she were in front of the source (at the right) and a lower number if behind the source (at the left). Thus, the frequency of the wave received by the observer appears higher where the motion of the source is toward the observer and lower where the motion of the source is away from the observer. Alternatively, the wavelength is shorter in the forward direction, resulting in a higher frequency or pitch, and longer in the backward direction, resulting in a lower frequency or pitch. This is consistent with the relation $S = f\lambda$ studied in Sec. 2.1. All sound waves travel at the same speed, so a high-frequency f implies short-wavelength λ, and vice versa. Notice also that in the direction perpendicular to the direction of travel, as the source passes directly by the observer, the spacing of the circular wave fronts is approximately the same as when there is no motion of the source. The frequency measured by a stationary observer as a source

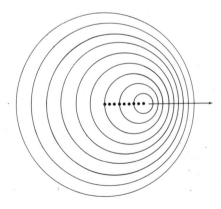

Figure 2–37   The dots show the location of a moving wave source at equal time intervals for a source moving to the right with a constant velocity. The circular wave fronts emitted by the source at each of these points are shown.

moves by is thus higher than normal when the source is moving toward the observer, the same as the normal frequency just as the source moves past the observer, and lower than normal when the source is moving away from the observer. By normal frequency we mean the frequency received when both source and observer are at rest.

Let us now study in more detail why the frequency appears higher when the source is moving toward the observer. Consider first a source emitting a series of equally spaced compressional waves toward an observer, as shown in Fig. 2–38. The compressions move uniformly through the air with the speed of sound and

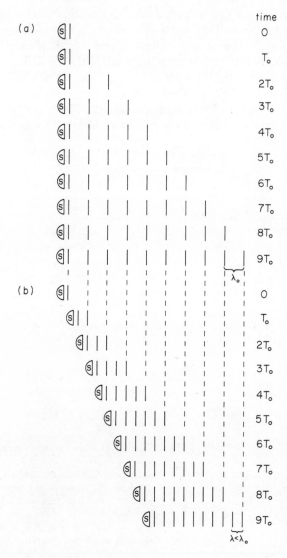

Figure 2–38 Part (a) shows compressions of frequency f_o emitted by the source and heard by the observer. Part (b) shows the sequence of compressions emitted at the same time intervals, but with motion of the source toward the observer with velocity $v_s = S/2$, resulting in a higher observed frequency.

create a constant frequency tone for the observer; the source emits the frequency f_0 and the observer hears the normal frequency f_0. Now suppose that the source is moving with a constant velocity of one-half the speed of sound to the right, while emitting compressions at the same frequency f_0, as shown in Fig. 2–38(b). Each successive compression is emitted closer to the previous one because of the motion of the source toward the observer, and the wave propagating through the air has a shorter wavelength, and thus a higher frequency (two times the normal frequency), when heard by the observer.

Now consider the case where the source is moving at a constant velocity of one-half the speed of sound away from the observer as illustrated in Fig. 2–39. We can work out the effect of the motion, as previously, to find that in this case the observer measures a frequency lower than f_0 (two-thirds of the normal frequency).

Next consider a stationary source and a moving observer as shown in Fig. 2–40. The velocity of the observer for this example is equal to the speed of sound so that, after one compression reaches the observer, it will take exactly one half-period for the next compression to be felt. During this time interval, both the observer and the next wave front will have moved one half-wavelength. Therefore, this particular velocity for the observer doubles the observed frequency. This can be read

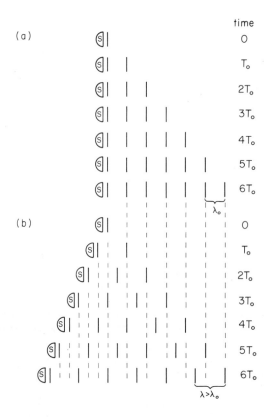

Figure 2–39   Part (a) shows compressions of frequency f_o emitted by the source and heard by the observer. Part (b) shows the sequence of compressions emitted at the same time intervals, but with motion of the source away from the observer with velocity $v_s = S/2$, resulting in a lower observed frequency.

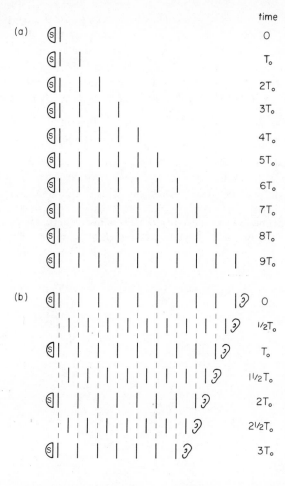

time

(a)

0

T_0

$2T_0$

$3T_0$

$4T_0$

$5T_0$

$6T_0$

$7T_0$

$8T_0$

$9T_0$

(b)

0

$1/2T_0$

T_0

$11/2T_0$

$2T_0$

$21/2T_0$

$3T_0$

Figure 2-40 Motion of the observer toward the source increases the observed frequency above the normal frequency f_0. Here the velocity of the observer (ear) is equal to the speed of sound, and the sequence of "pictures" of the wave in part (b) is taken at a rate of two per period.

from Fig. 2-41. Motion of the observer away from the source can be analyzed similarly.

The frequency f heard by the observer, as a function of source or observer velocity, is shown in Fig. 2-41 for a moving observer and Fig. 2-42 for a moving source. In both figures and in the equations shown, v_o and v_s, the observer and source velocities, respectively, are positive if the motion of the observer or source is toward the other and negative if the relative motion separates the source and observer. Figure 2-41 tells us that the faster the observer moves toward the source the higher the pitch heard, and the faster the observer moves away from the source the lower the pitch heard. Notice that if the observer moves away from the source faster than the speed of sound ($v_o = -S$) he or she outdistances the oncoming wave and therefore hears nothing ($f = 0$). As in the case of motion of the observer, relative motion of the source toward the observer increases the observed frequency, whereas relative motion away from the observer decreases the observed

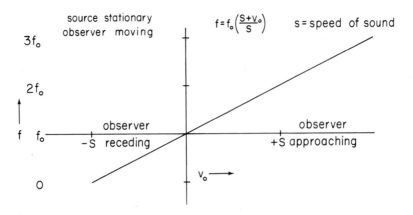

Figure 2–41 Frequency f heard by an observer moving with respect to the source with a velocity v_o listening to a tone of normal frequency f_o produced by a stationary source.

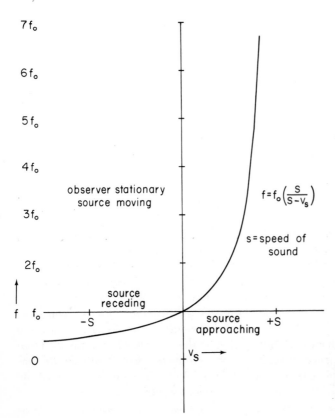

Figure 2–42 Frequency f heard by a stationary observer listening to a tone of frequency f_o produced by a source moving with velocity v_s.

frequency, as shown in Fig. 2–42. Notice that no matter how fast the source recedes from the observer the observer will still hear something (perhaps a very low frequency), but when the source approaches the observer at the speed of sound, *all* the waves produced by the source arrive at the observer at nearly the same time. This creates an extemely loud sound, since a large number of wave fronts are adding together at the same time and place. This phenomenon, called a *sonic boom*, will be discussed in the next section.

The Doppler effect can be used to determine the speed of underwater objects using sonar much as the same effect is used to determine the speed of cars on a highway with radar waves. A tone of frequency f_0 is emitted in the direction of an underwater object; the waves bounce off the object as shown in Fig. 2–43(a) if the object is stationary and as shown in Fig. 2–43(b) if the object is moving toward the source with a velocity of one-third of the velocity of sound in water. For both

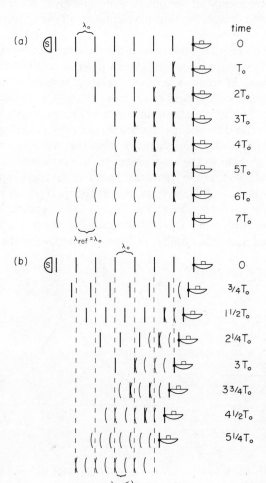

Figure 2–43 Underwater sound waves reflecting off a submarine at rest, as in part (a), return to the source with the same frequency. When the submarine is moving toward the source (here the submarine has a velocity of $S/3$), the frequency of the reflected wave is increased, as shown in part (b). Note that straight line wave fronts are moving to the right; curved wave fronts are moving to the left after having been reflected.

the source and object at rest, if the successive wave compressions are shown every full period, it is clear that they bounce off the object and return to the source with the same wavelength and, therefore, with the same frequency with which they started. For a moving object, the successive wave fronts are shown at time intervals of three-quarters of the initial period to simplify the geometry. If the object is moving toward the source with one-third of the velocity of sound in water, then in three-quarters of the initial period T_0 the wave will have gone three-fourths of one wavelength and the object one-quarter of one wavelength. At this time, the next wave compression will be reflected off the object. This sequence repeats itself as shown in the figure. Thus the reflected wavelength is less than the original wavelength, or, equivalently, the frequency of the reflected signal is higher than that of the original signal. In the particular case shown here, the wavelength is three-quarters of its initial value, which means that the frequency of the reflected wave is $\frac{4}{3}$ times the initial frequency. If the object is moving away from the source, a similar geometrical construction shows that the frequency of the reflected wave is decreased (or its wavelength increased).

It is important to draw specific attention to the difference between intensity (or amplitude) and frequency when discussing the Doppler effect. The Doppler effect deals only with changes in frequency due to relative motion of source and observer. Changes in intensity or loudness result from changes in the distance between source and observer (which clearly occur whenever there is relative motion between them) and are governed by the inverse square law. As an example to help distinguish between frequency and intensity effects, both the frequency and intensity of a train whistle that we hear as the train passes us at a crossing are plotted in Fig. 2–44. Figure 2–44(a) tells us that the frequency heard as the train approaches is higher than the normal frequency of the whistle. Just as the train passes by, at time t_x, the observed frequency is f_0. After time t_x, the observed frequency is lower than the normal frequency. While the train is directly approaching or moving away from us, its frequency is constant and either higher or lower, respectively, than the nor-

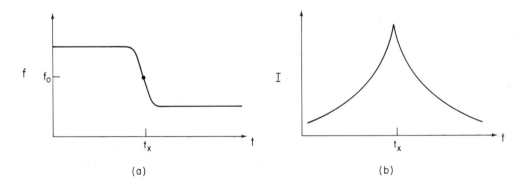

(a)   (b)

Figure 2–44 Frequency and intensity of the whistle of a train as the train passes a stationary observer; at a time t_x the train is closest to the observer.

mal frequency. This is because the Doppler effect is dependent only on the relative velocity; a constant relative velocity implies a constant frequency shift. Figure 2–44(b) tells us that as the train approaches its sound gets louder (according to the inverse square law), until it reaches its loudest level when it is crossing right in front of us. It then becomes softer as the distance between us and the train becomes greater. It is important to distinguish carefully frequency and intensity effects and to be aware of their origins.

2.6   Shock Waves and Sonic Booms

In this book we deal almost entirely with periodic continuous waves; however, for completeness we shall now discuss an example of nonperiodic waves that propagate like sound waves. A sound wave, as we have seen, is a disturbance of the air that propagates through the air as the molecules of the air effectively "bump" each other in a series of molecular collisions. In an explosion, say of a bomb or a firecracker, the air around the explosion is pushed out very rapidly by the force of the explosion. The *shock wave* produced is a single large *compression* that moves away from the explosion at the speed of sound. It is like a compression in a sound wave except that it is not periodic.

Another example of such a shock wave is the "sonic boom." Figure 2–45 shows the sequence of circular waves produced in a ripple tank by a source moving faster than the wave speed. The result of this high source speed is that the source out-distances all the circular waves it produced earlier. If the sequence of circles representing outgoing waves from the source is made very close together, the crests from adjacent groups of circular waves join together along lines on either side of the source, forming a V-shape, as shown in Fig. 2–46. A common example of this is the wake of a motorboat. If the boat were to move very slowly through the water, it would produce a series of circular waves as in Fig. 2–37. But since most boats move faster than the velocity of water waves, the characteristic V-shape results.

From a jet plane moving at a speed greater than the speed of sound, all the sound at a given instant of time (the roar of the motors, swish of air, and the like) is concentrated in a spherical wave that emanates from the plane at that time. The

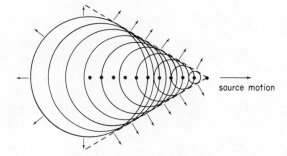

source motion

Figure 2–45 Circular waves from the sequence of points shown, where the source velocity exceeds that of the emitted wave. The dashed lines indicate the leading edge of the combined wave front.

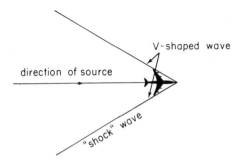

Figure 2-46 When the source velocity exceeds that of the speed of sound, the leading edges of the circular waves from the source combine to form a shock wave. The angle of the V decreases as the source velocity increases.

spherical wave fronts add together to form a cone-shaped surface called the *Mach wedge*, where the sounds originating during a longer time interval come together, creating a shock wave of large amplitude. This conical wave is a nonperiodic shock wave that can be very intense. In the case of a plane moving at a speed greater than the speed of sound, the effect is known as a *sonic boom*. The sonic boom follows the plane around, and does not occur only at the moment when the plane's speed becomes greater than the speed of sound, as is often believed. For this reason, it can be annoying or even dangerous to have aircraft continually flying at supersonic speeds over densely populated areas. By observing the circles in Fig. 2-45, we can see that, while the sound is "concentrated" along the edges of the cone creating an especially loud boom, sound from the plane continues to be heard at all points inside the cone. When a supersonic plane passes overhead, you therefore hear a large boom when the cone passes you, followed by a low-pitched Doppler-shifted roar.

In addition to explosions (such as bombs and firecrackers) and the sound from supersonic aircraft, nature has provided us with an excellent example of the sonic boom—thunder. When a lightning discharge occurs, the air is heated to extremely high temperatures and expands rapidly. In fact, this expansion may occur at a speed greater than the speed of sound in air and might therefore produce a shock wave or sonic boom. An electrical spark heats the air the same way. The rumbling of thunder results when the sound from a long lightning bolt (sometimes over 10 miles long) arrives at the observer over a long time interval, owing to the varying distance between the lightning bolt and observer. The duration of the thunder can be related to the length of the lightning path when the path is directed radially away from the observer, whereas a sharp shock is produced by a bolt that goes perpendicular to the line of sight. The crack of a cowboy's whip is generally believed to be a small sonic boom caused by the tip as it travels faster than the speed of sound.

2.7   Ultrasonics

Ultrasonics is the study of waves whose frequencies lie above the audible range, typically between 20 kHz and several megahertz (MHz). Ultrasonic waves are similar to sound waves in all respects except frequency.

Ultrasonic waves, like normal sound waves, reflect at any change in medium. Thus, they can be used to probe human internal organs instead of potentially more dangerous X-rays. Many types of bone and tissue structure can be observed using ultrasonic scanning, and a new area of medicine is developing to exploit such techniques.

Ultrasonic waves are used in conjunction with the principle of the Doppler effect in an important medical application, the ultrasonic fetal stethoscope. The dangers of X-rays have been discussed extensively in recent years; these dangers are particularly acute when a fetus in the first trimester of pregnancy is exposed to any type of high-energy electromagnetic radiation. Yet there are times when it is important to "look" into the uterus to obtain information concerning the growth of the fetus. This probing can often be done using ultrasonic waves, which, unlike X-radiation, do not damage human tissue, even in sensitive stages of embryonic development. The ultrasonic fetal stethoscope uses ultrasonic waves to observe the fetal heartbeat. Ultrasonic waves are passed through the mother's midsection, and reflections occur at any point where the type of tissue changes significantly. If there is a fetal heartbeat, the heart is continually expanding and contracting, thus changing its motion relative to the sound source and all the other organs internal to the mother and the fetus. Therefore, the frequency of the ultrasonic wave reflecting off the fetal heart will increase and decrease at the periodicity of the fetal heartbeat with some pattern characteristic of the heart motion. Not only is it possible to monitor fetal heart motion safely and effectively, but it is also possible to predict multiple births at a very early stage, long before the heart beat can be heard using a regular stethoscope.

Some types of burglar alarms use ultrasound in a similar way to detect motion in a store or warehouse at night. Any moving object will reflect waves from the ultrasonic source to create Doppler-shifted waves and waves of varying intensity, which are picked up by sensors that sound an alarm. Sometimes sensitive people can hear the sound (around 20 kHz) from such an alarm system if it is inadvertently left on during working hours. Because this sound can be extremely annoying or even painful, businesses are encouraged to turn these devices off when people are present, and in some states it is illegal to expose people to this sound. Ultrasonic burglar alarms for homes are also available.

The ability of ultrasound to be annoying can be applied for the extermination of rodents and insects. Very loud ultrasonic sources, when placed in a building and left on, will drive rodents away and will kill cockroaches. The intense ultrasound apparently disorients the insects, causing them to die from induced erratic behavior.

Ultrasonic waves can also be used to clean small objects such as watches and electronic components. The object is placed in a bath of water or some solvent, and an ultrasonic source vibrates the bath at a frequency typically between 20 and 60 kHz. Even firmly attached dirt can be loosened and removed in this way. Ultrasonic cleaning can be more efficient than mechanical or chemical means, because the waves can penetrate to otherwise inaccessible points.

EXERCISES

1. The frequency limits of the range of human hearing are from about 20 Hz to 20 kHz. The speed of sound is about 34,500 cm/sec. What are the wavelengths of these waves in cm? In m?

2. Draw a graph of a square wave of amplitude $A = 3$ V and period $T = 2$ ms. What is its frequency? If the speed of sound in some metal is 300,000 cm/sec, what is the wavelength of this wave in that metal?

3. In general, which would have a longer wavelength, a note from a violin or from a tuba? A piano and a flute each play 256-Hz tones. Which wave has the longer wavelength? How might the waves differ?

4. Compare and contrast transverse and longitudinal waves. Give examples of each. What particular type of wave does not need a medium? Describe the bell-in-vacuum experiment. What conclusions can be drawn from it?

5. Carefully draw and add together the following pairs of waves: (a) sine waves of equal amplitude with frequencies f and $2f$ beginning in phase; (b) similar waves beginning out of phase; (c) a wave of frequency f and amplitude A with a wave of frequency $3f$ and amplitude $A/3$ beginning in phase; (d) waves similar to those in part (c) except that they start out of phase.

6. Two sinusoidal tones of equal amplitude and with frequencies of 439 and 443 Hz are sounded simultaneously. Name and describe in detail (qualitatively and quantitatively) what you hear. Why is it important that the two waves have equal amplitudes for this effect to occur?

7. Describe qualitatively what happens to the sound heard by an observer moving with a constant speed past a stationary, steady source of sound, such as a car horn. What is the name of this effect?

8. (a) A jet plane flies overhead at twice the speed of sound. Describe, with the aid of a drawing, what you hear at the gound level.
   (b) Another jet flies overhead at four times the speed of sound. Draw its shock wave. Be sure to make clear any distinction between this shock wave and the one drawn in part (a). In which case does the shock wave travel faster?

9. If a jet flies directly away from you faster than the speed of sound, you can still hear the roar of its engines. But if you move away from a loud stationary jet at a speed greater than the speed of sound, you *cannot* hear the jet. Why?

10. Why can you hear things around corners? All other things being equal, which can we hear more readily around corners, French horns or flutes? Why?

11. Describe the Quincke's interference tube experiment. What principle is demonstrated?

3

Standing Waves
and the Overtone Series

We have studied the general properties of waves and some of the laws of physics that determine their behavior. We shall now study the properties of waves that are confined to a one-dimensional medium, such as transverse waves on a stretched string and sound waves in a narrow tube filled with air. After a discussion of how these waves behave at the end of such a one-dimensional medium and how waves add together to form standing waves, we shall see the special relationships among the waves that can exist in such a medium and how this naturally leads to the overtone series. This chapter will then conclude with a study of more complex and two-dimensional standing waves.

3.1 Transverse Standing Waves

A very special result occurs when two identical waves moving in opposite directions are added. Figure 3–1 shows segments of two such waves, along with the resultant sum wave, viewed at time intervals of one eighth-period. The figures represent a sequence of photographs at time intervals of one eighth-period showing a region two wavelengths wide. At time $t = 0$, when the two waves are in phase (that is, lined up peak to peak and trough to trough), the resultant is a wave of twice the original amplitude, as in (1) of Fig. 3–1, but with the same wavelength. One quarter-period later, at time $t = T/4$, each of the waves has moved one quarter-wavelength, and they are exactly out of phase, producing zero displace-

63

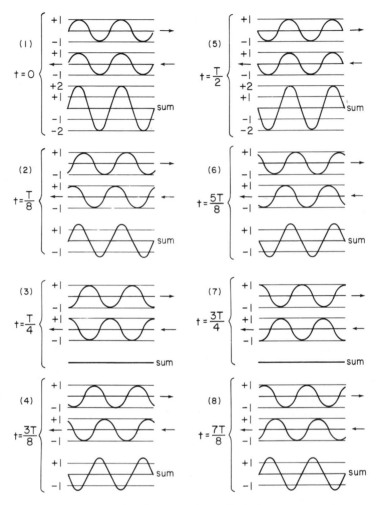

Figure 3–1   A two-wavelength section of two identical waves moving in opposite directions along a rope, at intervals of one eighth-period, showing the individual waves and the sum, the overall shape of the rope.

ment at all points, as shown in (3). One quarter-period after this, at time $t = T/2$, the two original waves are again in phase with each other, but the resultant wave, as shown in (5), is opposite in phase compared with the resultant wave at time $t = 0$. An additional quarter-period later, at time $t = 3T/4$, the waves are again out of phase, producing a zero displacement at all points, as shown in (7). Finally, after one full period, at time $t = T$, each wave has propagated one full wavelength in its respective direction, the resultant is the same as in (1), and the motion repeats.

What is surprising is the shape of the sum wave at one-eighth-period intervals between (1), (3), (5), and (7), as shown in (2), (4), (6), and (8) of Fig. 3–1. For these

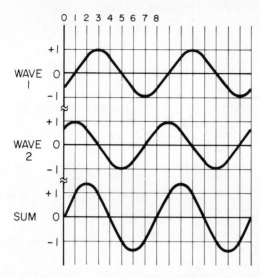

Figure 3-2 Expansion of drawing (2) of Fig. 3-1 showing details of how the two component waves are added.

cases, we must add the displacements of both waves at several points along the wave to obtain the resultant displacement at those points, and draw through the points a smooth curve that represents the shape of the resultant wave. This is illustrated in detail in Fig. 3-2, which is an expansion of (2) of Fig. 3-1. Adding the two original waves at the positions labeled 0 to 8 along the rope gives the results shown in Table 3-1. The resultant is a wave of the same wavelength as the two component waves, but whose amplitude is 1.4 times the amplitude of either individual wave, not double, as when the two original waves are in phase with each other. Figure 3-3 shows, superimposed on one graph, drawings of the resultant wave at times (1) through (8) of Fig. 3-1. The sum wave is *standing*, that is, progressing neither to the right nor to the left, but continually oscillating back and forth between configurations labeled (1) and (5). Its displacement varies between zero (no visible wave) and twice the amplitude of either of the two component waves. Figure 3-4 shows a graph of the motion of points (a) to (i) along the rope as a function of time. Points (a), (e), and (i), where there is no motion of the rope, are called *nodes* and are labeled N (remember *no d*isplacement). Points (c) and (g),

TABLE 3-1 **Addition of waves of Fig. 3-2[a]**

				Position along rope				
Displacement:	0	1	2	3	4	5	6	7	8
Wave 1	−0.7	0.0	+0.7	+1.0	+0.7	0.0	−0.7	−1.0	−0.7
Wave 2	+0.7	+1.0	+0.7	0.0	−0.7	−1.0	−0.7	0.0	+0.7
Sum	0.0	+1.0	+1.4	+1.0	0.0	−1.0	−1.4	−1.0	0.0

[a] The displacements of waves 1 and 2 are read from Fig. 3-2.

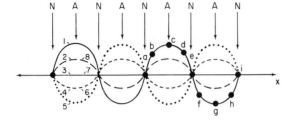

Figure 3–3 Consecutive "pictures" of two full wavelengths of the standing wave of Fig. 3–1, shown at a sequence of equal time intervals of one eighth-period, labeled 1 . . . 8.

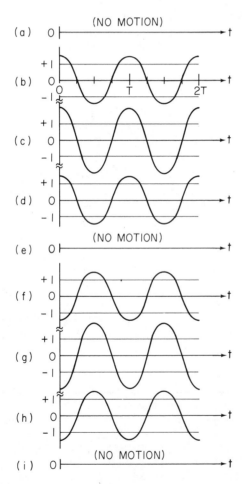

Figure 3–4 Motion of points a . . . i of the standing wave of Fig. 3–3, shown over a two-period interval beginning at $t = 0$.

where the transverse motion and displacement attain their maximum value, are called *antinodes* and are labeled *A*. One section of the standing wave between two nodes is called a *loop*. The length of each loop of the standing wave pattern is one half-wavelength of the original wave.

This discussion has involved only transverse waves in a rope or similar medium. Longitudinal standing waves can also be obtained in which the motion of the wave is along the one-dimensional medium. In later sections and chapters, we shall see that such standing waves are basic to the production of sound in many musical instruments.

3.2   *Resonance and the Overtone Series*

We have seen how two identical sinusoidal waves moving in opposite directions on a long rope combine to form a standing wave, and we have discussed the basic properties of such a standing wave. Now we shall look at what happens to waves at the end of a rope and then study the standing waves produced in a short section of such a medium.

Consider a single transverse pulse, consisting of one-half of an oscillation of period *T*, reflecting off the end of a rope. Figure 3–5 shows such a pulse for the case of a fixed end (we can think of it as attached to a wall and held motionless), as represented by the dot at the end of the rope in part (a), and the case of a free end (it can be moved by the wave) as in part (b). The "force" of the pulse hitting the fixed end of the rope causes the pulse to change from positive to negative, a 180° phase reversal. The wall pulls down on the end of the rope to ensure that there is no displacement at the end. This downward force can be thought of as generating a negative pulse traveling away from the end, the reflected pulse. No such phase reversal occurs when the pulse reflects off a free end, because there is no downward force exerted on the end of the rope.

Now let us expand to the case of three wavelengths of a sinusoidal wave reflecting off a fixed end, as shown sequentially in Fig. 3-6. The × marks the leading edge of the wave. Consecutive drawings are spaced at time intervals of one half-period so that the wave moves one half-wavelength between drawings. Both

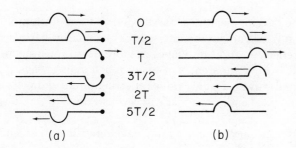

(a)   (b)

Figure 3-5 Reflection of a transverse pulse off a fixed end (a) and a free end (b) of a rope or coiled spring.

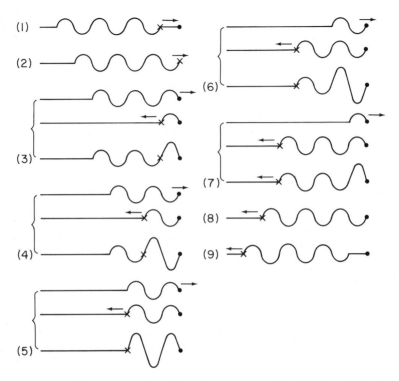

Figure 3-6 Reflection of a transverse pulse train, consisting of three periods of a sine wave, off a fixed end. The wave has a phase reversal upon reflection. The addition of incident and reflected waves produces a standing wave of twice the original amplitude. The × marks the leading edge of the incident wave as it reflects off the end.

incident and reflected waves are shown individually, followed by their sum, which is the overall shape of the rope. Notice the phase reversal (180° phase change) of the wave upon reflection; the first pulse of the incident wave is negative (drawing 1), whereas the first pulse of the reflected wave is positive (drawing 9). Also notice that the sum wave, obtained by adding the incident and reflected waves where they overlap, has twice the amplitude of the original wave shown in drawing (1) of Fig. 3–6.

Figure 3–7 shows the development of a standing wave by reflection of an infinite sinusoidal wave off a fixed end at time intervals of one quarter-period. Figure 3–8 shows the incident, reflected, and sum waves in more detail for drawings (5), (6), and (7) of Fig. 3–7. It is now apparent that to form a standing wave it is not necessary to have two initial waves; a single wave reflecting off an end will suffice. There must be a node at the fixed end, since that end is constrained to zero displacement at all times.

In fact, it is not even necessary to have a long wave reflecting off an end as in

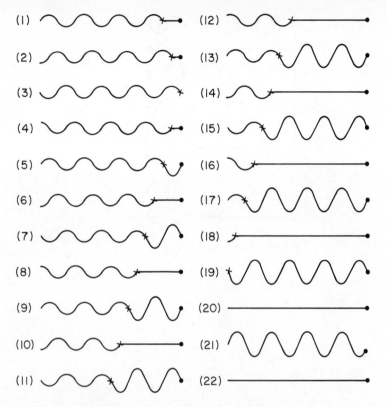

Figure 3-7   Reflection of a continuous transverse wave train off a fixed end, creating a standing wave, shown at intervals of one quarter-period. The × marks the progression of the first pulse of the train.

Fig. 3-7. If a long pulse is started in a stretched rope fixed at both ends, it may reflect off both ends to form a standing wave. Such a situation exists whenever a stringed instrument is plucked or bowed; plucking provides an initial pulse that dies rapidly, whereas bowing provides a continuous source of pulses.

What transverse standing waves can exist in a stretched rope fixed at both ends? The answer to this question can be readily obtained by observing that the constraints on any possible standing waves are (1) there must be a node at each end, since the ends are fixed, and (2) all the loops in the standing wave must be equal in amplitude and with length equal to one half-wavelength of some allowable standing wave. The two simplest possible standing waves in a stretched rope are illustrated in Fig. 3-9. The drawings at the left show one extremum position of the rope for each wave, and the arrows show the direction of the ensuing motion toward the opposite extremum, which is indicated by the dashed lines. The standing waves then continue to oscillate back and forth between these extremum positions. To the right of each standing wave is a drawing showing the extremum positions of each stand-

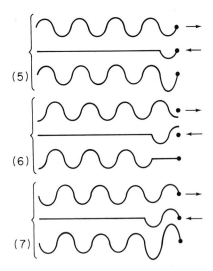

(5)

(6)

(7)

Figure 3–8 Incident, reflected, and sum waves for (5), (6), and (7) of Fig. 3–7.

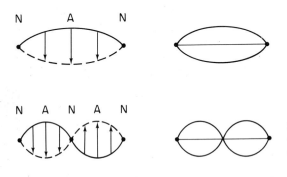

Figure 3–9 The two "simplest" standing waves on a stretched rope or wire, and their representations.

ing wave. This drawing is the graphical representation of the standing wave to its left; that is, by drawing such a graphical representation we mean to convey the existence of the standing wave that it represents. Figure 3–10 shows some "standing waves" that violate the preceding constraints and therefore cannot exist.

Figure 3–11 contains the graphical representations for the first six possible standing waves in a stretched rope, labeled by their *harmonic number N*, the meaning of which will soon become clear. The nodes must be equally spaced along the rope, with an antinode midway between any two successive nodes. Recalling that one loop of a standing wave is one half-wavelength, we observe that the first standing wave is one half-wavelength long, the second one full wavelength (of a different value) long, the third one and one-half wavelengths long, and so forth, as given in Table 3–2.

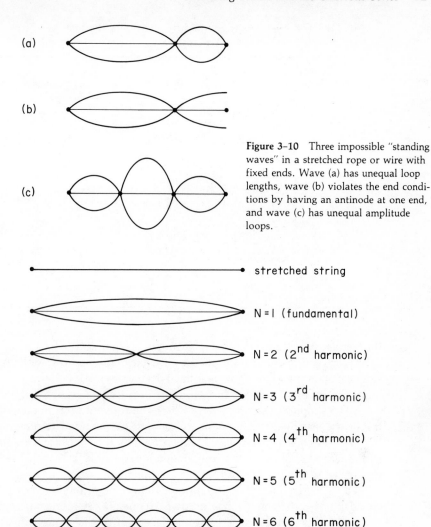

Figure 3–10 Three impossible "standing waves" in a stretched rope or wire with fixed ends. Wave (a) has unequal loop lengths, wave (b) violates the end conditions by having an antinode at one end, and wave (c) has unequal amplitude loops.

Figure 3–11 The representation of the first six possible standing waves in a stretched wire or rope.

In Sec. 2.1 we saw that the frequency and wavelength of a wave are related by the equation $v = f\lambda$, where v is the speed of propagation of the wave in the medium. In this case, v is the speed of the wave on the stretched rope, which is assumed to be constant for waves of all frequencies. We can now determine the frequencies of the various standing waves for a fixed length of rope, which are also given in Table 3–2. The harmonic number N, in addition to being equal to the number of loops in the standing wave, tells us the frequency of the standing wave;

TABLE 3-2   **Wavelengths and frequencies of the first six possible standing waves on a string of length** L

Harmonic number N	Wavelength	Frequency $(f = \frac{v}{\lambda})$
1	$\lambda_1 = 2L$	$f_1 = \dfrac{v}{2L}$
2	$\lambda_2 = L = \frac{1}{2}\lambda_1$	$f_2 = \dfrac{v}{L} = 2f^1$
3	$\lambda_3 = \frac{2}{3}L = \frac{1}{3}\lambda_1$	$f_3 = \dfrac{v}{\frac{2}{3}L} = 3f_1$
4	$\lambda_4 = \frac{L}{2} = \frac{1}{4}\lambda_1$	$f_4 = \dfrac{v}{\frac{1}{2}L} = 4f_1$
5	$\lambda_5 = \frac{2}{5}L = \frac{1}{5}\lambda_1$	$f_5 = \dfrac{v}{\frac{2}{5}L} = 5f_1$
6	$\lambda_6 = \frac{L}{3} = \frac{1}{6}\lambda_1$	$f_6 = \dfrac{v}{\frac{1}{3}L} = 6f$

the frequency of each successive harmonic is equal to the frequency of the simplest mode (or fundamental) multiplied by its harmonic number N. If, for example, the simplest standing wave in a rope has a frequency of 100 Hz when vibrating in one loop, the rope will vibrate in two loops at a frequency of 200 Hz, and so forth.

The simplest standing wave, where the rope vibrates in one loop at its lowest frequency, is called the *fundamental* or *first harmonic*. The other standing waves, called *harmonics* or *overtones,* are identified by their harmonic number N and have a frequency of N times the fundamental frequency. We shall use the term "harmonic" only when the frequencies of the higher standing waves are integrally related to the frequency of the fundamental. The different types of oscillation are called *modes;* each mode has its own frequency. A string may vibrate in many modes simultaneously, where the displacement of the string is just the sum of the displacements due to the individual modes.

Those readers with little or no musical training should read Appendix A before proceeding.

This complete set of sinusoidal standing waves with frequencies related by successive integers is known as the *overtone series.* The first ten musical notes of the overtone series whose fundamental frequency corresponds to the note G on the lowest line of the bass clef are shown in Fig. 3–12. The sequence of musical intervals of the overtone series is the same no matter what note is chosen as the fundamental. The intervals associated with the first eight harmonics of the overtone series are given in Table 3–3.

The overtone series is important because complex waves (those with shapes

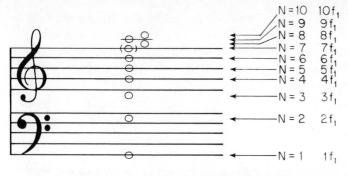

Figure 3-12 Notes on the musical staff having frequencies closest to the notes in the overtone series of G_2; each note is labeled by its harmonic number and the frequency of the corresponding harmonic.

TABLE 3-3 **Musical intervals between the fundamental and other notes of the overtone series**

N	f	Interval
1	f_1	Unison
2	$2f_1$	One octave
3	$3f_1$	One octave + one perfect fifth
4	$4f_1$	Two octaves
5	$5f_1$	Two octaves + one major third
6	$6f_1$	Two octaves + one perfect fifth
7	$7f_1$	Two octaves + one minor seventh
8	$8f_1$	Three octaves

other than pure waves) can be composed of combinations of waves whose frequencies are in the overtone series of the note whose fundamental frequency is the frequency of the complex wave. We shall study this in depth shortly.

A given musical interval, a type of "distance" between two pitches, is always obtained from two notes with a specific frequency ratio between them. For example, one octave (two notes with the same letter name) corresponds to a frequency ratio of 1 to 2. Thus, harmonics 1 and 2 form an octave, as do harmonics 2 and 4, 3 and 6, and 4 and 8. A true perfect fifth contains notes with a frequency ratio of 2 to 3, such as the ratio between harmonics 2 and 3 or harmonics 4 and 6. A true perfect fourth contains notes with a frequency ratio of 3 to 4, such as the ratio between harmonics 3 and 4 or harmonics 6 and 8. A true minor seventh has a ratio of 4 to 7, a true major third, 4 to 5, and a true minor third, 5 to 6. These intervals, with the exception of the minor seventh, are approximately the same on the piano, which is tuned in equal temperament.

Compared with the notes of the piano, the notes separated by octave intervals (N = 1, 2, 4, and 8) are the only notes of the overtone series that are perfectly in tune; harmonics 3, 5, and 6 are slightly different from the notes on the piano, while harmonic 7 is substantially lower than the note on the piano and is therefore enclosed by parentheses, as in Fig. 3–12. This does not imply that the notes of the overtone series are out of tune; they are not. Rather, the frequencies chosen for the notes on the piano represent a compromise in which some of the frequencies differ from those of the overtone series. The choice of keyboard note frequencies, called the *temperament,* is treated in Chap. 9.

3.3   *Mersenne's Laws*

The three physical laws governing the fundamental frequency of various stretched wires are called Mersenne's laws, after the French physicist and musical theorist Marin Mersenne (1588–1648). They relate the fundmental frequency f (or period T) to the linear mass W (mass per unit length) of the wire, the tension F in the wire, and the length L of the wire. A summary of the three laws is contained in Table 3–4. The symbol $\propto$ is read "is proportional to"; we cannot use an equals sign as the units on each side of the equation would not match. For example, $T \propto L$ means the period is proportional to the length of rope. If we make the rope twice as long, the period of the fundamental mode becomes twice its original value (the frequency becomes half of its original value), whereas if we make the rope one-third of its original length the period becomes one-third of its original value (the frequency becomes three times its original value). $T \propto 1/\sqrt{F}$ or $f \propto \sqrt{F}$ means that it requires four times as much tension to raise the frequency to twice its original value or, equivalently, decrease the period to half its original value; this corresponds to a change of one octave. $T \propto \sqrt{W}$ or $f \propto 1/\sqrt{W}$ means, for instance, if we make the mass per unit length four times as great, the period will double or, equivalently, the frequency will become one-half its original value. The laws therefore tell us that low frequencies are achieved using long, heavy wires with little tension, and, conversely, high frequencies are achieved by using short, thin wires with high tension. The difference between sizes and strings in the orchestral string family arises from application of a combination of these three laws; the strings on the cello, for exam-

TABLE 3-4   **Mersenne's laws for the fundamental frequency**
f or period T of a stretched string

	Period		Frequency	Constants
Law 1	$T \propto L$	or	$f \propto \dfrac{1}{L}$	W, F
Law 2	$T \propto \dfrac{1}{\sqrt{F}}$	or	$f \propto \sqrt{F}$	W, L
Law 3	$T \propto \sqrt{W}$	or	$f \propto \dfrac{1}{\sqrt{W}}$	L, F

W = weight per unit length of string
L = length of string
F = tension applied to string

ple, are longer, heavier, and held with less tension than those of the violin, and therefore yield a lower tone. Of course, it is not necessary to use all these laws in all cases to obtain any desired pitch.

The three laws can be summarized in a single formula:

$$f \propto \frac{1}{L}\sqrt{\frac{F}{W}}$$

An extreme illustration of Mersenne's laws involves the piano. An "ideal" piano would have all strings of the same type of wire under the same tension and obtain its frequency or pitch difference by changing only the length of the strings, thereby retaining similar tone quality over its entire range. Since a piano contains over seven full octaves of notes and each octave interval increases the frequency by a factor of 2, the frequency of the highest note (almost 4200 Hz) is over 128 times the frequency of the lowest note (about 27.5 Hz). Thus, according to Mersenne's first law, if the highest frequency string is 6 inches long, the lowest would be about 64 feet! Obviously some compromise must be made, and that compromise is to increase the mass per unit length of the lower strings by making them larger in diameter. Wrapping the center wire with an outer coil for the lowest strings increases the mass per unit length without greatly reducing flexibility. By obtaining the lower frequencies with a combination of increased length and increased mass per unit length, the tension in the strings can remain approximately constant over the entire piano to avoid unequal stress and warping of the frame.

3.4   *Longitudinal Standing Waves*

In Sec. 3.2 we discussed the nature of the transverse standing waves in a stretched rope or string. We shall now discuss the longitudinal analog to transverse standing waves in a string: standing sound waves in an air column or tube. We shall simultaneously develop a convenient transverse wave representation for the longitudinal standing waves in a tube, and observe the direct comparison between transverse standing waves in a string and standing sound waves in a tube. It will be helpful to make use of longitudinal waves in a long coiled spring (slinky spring) to visualize what is happening in a sound wave.

As in the case of transverse waves, longitudinal standing waves are formed when identical waves traveling in opposite directions are superimposed in the same medium. Figure 3–13 illustrates the motion resulting from such a situation. The columns of dots can be taken to represent either coils of a slinky spring or, for the case of a sound wave in air, successive layers of air molecules. Shown in the figure are the positions of the air layers at $t = 0$, $T/4$, $T/2$, and $3T/4$, where T is the period of the wave; these configurations are then repeated. The arrows show the extent of the motion of each layer from its equilibrium position, the average position each molecular layer would assume in the absence of any wave. Certain layers are totally motionless; these points are the nodes. Midway between any two nodes is a position where the motion of the air layer (or spring coil) is at a maximum, the antinodes. Since the nodes and antinodes are defined in terms of the velocity or displacement of the individual layers, they are sometimes called *velocity* or *displacement* nodes and antinodes.

Another quantity that can be used to describe locations along the wave is air pressure (which is analogous to the density of spacing of the spring coils). The air pressure is greater where the layers of air are closer together and smaller where they are farther apart. At a velocity antinode, the layers are moving rapidly back and forth together while maintaining about the same spacing, so the pressure remains approximately constant (no pressure fluctuation). At a velocity node, the spacing between the air layers changes, creating alternately high and low pressure. Thus a velocity node is an antinode for pressure fluctuation. Likewise, a velocity antinode is a node for pressure fluctuation (often called simply a *pressure* node). If the two original waves are sine waves, there will be sinusoidal variation of all the variables discussed: positions of layers, velocities, pressure at any point, and so on.

It is clear that to draw a longitudinal standing wave in the manner of Fig. 3–13(a) to (d) is very tedious. It is therefore helpful to make use of the transverse graphical representation shown in part (e). Both extrema in the corresponding transverse standing wave are drawn inside two parallel lines representing the tube. The nodes (N) are located at the crossover points and the antinodes (A) at the extreme positions. The maximum longitudinal velocity (or displacement) of the layer of molecules (or slinky spring layers) can be taken to be proportional to the distance between the two curves at any point along the standing wave.

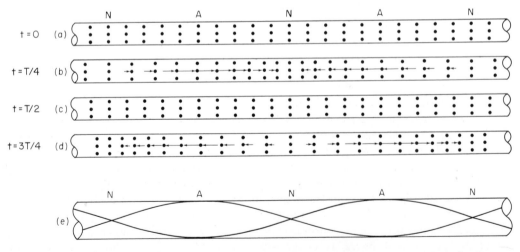

Figure 3–13 Motion of the layers of air molecules in a standing wave in a section of an infinite air tube, shown at four equal time intervals, (a) through (d), during one period. Arrows indicate displacement of molecular layers from their equilibrium positions. Drawing (e) is the transverse representation of this longitudinal standing wave.

We can verify that this molecular motion exists in the standing sound wave in an air column using a Kundt's tube, after August Adolph Kundt (1839–1894), which consists of a simple horizontal tube with a small amount of fine dust lying along the bottom. Creation of a standing wave in the tube will result in violent motion of the dust at an antinode, where the molecular layers are in rapid motion, alternating with quiescent regions at the nodes, where there is very little molecular motion due to the wave. The dust is swept into regions where the motion of the air layers is minimum. Typical dust patterns created by a standing sound wave in a Kundt's tube are shown in Fig. 3–14 in side view. It is the overall structure of nodes and antinodes that marks the loops of the standing wave. Small closely spaced striations are clearly visible, particularly at all points away from the nodes. These striations are kept in continual motion by the vibrations of the air molecules. Figure 3–14(a) is associated with a wave of higher frequency or shorter wavelength than Fig. 3–14(b), resulting in a shorter distance between nodes in (a). The period of the oscillation for (a) is shorter than that of (b), so the total travel distance for molecular layers is less, resulting in a smaller spacing between the striations in (a). Figure 3–14(a) shows two loops or one full wavelength; Fig. 3–14(b) shows one loop or one half-wavelength.

Of course, the most interesting standing waves are those in finite tubes. To describe such standing waves fully, we must first investigate how waves reflect at an end of the medium, as we did in the case of transverse standing waves.

Consider first a compressional pulse traveling down a slinky spring toward either a fixed end or a free end, as shown sequentially in Fig. 3–15(a) and (b), respectively. When the compression reaches the fixed end, it will reflect back as a compression. This is because when the coils adjacent to the wall are compressed and try to expand outward, they can do so only back in the direction of the spring; the wall does not allow them to continue to move in its direction. However, when compression reaches a free end, its motion causes the spring to extend, forming a rarefaction; it is this rarefaction that then travels back down the spring as the reflected pulse. Thus, unlike the case of transverse waves, for longitudinal waves in a slinky spring (and longitidunal waves in general), there is a phase reversal (in

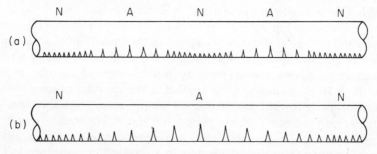

Figure 3–14 Dust patterns created by standing sound waves in a Kundt's tube. Velocity nodes and antinodes are indicated. The frequency of wave (a) is twice that of (b), since its wavelength is one-half. The individual ridges result from the inherent vibrations of the air molecules.

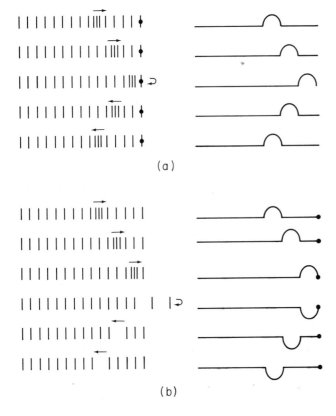

Figure 3-15 Reflection of a compressional pulse in a slinky spring off a fixed end (a) and a free end (b). There is a phase change in pressure or density upon reflection from a free end. The transverse representation of each is shown on the right.

pressure) at a free end and no phase reversal at a fixed end. A transverse representation of the pulses is given to the right of each drawing in Fig. 3–15, illustrating the described behavior in the reflection of a pulse. Notice that there is a phase reversal in pressure (compression to rarefaction, or vice versa) when the sound pulse reflects at an open end, but there is no phase reversal in pressure when it reflects at a closed end. This behavior is summarized in Table 3–5.

The following experiment shows the analogy between the open end of an air tube and the free end of a slinky spring or the closed end of an air tube and the fixed end of a slinky spring. A loudspeaker is placed at the end of a tube 120 cm long with a microphone inserted partway into the center of the tube, as illustrated in Fig. 3–16. A 10-Hz square wave is applied to the speaker, causing the speaker cone to move rapidly toward the tube, then rapidly away from the tube, with 50 ms between each motion (100-ms period for the 10-Hz square wave). Whenever the speaker moves toward the tube, a compression is formed; rapid motion of the speaker away from the tube results in a rarefaction, as shown in Fig. 3–17.

It takes about 2 ms for the pulse to travel 60 cm to the microphone, 2 ms more for it to reach end B, 2 ms more for the reflected pulse to travel back to the microphone, and so on. Thus any compressional pulse produced by the speaker can

TABLE 3-5 **Occurrence of a phase change when a pulse reflects off the end of the wave medium.**

	TRANSVERSE Displacement	LONGITUDINAL Pressure	
Fixed end	YES	NO	Closed end
Free end	NO	YES	Open end

reflect back and forth in the tube many times before the next pulse is produced by the speaker 50 ms later.

Figure 3–18 shows the response of the microphone as a single pulse reflects back and forth in the tube. The pulse is produced at time $t = 0$, and 2 ms later a compression passes the microphone. This compression reaches end B 2 ms after passing the microphone, is reflected with a phase reversal in pressure, and returns to the microphone 2 ms later as a rarefaction. Four milliseconds later, after reflec-

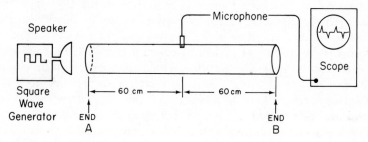

Figure 3-16 Apparatus for investigation of reflection of sound pulses off open and closed ends of air columns.

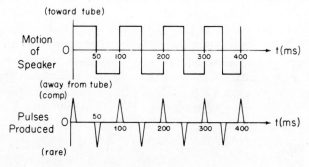

Figure 3-17 Application of 10-Hz square wave motion of the speaker of Fig. 3–16 produces a sequence of alternating compressional and rarefactional pulses.

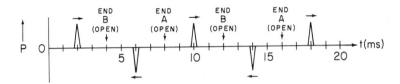

Figure 3-18 Pulses arriving at the microphone after reflection off ends of tube for experiment of Fig. 3-16, with both ends open. At each open end the pulse reflects with a phase reversal in pressure. The pulse began at end *A* at time *t* = 0. Arrows indicate the direction the pulse is moving past the microphone.

tion from end A with another phase reversal, the pulse again passes the microphone as a compression.

If end B of the tube is closed, there will be no phase reversal upon reflection from that end. Thus the first reflection from end B will return as a compression, as shown in Fig. 3-19. In this case, when the pulse reflects from the open end (end A), there will be a phase reversal, and when it reflects from the closed end (end B), there will be no phase reversal.

Consider reflections of longitudinal waves under both types of end conditions. At an open end, there will be motion of the air layers, as the air is free to move there. Similarly, coils move at the free end of a slinky spring. Imagine a compressional pulse traveling along a spring and reaching a free end. The compression forces the spring to extend, a rarefaction forms at the end, and this rarefaction propagates back along the spring. Thus a phase reversal (in coil density) occurs at the free end of a spring. There is, therefore, a pressure node and a velocity (or displacement) antinode at the free end; the initial compression is always canceled there by the reflected rarefaction. Likewise, there is a pressure node and a velocity (or displacement) antinode at the open end of an air tube. In short, there is a pressure node and a velocity antinode at a free end or an open end for a longitudinal wave, and such waves undergo a phase reversal (in pressure) upon reflection.

Conversely, at a fixed end of a slinky spring or at a closed end of an air col-

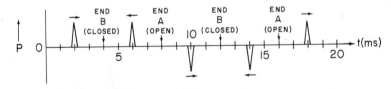

Figure 3-19 Pulses arriving at microphone after reflection off ends of tube for experiment of Fig. 3-16, with end *A* open and end *B* closed. When a pulse is reflected by an open end, there is a phase reversal in pressure; when a pulse is reflected by a closed end, there is no phase reversal in pressure. The pulse began at end *A* at time *t* = 0. Arrows indicate the direction the pulse is moving past the microphone.

umn there can be no motion of the layers. In fact, a thin wall could be put across the tube in Fig. 3-13 at any nodal point, and the standing wave could still exist the same as before. In that case, the standing wave would be formed from the original wave and its reflection. Thus, at a closed end of an air column or a fixed end of a slinky spring there must always be a velocity (or displacement) node or a pressure antinode. Table 3-6 summarizes the standing-wave nodal and antinodal end configurations for the cases of transverse waves, slinky spring longitudinal waves, and sound waves in tubes.

Of course, one obtains reflections from either end, so, as in the case of a standing wave in a wire discussed previously, one can set up standing waves in air tubes subject to the proper end constraints. The first six such standing waves for an open tube (open at both ends) are shown in Fig. 3-20, using the velocity (or displacement) diagrams of Fig. 3-13. The open tube resonates at the fundamental frequency when the tube is one half-wavelength long, with an antinode at each end and a node in the middle. Successive harmonics each add an additional loop to the standing-wave pattern. As in the case of the rope, the nodes and antinodes must be equally spaced along the tube.

The end constraints placed on the possible standing waves in a closed tube (open at one end and closed at one end) allow only the formation of odd harmonics in a closed tube, as shown in Fig. 3-20. For the same fundamental frequency the closed tube is half the length of the open tube; the open tube is one half-wavelength long and the closed tube one quarter-wavelength long for their respective fundamental frequencies.

Table 3-7 summarizes the relationships between harmonic number, frequency, wavelength, and length of tube for the first six harmonics of open and closed tubes, respectively. If we choose the length of the open tube L_o twice that of the closed tube L_c, their fundamental frequencies f_1 will be the same.

These ideas are directly applicable in determining the fundamental frequencies of wind instruments and the overtone (or harmonic) structure of their sounds. On any wind instrument, the bell end acts as an open end. An embouchure hole, such as that on the orchestral flute, or a fipple, such as on the recorder family, is also an

TABLE 3-6 **End configuration for standing waves in a one-dimensional medium**

| | TRANSVERSE | LONGITUDINAL | | |
| | | | Velocity or | |
	Displacement	Pressure	displacement	
Fixed end	NODE	ANTINODE	NODE	Closed end
Free end	ANTINODE	NODE	ANTINODE	Open end

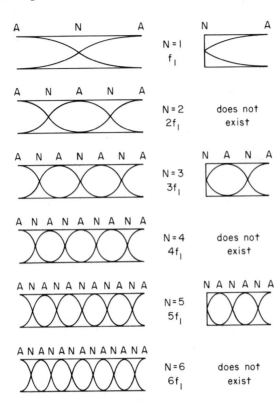

Figure 3-20 Transverse representations of velocity patterns for standing sound waves in open and closed tubes. Air displacement (or velocity) is graphed here. A closed end is a node just as a fixed end of a string is a node. (Compare with Fig. 3–11.) A closed tube of a given length and an open tube of double this length have the same fundamental frequency.

open end. Thus, these instruments can support standing waves of all harmonics. The reed end of a reed instrument (such as the clarinet) acts as a closed end, because the reed is slapped closed most of the time; it opens at regular time intervals to allow short bursts of air to enter the instrument. Because the clarinet has a cylindrical bore (the width of the air tube is constant along most of its length) and one closed end, low-numbered odd harmonics dominate, and it behaves acoustically like a closed tube, as shown in Fig. 4–19. Although the oboe, bassoon, and saxophone families are reed instruments, they have conical-shaped bores and, for reasons discussed in Chap. 10, the conical bore allows all harmonics (both even and odd) to exist. Among the reed instruments, the clarinet family is unique in that it is the only one with a cylindrical bore, which not only gives mainly odd harmonic structure, but also causes clarinets to overblow at the musical interval of a twelfth. By blowing harder into the instrument, or using a register key, one can excite the next possible higher harmonic. For the clarinet, the next allowable harmonic has a frequency three times the fundamental frequency and is thus an octave and a fifth above the note played normally (see Appendix A). The other woodwinds overblow at one octave interval. Whether the reed on a woodwind instrument is single

TABLE 3–7 **Wavelength and frequency of standing waves in open and closed tubes of length L_o and L_c, respectively. The fundamental frequencies f_1 are the same if $L_o = 2L_c$.**

Harmonic number	Open tube		Closed tube	
	Wavelength	Frequency	Wavelength	Frequency
1	$\lambda_1 = 2L_o$	$f_1 = \dfrac{S}{2L_o}$	$\lambda_1 = 4L_c$	$f_1 = \dfrac{S}{4L_c}$
2	$\lambda_2 = L_o = \dfrac{1}{2}\lambda_1$	$f_2 = \dfrac{S}{L_o} = 2f_1$	—	—
3	$\lambda_3 = \dfrac{2}{3}L_o = \dfrac{1}{3}\lambda_1$	$f_3 = \dfrac{S}{\frac{2}{3}L_o} = 3f_1$	$\lambda_3 = \dfrac{4}{3}L_c$	$f_3 = \dfrac{S}{\frac{4}{3}L_c} = 3f_1$
4	$\lambda_4 = \dfrac{1}{2}L_o = \dfrac{1}{4}\lambda_1$	$f_4 = \dfrac{S}{\frac{1}{2}L_o} = 4f_1$	—	—
5	$\lambda_5 = \dfrac{2}{5}L_o = \dfrac{1}{5}\lambda_1$	$f_5 = \dfrac{S}{\frac{2}{5}L_o} = 5f_1$	$\lambda_5 = \dfrac{4}{5}L_c$	$f_5 = \dfrac{S}{\frac{4}{5}L_c} = 5f_1$
6	$\lambda_6 = \dfrac{1}{3}L_o = \dfrac{1}{6}\lambda_1$	$f_6 = \dfrac{S}{\frac{1}{3}L_o} = 6f_1$	—	—

(clarinet and saxophone families) or double (oboe and bassoon families) is irrelevant in determining which harmonics of standing waves are possible in the instruments. The only relevant factors are the shape of the bore (cylindrical or conical) and the existence of a node at a closed end for a cylindrical tube; however, the style of reed helps determine the relative *amplitudes* of possible harmonics.

Brass instruments are blown by buzzing the player's lips directly into a mouthpiece; such a mouthpiece is also a closed end. Here, again, the bore of the instrument as well as complicated effects due to mouthpiece and bell shape allow all harmonics to exist in these instruments.

Brass and woodwind instrument players are familiar with the following effect: As they play, their instruments warm up and rise in pitch. This rise in pitch can be explained using several of the principles described previously. The velocity of sound is greater in warm air than in cooler air. As the air column in the instrument gets warmer, the frequencies of the fundamental and all harmonics increase. This is because the wavelength of each standing wave, related to the length of the instrument, remains the same while the velocity of sound in the instrument increases. By the formula $f = S/\lambda$, we can see that, if S increases and λ stays the same, the frequency f must increase.

Thus far we have studied one-dimensional standing waves in strings and air columns. Very important standing waves exist in one-dimensional solids such as tuning bars or xylophone bars, and multidimensional standing waves become important in a number of applications.

Tuned percussion instruments, such as chimes, bells, and timpani, are unique in that their resonant frequencies are not related by integers; that is, the resonances do not have frequencies of exactly N times the fundamental frequency. Several modes of oscillation are given in Fig. 3–21 for a stretched circular membrane such as a drumhead or tympani. The frequency ratios differ greatly from integral values, as will be discussed in Chap. 14. In addition, the particular frequency that we hear as the "pitch" of the timpani is not the lowest one produced, but usually the second lowest. The lowest is heard as a sort of dull low-frequency thud that dies away quickly.

A Chladni plate, named after the German acoustician Ernst Florens Friedrich Chladni (1756–1827), is a thin regularly shaped metal plate that can be excited to produce two-dimensional standing waves by drawing a violin bow across one edge or by some other, usually electrical, means. If salt or sand is sprinkled onto the vibrating plate, it will be jostled at the antinodal regions and collect along the nodal lines. The patterns thus produced are often very beautiful and can change radically if the excitation frequency is changed slightly. Some patterns are shown in Fig. 3–22 for circular and square plates; the lines drawn correspond to nodal lines. The tap tone, the tone produced by gently rapping with a knuckle the front or back plate of a stringed instrument like the violin, has a standing wave similar to these Chladni figures. The wood forming the plates of a violin must be carefully shaped so that the tap tone will be at exactly the right frequency to strengthen the notes in a specific frequency range of the instrument. The "singing" of a high-quality crystal wine glass when a moistened finger is rubbed around its edge is due to the existence of an audio-frequency standing wave in the glass, which is excited by the rubbing. The fact that a good wine glass can be broken by a loud sound is evidence that a mode of oscillation exists in the audible range. An intense sound of that frequency can drive the standing wave to an amplitude greater than the elastic limit of the crystal. This effect is similar to the creation of standing sound waves around the in-

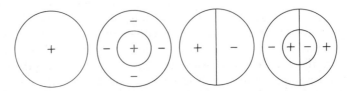

Figure 3-21 Some vibrational modes in a circular drumhead. Nodal lines are indicated, along with the relative phase of various parts of the drumhead, indicated by + (up) and − (down).

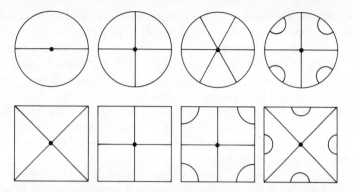

Figure 3-22   Standing wave patterns in circular and square Chladni plates, supported in the center. Nodal lines are indicated.

side walls of a circular room, such as a large cathedral dome. Walking around the edge of the dome, one hears alternating loud (antinodal) and soft (nodal) regions when a constant musical tone is played by an instrument near the edge of the dome.

The collapse of the Tacoma Narrows Bridge, as discussed in Chap. 1, is clearly a case of resonance. We now see, using the vocabulary developed here, that a certain torsional or twisting mode of oscillation was driven in a complex manner by the wind, creating a standing wave as shown in Fig. 3–23. The amplitude of the oscillation eventually exceeded the elastic limit of the materials in the bridge, thereby leading to its collapse.

Soldiers marching across small bridges break step to avoid creating large periodic oscillations that might drive a resonance in the bridge and cause it to collapse.

A recent medical technique uses resonances to check the healing of a bone in a

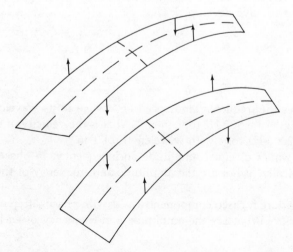

Figure 3-23   Oscillation mode of the center span of the Tacoma Narrows Bridge. Ends remain fixed; nodal lines are dashed.

cast without the use of possibly dangerous X-rays and without removing the cast. A bone of some length, when in one piece, possesses a set of resonant frequencies characteristic of its size, material, and the way in which it is connected to neighboring bones. Often these frequencies are in the ultrasonic region. When waves of the appropriate frequency range are focused on the bone, absorption of the waves can be observed at the resonant frequencies of the bone. If the bone is broken, it resonates differently with a new set of frequencies. By observing changes in the resonant frequencies of the bone, it is possible to monitor its healing. Since sound waves pass through any type of cast, this procedure can be carried out easily without removal of the cast.

Resonances in air tubes can also be used to check for holes and discontinuities in inaccessible pipes in nuclear reactors, oil refineries, chemical plants, and other industrial situations. By observing the resonances in a long tube with sound waves and repeating the process over long time intervals, one can observe changes in the resonances with time, which indicate deterioration of the pipe. Long-wavelength (low-frequency) waves are used because they diffract easily and therefore bend with the pipe. If a hole appears in the pipe, this point will then act as an open end, creating a velocity antinode at that point and thereby changing the resonant frequencies of the pipe. Using this technique, holes as small as 2 mm diameter in pipes up to 10 m long have been detected.

In the Renaissance, choral works in minor keys would often end in a major key, the third of the final chord being raised one half-step to make a major rather than a minor chord. The note two octaves plus a major third above the bass note in the final chord is present in the overtone series of the bass note. If the minor third, one half-step below the harmonic of the bass note, is being sung, these two notes will produce beating. This becomes particularly evident when the sound reverberates in a large room or cathedral. By replacing the minor third with a major third in the final chord, this beating is eliminated. This major third at the end of a piece in a minor key is called the *tierce de Picardi,* or Picardi third, after the town in France where it is believed to have developed.

EXERCISES

1. What formula relates the frequency and speed of a wave to its wavelength? What is the wavelength of a 500-Hz tone in air? What is the velocity of sound in a gas that has a 2000-Hz tone with wavelength of 1 m?

2. (a) Draw two sine waves of equal amplitude and frequency, in phase, and add them graphically. What are the amplitude and frequency of the sum wave?

   (b) Repeat this procedure for two component waves with phase differences of 45°, 90°, and 180°. What are the amplitude and frequency of each sum wave?

(c) Draw a square wave of amplitude $A = 1$ V and a square wave of amplitude $A = 2$ V with twice the frequency of the first wave, beginning in phase. Add the two waves. What is the frequency and amplitude of the sum wave? Add similar waves with a different initial phase relationship.

3. (a) What are the first four resonant frequencies of an open tube 345 cm long? Draw the velocity (or displacement) standing waves for each, and label the nodes N and the antinodes A. Label each mode by its harmonic number.

 (b) What are the first three resonant frequencies of a closed tube 690 cm long? Draw the standing waves for each, and label the nodes and antinodes. Label each mode by its harmonic number.

4. What is a mode?

5. Which type of tube, open or closed, can never have a node in the exact center?

6. Why does the pitch of a wind instrument rise when it warms up?

7. Longitudinal standing waves can be produced in a 2-m metal rod by holding it at some point with one hand and stroking it with the other hand, causing it to oscillate with antinodes at each end. Why are the ends antinodes and not nodes? When the rod is pinched at the center while exciting the rod, the frequency emitted is 2500 Hz. (a) What is the speed of sound in the metal if this is the fundamental mode? (b) Where should the rod be pinched to excite the next possible standing wave? What is its frequency?

8. How long should a closed tube (in air) be such that its fundamental frequency is 100 Hz? What is the frequency of the next possible standing wave?

9. A closed tube is 50 cm long. What is the length of an open tube that has the same fundamental frequency? What is this frequency? These two tubes are placed in a gas with a speed of sound greater than the speed of sound in air. Do the frequencies of their fundamentals remain equal to each other? Do they increase or decrease?

10. The third possible standing wave in a closed tube has frequency of 1500 Hz. What is the length of the tube?

11. The frequency of the fundamental of a certain tube in air is 200 Hz. The frequency of the fundamental is 250 Hz when the tube is placed in an unknown gas. What is the speed of sound in this gas? Do you have to know what type of tube was used (open or closed)?

12. A certain stretched string has a fundamental frequency of 100 Hz. What must be done to its (a) length, (b) tension, or (c) weight per unit length, individually, to raise the fundamental frequency to 200 Hz?

13. A certain stretched string has a fundamental frequency of 175 Hz. What is the new frequency if one (a) decreases the tension by a factor of 4, (b) doubles the length, or (c) decreases the weight per unit length by a factor of 9?

14. Draw a musical staff and write out the notes corresponding to the overtone series of G_2 (low G on the bass clef). Label each note by its harmonic number and frequency, if the G_2 has a frequency of about 100 Hz. Which of these notes are resonances in an open tube whose fundamental note is G_2? Which of

these notes are resonances in a closed tube whose fundamental note is G_2? Which of these notes are in the overtone series of G_3, the note one octave above G_2?

15. Certain rope bridges can oscillate in the wind or as a result of the motion of people walking on them. In general, the frequency of oscillation is much lower (for example, 1 Hz) than that of a stretched string. Explain this using Mersenne's laws.

16. Perform the graphical additions for Fig. 3–1, parts (2), (4), (6), and (8).

4

Analysis
and Synthesis
of Complex Waves

We have seen that the various notes of the overtone series can be produced as resonances in stretched strings, open tubes, and closed tubes. Now we shall consider how these harmonics are combined to form a single complex wave. The process of combining harmonics to form a complex wave is called *Fourier synthesis*, after the French mathematician Jean Baptiste Joseph Fourier (1768-1830). The inverse process, that of determining the harmonic content of complex waves, is called *Fourier analysis.* At the end of this chapter, we shall study various factors affecting the timbre, or quality of a musical tone, which is related to the harmonic content of complex waves.

4.1 *Synthesis of Complex Waves*

Consider the examples of wave addition given in Fig. 4-1. Shown are the fundamental ($N = 1$), the second harmonic ($N = 2$), and the algebraic sum of these two (first and second) harmonics. These graphs are generally taken to represent the wave shape in time, as displayed on an oscilloscope, but could also represent the instantaneous shape of a standing wave in a freely vibrating string or the "shape" of a sound wave traveling through space. The difference between the sum wave in cases (a) and (b) is due solely to the change in relative phase between the two component harmonics. Figure 4-2 shows the synthesis of a complex wave from equal amplitudes of first and third harmonics; again the difference in the two resultant

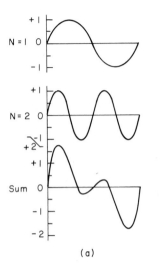
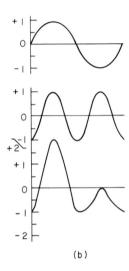

Figure 4-1 Fourier synthesis of complex waves from equal amplitudes of fundamental and second harmonic. Parts (a) and (b) differ in the phase of the second harmonic.

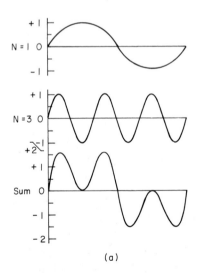
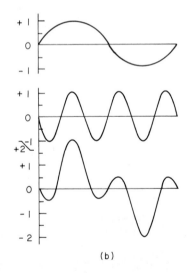

Figure 4-2 Fourier synthesis of complex waves from equal amplitudes of fundamental and third harmonic. Parts (a) and (b) differ in the phase of the third harmonic.

wave shapes is due to the difference of the relative phase between the two components. Figure 4–3 shows complex waves formed from first and second harmonics with different amplitude relationships; in both cases the relative phase of the second harmonic with respect to the fundamental is the same.

We can now see that the shape of a complex wave, a wave formed by the addition of two or more harmonics, is determined by (1) the number and relative amplitudes of the component harmonics and (2) the phases of the higher harmonics relative to the fundamental. It is interesting to listen to various complex tones formed from two harmonics while changing the amplitude and/or phase of the higher harmonic relative to that of the fundamental. The tone quality or *timbre* is clearly affected by moderate changes in the amplitude of the higher harmonic, but hardly affected at all by rather large changes in the relative phases of the two harmonics. This property of the ear (summarized as Ohm's law of hearing) is one of the basic features of the place theory of hearing and will be discussed in Chap. 6.

The *standard* complex waves discussed in Chap. 1 (triangular, square, sawtooth, and pulse train) can be synthesized from the notes of the overtone series as shown in Figs. 4–4 through 4–7. Each figure contains at the left the sequence of harmonics present in the complex wave with the appropriate amplitude and phase, and at the right the complex wave formed by the addition of each successive component. Notice that with the addition of each higher harmonic the complex wave looks more like the standard shape shown at the top. For reference, the relative amplitudes of the harmonics of each of the standard waves are given in Table 4–1. This table shows no relative phases, which must be chosen appropriately to pro-

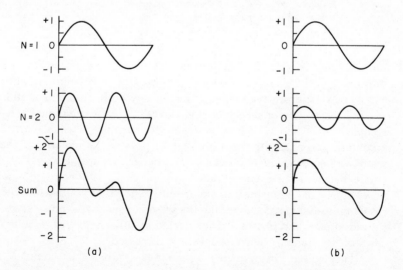

Figure 4–3 Fourier synthesis of complex waves from fundamental and second harmonic. In part (a) the amplitude of the fundamental and second harmonic are equal; in part (b) the amplitude of the second harmonic is half that of the fundamental.

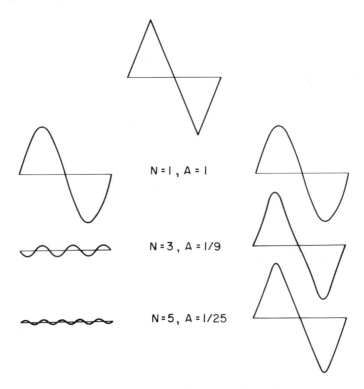

N = 1 , A = 1

N = 3 , A = 1/9

N = 5 , A = 1/25

Figure 4-4   Fourier synthesis of a triangular wave. At the left are the successive harmonics; at the right are the sum waves including each successive harmonic. The graph at the top is the wave being synthesized.

duce the desired complex wave shape. However, the amplitudes are far more important than the phases in determining the timbre, or sound quality, of the wave.

The standard wave shapes (triangle, sawtooth, square, and pulse train) can be synthesized by combinations of sine waves of (1) the fundamental frequency of the complex wave and (2) some or all of the harmonics, each with the appropriate amplitude and phase. In general, it can be proved mathematically that *any* periodic wave with some frequency f can be synthesized from sine waves of the frequency f and its harmonics, with amplitudes and phases determined by the shape of the complex wave. This mathematical statement is known as *Fourier's theorem*. In general, for complicated wave shapes like the wave forms of real musical instrument tones, the amplitudes and phases are not related by any simple formula, as they are in the cases of the standard wave shapes just discussed.

Throughout the sequence of Fourier syntheses of complex waves, it was necessary to draw only one period of the complex wave and a number of periods of each harmonic equal to its harmonic number. After one period, the complex wave repeats itself, in the same manner as do the fundamental and each of the harmonics.

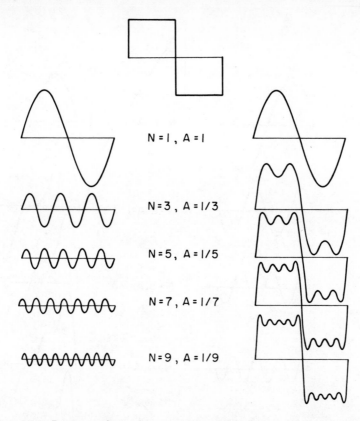

Figure 4-5 Fourier synthesis of a square wave. At the left are the successive harmonics; at the right are the sum waves including each successive harmonic. The graph at the top is the wave being synthesized.

TABLE 4-1 Relative amplitudes of harmonics present in the standard waves

Wave	Harmonic amplitudes
	$N = $ 1, 2, 3, 4, 5, 6, 7, 8, 9, 10 . . .
Sine	1, 0, 0, 0, 0, 0, 0, 0, 0, 0, . . . , $N = 1$ only
Triangle	1, 0, $\frac{1}{9}$, 0, $\frac{1}{25}$, 0, $\frac{1}{49}$, 0, $\frac{1}{81}$, 0, . . . , $\frac{1}{N^2}$ for odd N
Square	1, 0, $\frac{1}{3}$, 0, $\frac{1}{5}$, 0, $\frac{1}{7}$, 0, $\frac{1}{9}$, 0, . . . , $\frac{1}{N}$ for odd N
Sawtooth (ramp)	1, $\frac{1}{2}$, $\frac{1}{3}$, $\frac{1}{4}$, $\frac{1}{5}$, $\frac{1}{6}$, $\frac{1}{7}$, $\frac{1}{8}$, $\frac{1}{9}$, $\frac{1}{10}$, . . . , $\frac{1}{N}$ for all N
Pulse train	1, 1, 1, 1, 1, 1, 1, 1, 1, 1, . . . , equal for all N

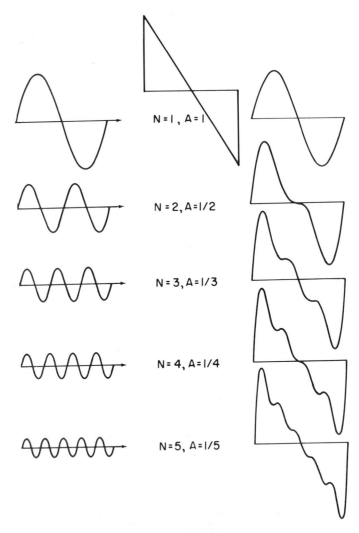

Figure 4-6 Fourier synthesis of a sawtooth wave. At the left are the successive harmonics; at the right are the sum waves including each successive harmonic. The graph at the top is the wave being synthesized.

As a rule, therefore, any wave containing harmonics of some fundamental frequency f will repeat itself with a periodicity $T = 1/f$. This is true even if the wave has no fundamental in its harmonic structure, as shown in Fig. 4-8. In these cases, although the fundamental is missing, the complex wave is periodic with the fundamental frequency because both frequencies present in these complex waves are harmonics of the fundamental. Furthermore, owing to inherent properties of the ear to be discussed in Chap. 6, the pitch we actually hear is that of the fundamental,

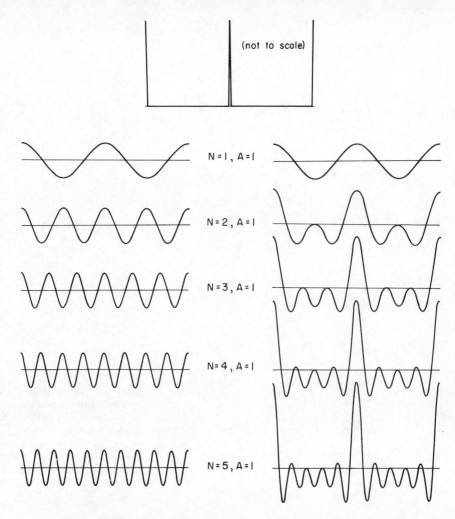

Figure 4–7 Fourier synthesis of a pulse train. At the left are the successive harmonics; at the right are the sum waves including each successive harmonic. The graph at the top is the wave being synthesized.

although the tone quality of these notes is different when the fundamental is actually present with some significant amplitude.

4.2 Fourier Analysis and Fourier Spectra

We saw in the last section that any periodic wave can be synthesized from sine waves of the frequencies of the harmonics of the fundamental frequency, which is

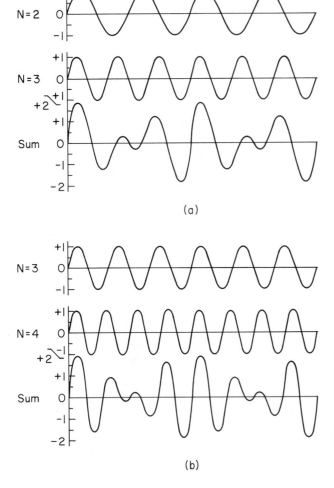

Figure 4-8 Fourier synthesis of complex waves from equal amplitudes of (a) second and third harmonics, and (b) third and fourth harmonics. One period of each complex wave is shown. The periodicity of both complex waves is that of the fundamental, although the fundamental is missing from the synthesis.

the frequency of the complex periodic wave itself. Because the tone quality of the complex wave is determined primarily by the amplitudes of the harmonics present, it is useful to display the harmonic content graphically. A graph called the *Fourier spectrum* is a useful means for displaying the harmonic content of complex periodic waves.

Figure 4-9 shows the Fourier spectra of two pure tones or sinusoidal waves, where the first (a) has a frequency of f_1 and an amplitude of 1, and the second (b) has a frequency of $3f_1$ and an amplitude of 0.5. These graphs illustrate the obvious fact that a sine wave is made up of just one sine wave, namely, one with its own frequency. The amplitude is in arbitrary units; what is important is only the relative amplitudes of the harmonics with respect to the fundamental and to each other. In each case shown in Fig. 4-9, the frequency of the sinusoidal wave is its fundamental

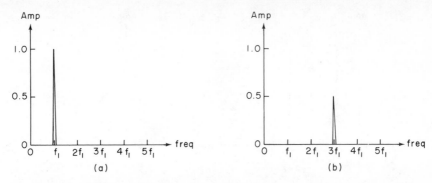

Figure 4-9 Fourier spectra of pure tones of (a) amplitude $A = 1$ and frequency $f = f_1$, and (b) amplitude $A = 1/2$ and frequency $f = 3f_1$.

(f_1 and $3f_1$, respectively). Shown in Fig. 4–10 are Fourier spectra of the complex waves of Fig. 4–1. The graph simply presents the information that in the complex waves of Fig. 4–1 there are two harmonics present, $N = 1$ (frequency f_1) and $N = 2$ (frequency $2f_1$), both equal in amplitude. Additional information, that is, the relative phases between the fundamental and the second harmonic, is missing, as in all similar Fourier spectra graphs. Only the amplitudes of the harmonics are shown. Figure 4–11 shows the Fourier spectra of the two complex waves of Fig. 4–2. Again, because no phase data are presented in this format, the spectra are identical, showing that both complex waves contain equal amplitudes ($A = 1$) of first and third harmonics.

Figure 4–12 shows the Fourier spectra for the waves in Fig. 4–3. In this case, the graphs of the Fourier spectra reflect the differences in the amplitudes of the second harmonic between the two waves. Figure 4–13(a) and (b) show the Fourier spectra of the waves of Fig. 4–8(a) and (b), respectively. In these cases, the fun-

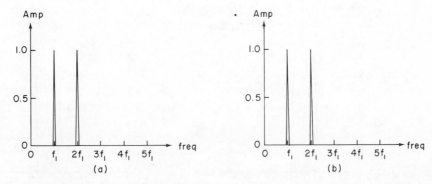

Figure 4-10 Fourier spectra of the complex waves of Fig. 4–1(a) and (b), respectively. Both contain equal amplitudes of the fundamental and second harmonic, but their shapes differ because the phase of the second harmonic relative to that of the fundamental is different in the two cases.

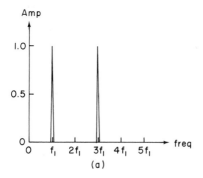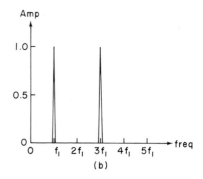

Figure 4-11 Fourier spectra of the complex waves of Fig. 4-2(a) and (b), respectively. Both contain equal amplitudes of the fundamental and third harmonic, but their shapes differ because the phase of the third harmonic relative to that of the fundamental is different in the two cases.

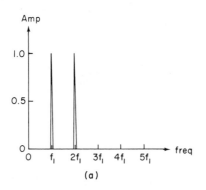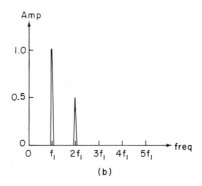

Figure 4-12 Fourier spectra of the complex waves of Fig. 4-3(a) and (b), respectively. Both contain the fundamental and second harmonic, but the amplitude of the second harmonic is halved in graph (b).

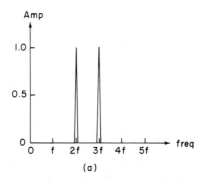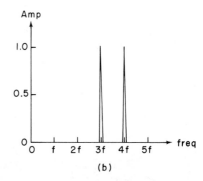

Figure 4-13 Fourier spectra of the complex waves of Fig. 4-8(a) and (b), respectively.

damental frequency is absent in the complex wave and therefore does not appear as a peak in the Fourier spectrum graph. This does not violate Fourier's theorem; in this case, the amplitude of the fundamental is 0.

The Fourier spectra for the standard waves, whose harmonic amplitudes are tabulated in Table 4-1, are shown in Figs. 4-14 through 4-17. These Fourier spectra present graphically the amplitudes of the harmonics present in the standard complex waves. The wave shape uniquely determines the Fourier spectrum, but because there is no phase information in the Fourier spectrum, the spectrum cannot uniquely determine the wave shape. In other words, as we have seen, many wave

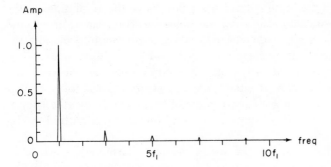

Figure 4-14 Fourier spectrum of a triangular wave of frequency f_1 up to the tenth harmonic.

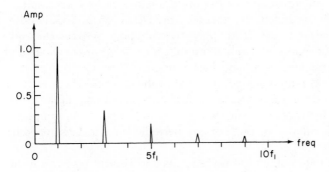

Figure 4-15 Fourier spectrum of a square wave of frequency f_1 up to the tenth harmonic.

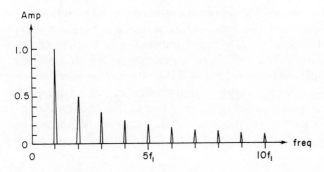

Figure 4-16 Fourier spectrum of a sawtooth wave of frequency f_1 up to the tenth harmonic.

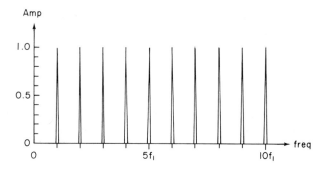

Figure 4-17 Fourier spectrum of a pulse train of a frequency of f_1 up to the tenth harmonic.

shapes may correspond to the same Fourier spectrum. Since the amplitudes of the harmonics have much more influence than the phases in determining the sound quality or timbre, the Fourier spectrum tells us the most important information pertinent to the timbre of the sound wave.

We can use an electronic device called a *Fourier analyzer,* or *spectrum analyzer,* to determine the harmonic content of any arbitrary musical tone or other wave form. For example, if we put a square wave of frequency f_1 into the Fourier analyzer, it will produce the Fourier spectrum of the square wave, plotted in Fig. 4–15.

A very general correlation can be made between harmonic structure as observed in the Fourier spectrum and sound quality or timbre. A simple sinusoidal wave (one harmonic, the fundamental) sounds pure or plain. On the other hand, as large-amplitude harmonics are added, the tone becomes richer, as in the sawtooth, or even raspy, as in the extreme case of the pulse train. Sounds between these extremes of harmonic content sound relatively more simple or rich. Square waves and other waves that have large-amplitude odd harmonics and little or no even harmonics, produce a hollow or woody sound, like that of the clarinet. The triangular wave, containing only odd harmonics but with very small amplitudes, lies somewhere between the sine wave and square wave, with the pure tone simplicity and a slight woodiness of sound. The pitch of any periodic wave is determined by its fundamental frequency.

Figures 4–18 through 4–21 show the wave forms and Fourier spectra for different notes played on a recorder, a clarinet, a violin, and a krummhorn. The recorder tone is very simple, almost like a pure sinusoidal tone, as seen in its wave form, Fig. 4–18(a). Its Fourier spectrum, Fig. 4–18(b), is therefore very simple, consisting of the fundamental and only a few higher harmonics with relatively small amplitudes. On the other hand, the krummhorn has an extremely rich, reedy tone, and possesses a very complex waveform, Fig. 4–21(a). Its Fourier spectrum, Fig. 4–21(b), therefore contains a large number of harmonics (over 40 can be easily counted), some of which have amplitudes larger than that of the fundamental. Even if the amplitudes of the harmonics are much larger than that of the fundamental, the note sounds at the pitch of its fundamental since all the frequencies are har-

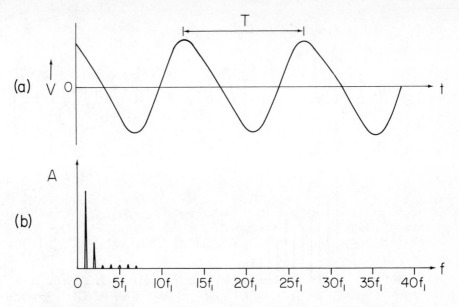

Figure 4–18 Wave form and Fourier spectrum of the note C = 523.25 Hz played on an alto recorder.

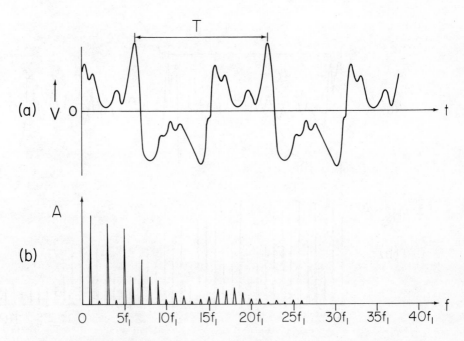

Figure 4–19 Wave form and Fourier spectrum of the note B = 233.08 Hz played on a clarinet.

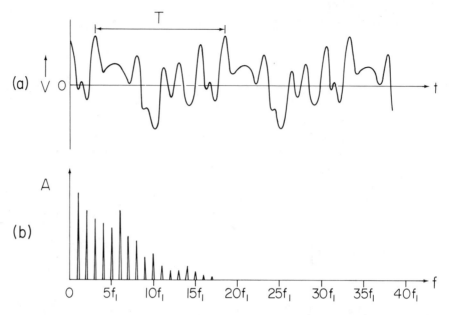

Figure 4-20 Wave form and Fourier spectrum of the note B$^\flat$ = 493.88 Hz played on a violin.

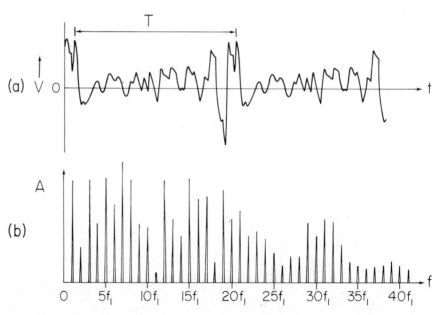

Figure 4-21 Wave form and Fourier spectrum of the note G = 196.00 Hz played on a krummhorn.

monics of that fundamental. This would be true even if the fundamental were missing.

The clarinet and the violin spectra and wave forms lie between these two extremes in complexity of wave shape (richness in timbre) and abundance of large-amplitude overtones. The Fourier spectrum of the clarinet note (Fig. 4–19(b)) has large-amplitude low-numbered odd harmonics, giving it a tone similar to that of the square wave. The sound curve of the clarinet note does not look like a square wave because the amplitudes and phases of the harmonics are different from those of a square wave. The violin has a medium-complex sound curve, as shown in Fig. 4–20(a), and therefore an intermediate number of harmonics in its Fourier spectrum, as seen in Fig. 4–20(b).

From this brief study of instrument sounds and their Fourier spectra we see that there is obviously a correlation between tones that sound pure (or simple) and their simple Fourier spectra, which contain few or no large-amplitude harmonics. There is also a correlation between complex, rich tones and their Fourier spectra, which contain a large number of large-amplitude harmonics. It also appears that tones containing certain types of harmonic stucture, characterized, for example, by odd harmonics only, can be identified by a particular timbre. However, one can deal in broad generalities only. As we shall see in the next section, any attempt to correlate Fourier spectra conclusively with instrument timbre breaks down when applied to details of subtle differences between instruments, even when the tonal variation is "obvious" to our ears.

4.3   *Analysis of Tone Quality*

We have studied, in a general way, the effect of harmonic content on the timbre of a musical tone. The explanation of different musical tone quality in terms of harmonic structure is limited, and we will discuss next some of the other factors that affect tone quality.

Amplitudes of harmonics   As discussed previously, these data are basic but have limited use in explaining details.

Attack and decay transients   When a piano note is struck, a note on a wind instrument attacked by tonguing, a violin bow attack started, or any other musical sound begun, unique sounds occur that last for very short time periods. These sounds are called *attack transients* and may possess harmonic content very different from the steady note that follows after the attack is completed. It has been shown that attack transients are very important in determining how one perceives the tone quality of an instrument; the auditory system responds to rapidly changing frequencies present in the attack, and this response affects the sound quality associated with that tone.

Several experiments illustrate this effect as follows. Tape recordings are made with different instruments attacking and holding the same musical note. The tape is then cut and spliced so that after the attack of one instrument the sustained note of another is heard. For example, the sustained tone of a violin can be chosen to follow the attack of an oboe. When the tape is played, most people will identify the sound as that of an oboe. In fact, the steady tones of most band and orchestral instruments are similar enough in their harmonic structure that the substitution of the attack of a second instrument will result in identification of the steady note as that of the instrument whose attack was used.

Another example of the effect of attack transients can be illustrated by playing a tape of piano music backward. Absence of the usual attack transients in the playback plus the unnatural increase in loudness results in a sound very different from that of a piano.

The *decay transients* of a musical tone also may have important effects on tone quality. This is especially true for some percussion instruments and plucked strings. These instruments have a common property that the vibration amplitude continually decreases after the initial attack.

Finally, it should be added that the subject of attack transients is of considerable contemporary research interest. It takes extremely costly and sophisticated electronic equipment, including high-speed computers, to analyze attack transients and to reproduce common instrumental or vocal transients.

Inharmonicities Thus far we have primarily studied only sounds that are composed of integrally related frequency components, that is, sounds whose overtone frequencies are the harmonic number times the fundamental frequency. Transients are, in general, not composed of simple harmonics only. Several exceptions to this basic rule are of importance in a discussion of tone quality.

In Chap. 1 we learned that we need a linear restoring force for SHM. For a vibrating piano string, this force results from the tension applied to the string; the tension tends to force the string back to its equilibrium position, that is, straight. There is another force also tending to make the string straight; it is due to the inherent stiffness of the metal strings. Even if there is no tension in the string, it will, when bent, tend to become straight. When the stretched string is vibrating in its fundamental mode, the effect of the stiffness is negligible; the string does not have any sharp bends. For higher harmonics, however, the bending becomes more pronounced because of the shorter wavelengths. When the piano string is played, it vibrates in many possible modes simultaneously. Because of the extra restoring force due to the stiffness of the string, each successive harmonic becomes higher in frequency with respect to the ideal value of Nf_1. The string stiffness causes the harmonics of the string to deviate from an exact, integral relationship of f, $2f$, $3f$, and so on. This is called *inharmonicity*, and it is more extreme in treble strings than bass strings because a given displacement of a short string produces a greater bending than in a long string. In fact, in most cases the sixteenth harmonic of a piano note (up four octaves from the fundamental) is about one half-step higher in pitch than the exact harmonic, a significant amount. The high notes on a piano are usually tuned slightly high and the low notes tuned slightly low to eliminate beats and to provide a better relationship between the harmonics of adjacent octaves. Piano tone quality is believed to be created by a combination of its unique attack and decay transients, the inharmonicities of the overtones of the strings, and the detailed tuning relationship between the strings tuned to the same note.

Wind instruments with a conical or tapered bore, as differentiated from a cylindrical bore, may have nonintegral harmonics. For some conical wind instruments, the inharmonicity is greater than that of the piano. In general, wind instruments with the conical flare largest at the blown end (such as the recorder or baroque flute) have harmonics that are low in frequency relative to the integral harmonics, whereas wind instruments with the conical flare largest at the bell end (such as the contemporary oboe, bassoon, and saxophone families) possess relatively higher harmonics.

In the cases of the piano and conical-bore wind instruments, the inharmonicities are generally not obvious to the listener, but can be demonstrated using electronic equipment. Even an untrained listener can hear the inharmonicity in some tuned percussion instruments, such as chimes. The frequencies of the overtones of such systems are not even close to bearing an integral relationship. The "note" we hear being "played" by a tuned drum like the tympani is usually the *second* lowest frequency standing wave, or resonance, in the system, not the fundamental. The fundamental of the chime can often be heard as a sort of dull thud at an intensity and frequency well below that of the note the instrument is sounding.

Two features are important in understanding and electrically reproducing (say on a synthesizer) tuned percussion sounds, such as those of a chime. First, and possibly most obvious, is the sudden attack when the chime is struck, followed by the slow decay of the sound. Second is a unique set of nonharmonic overtones, including the discernible fundamental *below* the "pitch" of the bell.

The tabla, an Asian Indian drum often played as accompaniment to the sitar, is unique among tuned drums. Its drumhead is specially weighted to force several of the resonances to have an integral relationship and to remove the loud low fundamental sound. As a result, the tabla has an especially clear, ringing tone with a well-defined pitch.

Further discussion of inharmonicities of tuned percussion sounds will be given when we study the electronic synthesizer and, in particular, the ring modulator.

Formants   Even after the substantial studies that have already been done on harmonic content, transients, and inharmonicity, there are significant problems with respect to tone differences between various instruments that remain unsolved. One of the procedures physicists like to use in explaining physical phenomena is to stand back, view the phenomenon from some distance, and draw general conclusions from the data. For example, suppose a series of Fourier spectra are plotted for every note on some instrument, and all harmonics that lie within a certain frequency region are found to be emphasized relative to the other harmonics. Such a frequency region is called a *formant region* of the instrument. A formant region can exist even if there is no other correlation or apparent similarity between the Fourier spectra of any groups of notes. In that case, the formant alone may be responsible for the tone quality.

Such a formant is believed by some to be responsible for the tone of the bassoon, where there is no apparent similarity between the Fourier spectra of different bassoon notes, other than an increase in amplitude of the harmonics in a high-frequency region. The resonator on an English horn may produce a formant at a high frequency that might contribute to the unique plaintive sound of the instrument.

Each human voice attains its unique tone quality (either speaking or singing) from the particular resonant frequencies of the three *resonant cavities*, the larynx, the mouth, and the nasal cavity. The ranges or groups of frequencies emphasized in each individual's voice, and therefore the vocal quality, are largely dependent upon details of the size and shape of these cavities, as well as the workings of the vocal folds. One can therefore view the particular groups of emphasized frequencies as formants leading to the unique quality of each individual's voice. Furthermore, each vowel sound has a particular character, which is determined by the resonant cavities that dominate the sound production or share in production of the sound, such as the nasal cavity for long "e," the mouth for "oo," and the throat for "uh." We can think of certain vocal formants being responsible for these vowel sounds, independent of the frequency or pitch at which the sound is uttered. We shall return to the human voice in Chap. 6.

Vibrato   Vibrato (periodic changes in the pitch of a musical tone) also adds a distinctive flavor to the tone. The vibrato of a singer's voice, for example, aids significantly in distinguishing the voice from other musical sounds. The term "vibrato" in general use refers not only to periodic changes in pitch, but also to periodic changes in amplitude, which should more correctly be called *tremolo*. The "diaphragm vibrato" of a flute player is close to pure tremolo; the vibrato obtained when a trombone player wiggles the slide in and out is almost a pure pitch vibrato. Singing vibrato is actually a mixture of true vibrato and tremolo. Vibrato on a violin or other string instrument is close to pure pitch vibrato.

Chorus effect   When many identical instruments play the same part in unison, the sound attains a quality considerably different from that of any of the individual instruments. This phenomenon is called the *chorus effect*. This is due to the superposition of many similar tones with random relative phases and slightly different frequencies and tone qualities. It is particularly evident in the string sections of a symphony orchestra. The sound of the solo violin playing in a concerto is different from the sound of the whole violin section that accompanies it, and the two are therefore clearly distinguishable.

4.4 Resonance Curves and Musical Sound Production

There are two fundamental steps in the production of a musical sound: (1) production of actual physical vibrations and (2) use of resonators to increase the amplitude of certain overtones. In this section, we shall see how the spectrum of the initial vibrations and the resonator characteristics determine the sound.

An elementary optics experiment is to pass a beam of white light (say from the sun or a very hot lamp) through a prism and observe that the white light is separated into its component colors, as shown in Fig. 4–22. The reverse process can also be demonstrated: Superposition of all the colors in the visible region results in white light. The audio analog to white light is white noise, which consists of all the frequencies of the audible spectrum superimposed with equal intensities; the graph of white-noise intensity versus frequency is shown in Fig. 4–23. This curve (line) shows clearly that all frequencies are present and all have the same intensity. White noise sounds like static on a radio or perhaps like rushing water. It is produced electronically in a synthesizer white-noise generator; a similar sound results from blowing onto a flute mouthpiece or occurs when the air rushes out of the windway of a recorder and hits the sharp, wooden wedge near the mouthpiece.

A variation on white noise is called *colored noise* or *filtered noise*. This is a general label for any noise that does not have all audio frequencies represented in nearly equal intensities, but instead has a very loosely defined pitch center caused

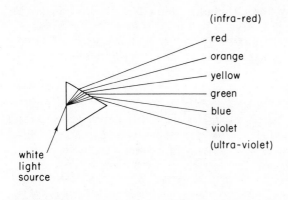

Figure 4–22 Separation of white light into its component colors using a prism.

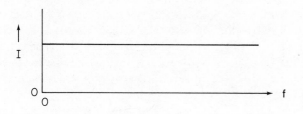

Figure 4–23 Spectrum of white noise.

by emphasis of some frequency range or ranges. Colored noise, like white noise, has a continuum of frequencies present, rather than only the frequencies in some overtone series or some other set of particular frequencies. The Fourier spectrum of one type of colored noise is shown in Fig. 4–24. This noise would sound like a very fuzzy whistling tone of frequency f_0. An example of colored noise would be the howling wind, which has in its spectrum virtually the entire audible range, with an emphasis on the frequency region around the frequency of the howling tone. This region is, in general, rather poorly defined. It should be emphasized that the graph of Fig. 4–24 illustrates only one type of colored noise, and that the term "colored noise" applies in general to steady-state nonmusical sounds that contain a continuum of frequencies in unequal intensities.

In electrical and audio engineering usage, the term "pink noise" means specifically noise with the same energy content in each octave of musical interval. This is equivalent to noise whose power per unit frequency drops off with frequency at a rate of 3 decibels per octave, a close approximation to the normal distribution of energy in orchestral and much other music. (We shall discuss the decibel intensity scale in Chap. 6.) This type of noise is useful in some tests of audio components.

White noise and colored noise are not generally classified as musical sounds. To produce a musical tone from white noise, a device is needed that will suppress most frequencies and strongly resonate those few that are to make up the desired musical tone.

The simplest such device, the Helmholtz resonator, named after the famous German physicist Hermann von Helmholtz (1821–1894), is illustrated in Fig. 4–25.

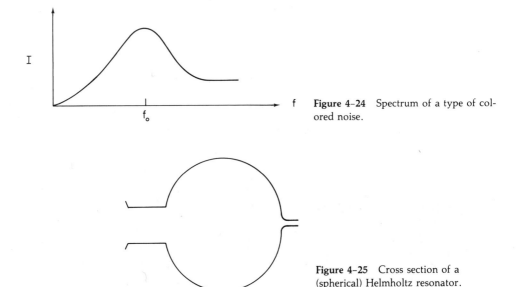

Figure 4–24 Spectrum of a type of colored noise.

Figure 4–25 Cross section of a (spherical) Helmholtz resonator.

It is a simple spherical cavity with a wide mouth, like the resonator of Fig. 4–25, with or without a small nipple on the side opposite the mouth. If a burst of air is blown into the cavity through the large neck, it will compress the air inside the cavity, which will then expand and rush outward, creating a low pressure inside the cavity. Outside pressure will then force the air back into the cavity at greater than atmospheric pressure. In this way the pressure in the cavity rapidly rises above, then drops below, the normal equilibrium atmospheric pressure. In fact, this resonance condition occurs for any such resonator at a frequency determined by the size and shape of the resonator. The small nipple on the right side of the Helmholtz resonator does not change the resonant frequency, but is inserted into the ear to allow the listener to hear the resonance. Helmholtz used a large set of these resonators in the nineteenth century to verify the existence of harmonics in complex tones, before the era of electronic Fourier spectrum analyzers. By holding successively smaller (higher-frequency) resonators to his ear with a musical note playing into the large opening, Helmholtz heard an increase in the amplitude of any frequency that was present in the harmonic structure of the instrument. Thus, he could roughly determine the Fourier spectrum of the note.

The Helmholtz resonator is very important because it possesses a single, low, isolated resonant frequency, as opposed to stretched wires, open and closed tubes, and other complex vibrating systems, which possess multiple closely spaced resonant frequencies. When a Helmholtz resonator is excited by white noise, the resonator causes an increase in amplitude of the very narrow band of frequencies near to its resonant frequency. We can perform an experiment to demonstrate this resonant property using the equipment shown in Fig. 4–26, where, instead of the ear, a sound probe and oscilloscope measure the resonant response. Instead of using a white-noise generator, we use a sine-wave generator, which produces all frequencies with the same amplitude, one at a time. As the frequency of the sine-wave generator is swept slowly across the audible range, the amplitude of the signal picked up by the sound probe will reach a maximum at the resonant frequency of the Helmholtz resonator and then decrease. The graph of this amplitude versus frequency is called the *resonance curve* for this Helmholtz resonator; it is shown in Fig. 4–27. The graph clearly shows a single rather narrow resonance that occurs at

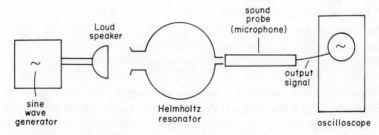

Figure 4–26 Equipment used to obtain the resonance curve of a Helmholtz resonator.

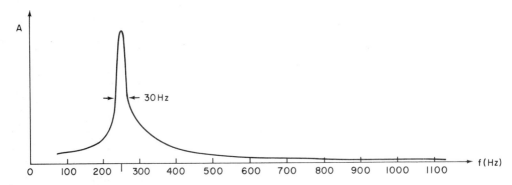

Figure 4-27   Resonance curve of a Helmholtz resonator whose resonant frequency is 250 Hz. The Helmholtz resonator has a single, narrow low-frequency peak with no other resonance structure below about nine times its fundamental frequency.

the frequency that is produced by blowing across the opening of the resonator, just as one would blow across the mouth of a soda bottle.

In addition to its importance in the historical development of acoustics, the Helmholtz resonator has important contemporary application to string instruments. The air cavity within the body of any string instrument acts as a type of Helmholtz resonator, creating emphasis of notes near its resonant frequency. In the case of the contemporary violin family, placement of the frequency of this "air resonance" is extremely important in producing uniform tone quality and loudness for all the notes near the low-frequency range of the instrument. The bottles of a bottle or jug band, particularly the bass bottles, also behave like Helmholtz resonators. In Chap. 7 we shall see that certain loudspeaker enclosures behave like Helmholtz resonators.

Drawing the resonance curves for other resonant systems is a convenient way to summarize the basic acoustical property of the system, that is, how the system responds to a stimulus at any frequency in the audible range. Figure 4–28 shows the resonance curve for an open tube. This graph tells us one thing that we already know, that an open tube resonates at all harmonics of the fundamental frequency; it also tells us how well it resonates at each of the harmonics, which is important in determining the amplitudes of the various harmonics and therefore the tone quality in this "instrument." The closed tube can support only odd harmonics, as shown in the resonance curve of Fig. 4–29. In both of these cases, the amplitudes, frequencies, and widths of the resonances are determined by the length and diameter of the tube. Resonance curves for real musical instruments tell us what overtones are possible in the sound of the instrument. Note that the resonance curve and the Fourier-spectrum curves are not the same, although they may be plotted on the same graph. The resonance curve tells which harmonics could be present in the tone of an instrument, if they were excited, whereas the Fourier spectrum gives the actual harmonic structure of a musical tone. The resonance curve describes the physical system; the Fourier spectrum describes the tone emitted.

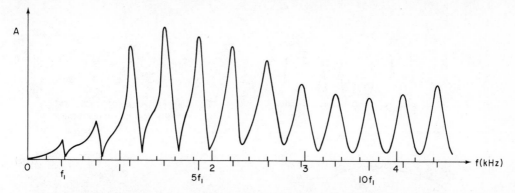

Figure 4-28 Resonance curve for an open tube whose resonant frequency is about 370 Hz. Harmonics of the fundamental frequency f_1 are indicated by the marks below the horizontal axis. All harmonics are present.

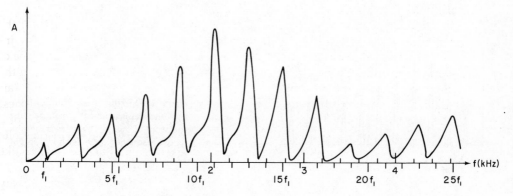

Figure 4-29 Resonance curve for a closed tube whose length is the same as that of the open tube of Fig. 4-28, giving a fundamental frequency of about 185 Hz. Harmonics of the fundamental frequency f_1 are indicated by the marks below the horizontal axis. Only odd harmonics are present.

As a further example illustrating these concepts, consider the hypothetical musical instrument whose resonance curve is shown in Fig. 4-30. Because all harmonics of the fundamental frequency f_1 are present in the resonance curve, this instrument is a member of the open-tube type. Now suppose that this instrument is excited by the noise spectrum given in Fig. 4-31. The strength of each harmonic present in the complex tone produced by the instrument will be a sort of product of the amplitude of the resonance curve at that frequency and the amplitude of the noise spectrum driving that frequency component, as shown in Fig. 4-32. Because of the complicated interaction between the air column and the noise source, this product will only approximate the actual resulting spectrum. The harmonics in the

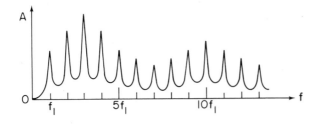

Figure 4–30 Resonance curve of a hypothetical music instrument.

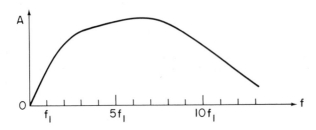

Figure 4–31 Noise spectrum produced by the hypothetical musical instrument whose resonance curve is shown in Fig. 4–30.

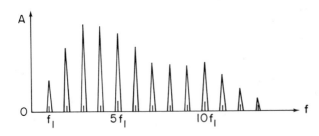

Figure 4–32 Fourier spectrum of the hypothetical instrument when the noise spectrum of Fig. 4–31 excites the note whose resonance curve is shown in Fig. 4–30.

resulting spectrum tend to be narrower and of higher amplitude than the simple product would predict, because there is always some interaction of the oscillating air column with the noise source. For example, both the noise spectrum and resonance curve amplitudes are greater at $3f_1$ than at $2f_1$ or f_1. Consequently, the amplitude of the Fourier spectrum is quite a bit larger at $3f_1$ than at $2f_1$ or f_1. Even though the noise spectrum between $7f_1$ and $10f_1$ is falling in amplitude, the increase in the amplitude of those frequencies in the resonance curve results in the amplitudes of components of the Fourier spectrum being approximately equal. In general, if either the noise spectrum or the resonance curve is zero at some harmonic frequency, the Fourier component at that frequency will be absent or have zero amplitude. While this procedure is neither quantitative nor rigorous, it does illustrate the qualitative relationship between the resonance curve, the noise spectrum, and the Fourier spectrum.

In reality, the production of sound by exciting resonances in musical instruments is more complicated. Even in the "simple" case just discussed, there is a *feedback* mechanism by which the harmonics present are strengthened in intensity and narrowed in frequency. This process happens very quickly and becomes part of the attack transient. The spectrum of the sound source can be affected by the standing wave. A more complete treatment of this topic will be given in Chap. 10, when we discuss the woodwind instruments.

Another illustration of these concepts involves the sound produced when water is poured into a cylinder, illustrated in Fig. 4-33. The sound of the gurgling water is nearly white noise, containing almost all frequencies of the audible spectrum, as graphed in Fig. 4-34. The cylinder is a closed tube, with water forming the closed end, and at some time (say when the water is as in Fig. 4-33) has a resonance curve consisting of odd harmonics of its fundamental f_0, as in Fig. 4-29. The sound produced by the gurgling noise with resonances from the closed tube superimposed on the noise is shown in Fig. 4-35. As water fills the cylinder, the noise remains roughly the same, but f_0 and all its odd harmonics increase in frequency because the length of the resonant air column decreases. The overall sound consists of a resonably well-defined pitch center, composed of a fundamental and some (odd)

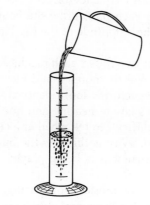

Figure 4-33 Experiment to analyze sound of water pouring into cylinder.

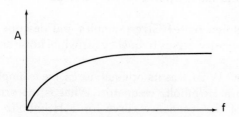

Figure 4-34 Noise spectrum of gurgling water as it is poured into cylinder.

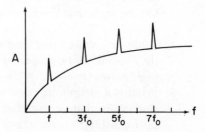

Figure 4-35 Complete sound spectrum of water pouring into cylinder at some instant of time. Spectrum consists of closed tube harmonics superimposed on gurgling noise.

harmonics superimposed upon a continuous wooshing, gurgling sound. As the water fills the cylinder, all the wavelengths of the resonances decrease, and the pitch rises.

EXERCISES

1. Draw five identical pairs of graph axes numbered 1 to 5; one of each pair should have its horizontal axis labeled in milliseconds, while the second should have its horizontal axis divided into segments of 100-Hz frequency.
    (a) On the first of each pair draw the following waves:
    (1) sine wave with period of 10 ms;
    (2) triangle wave with frequency of 200 Hz;
    (3) square wave with period of 20 ms;
    (4) sawtooth wave with period of 10 ms,
    (5) pulse train with frequency of 300 Hz.
    (b) On the second of each pair, draw the Fourier spectrum of each of these waves. Assume the amplitude of the fundamental frequency is 1.0 in each case.

2. The Fourier spectra of two notes, each produced by a different musical instrument, are identical. Will the notes sound identical, similar, or possibly very different? Will the waveforms look identical, similar, or possibly very different? What is Ohm's law of hearing and how can it be applied to this exercise? It is sometimes difficult to tell whether a certain note is being played on a violin, viola, or cello. Why is this? As the music continues, you can finally distinguish which instrument is being played. What might aid in this identification?

3. List six characteristics of musical sounds that have a bearing on their tone quality. State what each means, and give some examples where each might be of use. Why does one loud violin sound different from ten violins playing in unison such that the total sound intensity level is the same as for the single violin?

4. What are white noise and colored noise? Give examples and describe how they sound. Draw graphs of intensity versus frequency for white noise and for pink noise.

5. What is a Helmholtz resonator? What was its original use? Give examples of resonators that are similar to a Helmholtz resonator. What is the primary distinguishing characteristic of the resonance curve for a Helmholtz resonator?

6. (a) The clarinet is a cylindrical reed instrument whose reed end acts acoustically like a closed end. What harmonics might one expect to be emphasized in the resonance curve of the clarinet?

(b) The flute has the mouthpiece hole at one end and finger holes (or open tube) at the other end. What harmonics might one expect to be emphasized in the resonance curve of the flute?

(c) A flute player plays a note. Then, while keeping her fingers in the same positions, she blows harder into the mouthpiece. The note sounded is one octave higher than the first one. Why is it an octave higher?

(d) When a clarinetist blows extra hard, while keeping his fingers in the same position, he "overblows" at an octave and a fifth. Why an octave and a fifth?

7. What is a formant? The human vocal system acts like a closed tube about 17.25 cm long. What are the frequencies of the first two formant regions?

8. Discuss the basic relationship between the resonance curve and the Fourier spectrum for a note produced on a wind instrument. How is the applied noise spectrum and the resonance curve used explicitly to determine the Fourier spectrum of the note produced? Draw two different noise spectrums and two different resonance curves. Draw the Fourier spectra of the four possible notes produced by using each of the two noise spectra with each of the resonance curves.

REFERENCES

These references apply to Chaps. 1 through 4.

APEL, WILLI. *Harvard Dictionary of Music.* 2nd ed. Cambridge, Mass.: Harvard University Press, 1973.

A classic dictionary of music, with emphasis on historical aspects.

BACKUS, JOHN. *The Acoustical Foundations of Music.* 2nd ed. New York: W. W. Norton & Company, Inc., 1977.

BENADE, ARTHUR, H. *Fundamentals of Musical Acoustics.* New York: Oxford University Press, Inc., 1976.

These two books are the classics in the field of musical acoustics. Backus is somewhat more general and at a less sophisticated mathematical and physical level; Benade is much more detailed and at times requires considerable sophistication on the part of the reader. Both are excellent books.

BENADE, ARTHUR H. *Horns, Strings, and Harmony.* Garden City, N.Y.: Doubleday & Company, Inc. 1960.

Covers many aspects of acoustics at the level of advanced high school or nontechnical college students.

BLOM, ERIC, ed. *Grove's Dictionary of Music and Musicians.* 5th ed. New York: St. Martin's Press, Inc., 1973.

A ten-volume set containing a wealth of information on instruments, acoustics, and other areas related to music at the level of the nontechnical reader.

GIANCOLI, D. *Physics: Principles with Applications.* Englewood Cliffs, N.J.: Prentice-Hall, Inc., 1979.

Fundamental principles of physics, including waves and sound, for the undergraduate.

HALLIDAY, DAVID, AND ROBERT RESNICK. *Physics, Part I and II.* Combined 3rd ed. New York: John Wiley & Sons, Inc., 1966.

Popular physics text for students with mathematical and technical backgrounds.

HELMHOLTZ, HERMANN VON. *On the Sensations of Tone.* New York: Dover Publications, Inc., 1954.

An English translation of the original German classic by one of the founding fathers of acoustics. This book gives a unique historical perspective to many aspects of acoustics and is to a great extent understandable by the general reader.

KARZMARK, C. J., AND FEARGHUS O'FOGHLUDHA. "Introducing the AAPM." *Physics Today* 26 (November 1973): 39.

Discusses activities of the American Association of Physicists in Medicine, including recent ultrasonic applications.

LINDSAY, R. BRUCE. *Acoustics: Historical and Philosophical Development.* Stroudsburg, Pa.: Dowden, Hutchinson & Ross, Inc., 1976.

A large collection of papers on acoustics dating from the time of Aristotle to the turn of the twentieth century, including many of the historically important authors from the nineteenth century.

OLSON, HARRY F. *Musical Engineering—An Engineering Treatment of the Interrelated Subjects of Speech, Music, Musical Instruments, Acoustics, Sound Reproduction, and Hearing.* New York: McGraw-Hill Book Company, 1952.

This is a classic book in the field of engineering applications of acoustics, written for the mathematically more advanced student.

ROEDERER, JUAN G. *Introduction to the Physics and Psychophysics of Music.* 2nd ed. New York: Springer-Verlag, New York, Inc., 1975.

An excellent, basic introduction to psychoacoustics, written for a slightly more mathematical audience than the present text.

ROSSING, THOMAS D. "Resource Letter MA-1: Musical Acoustics." *American Journal of Physics* 43 (1975): 944.

This is one of many AJP resource letters dealing with various topics in physics, containing a good list of primary reference material.

RYAN, PATRICK. "Now Just Ultrasound them Varmints to Death." *Smithsonian Magazine* 10, no. 10,: 144.

Short discussion of ultrasound as a method for exterminating rodents and insects.

SACHS, CURT. *The History of Musical Instruments.* New York: W. W. Norton & Company, Inc., 1940.

This is possibly the most complete book on the history of instruments, with many good drawings and diagrams.

SAUNDERS, FREDERICK A. "Physics and Music." *Scientific American,* July 1948.

This is an early modern survey article on acoustics, presaging the contemporary interest in musical acoustics.

SAVAGE, WILLIAM R. *Problems for Musical Acoustics.* New York: Oxford University Press, Inc., 1977.

Many problems, with some discussion, at or near the level of this book, with emphasis on music applications.

SETO, WILLIAM W. *Theory and Problems of Acoustics* (Shaum Outline Series). New York: McGraw-Hill Book Company, 1971.

Contains theory plus 245 solved problems dealing with many aspects of waves and acoustics; mathematical level is generally well above that of this text.

Acoustical References

BERG, RICHARD E., AND DAVID G. STORK. *Demonstrations in Acoustics.* College Park, Md: University of Maryland, Department of Physics and Astronomy, Lecture Demonstration Facility, 1980.

This is a set of four 1-hour color videotapes designed as a companion to this text. Many of the experiments described in the present text are performed and brief explanations given.

The Science of Sound, Folkways Record Album No. FX6136, Descriptive Literature by Bell Telephone Laboratories.

This is an excellent recording, covering many areas of acoustics, which can be readily used in class lectures. The records contain short segments illustrating such topics as the overtone series, tone quality, filtering, distortion, reverberation, the Doppler effect, and others.

Several periodicals deal in part with topics in acoustics:

Journal of Acoustical Society of America (JASA)
Acustica
Physics Today
American Journal of Physics (AJP)
The Physics Teacher

The first two journals contain primarily research papers of researchers in areas of acoustics written to and for each other and are therefore likely to be too difficult for the general reader. *Physics Today* and the *American Journal of Physics* contain a mixture of types of articles, some of which are appropriate for the nonmathematically sophisticated reader. *The Physics Teacher* is used heavily by high-school teachers, and as such contains a high proportion of articles readable by nontechnical individuals.

5

Electronic Music and Synthesizers

Electronic synthesizers are used in many settings: serious contemporary music, light background music, jazz, and rock and roll. Even some classic works of Bach, Beethoven, and others have been transcribed for synthesizer. As advances in electronics allow production of portable, reasonably inexpensive synthesizers, and as musical tastes continue to change, the synthesizer will play an ever increasing role in music.

In this chapter, we shall describe the important synthesizer components and their operation and then the accomplishments and limitations of the synthesizer. Finally, we shall discuss the technique of sound-on-sound recording, by which most electronic music recordings are produced. We begin with a discussion of the basic concepts of modulation and combination of waves, concepts necessary for an understanding of the synthesizer.

5.1   Combination of Waves and Modulation

It is often desirable to combine two or more waves to obtain a special sound or effect; modulation (unlike simple addition) includes several very specific techniques for combining two waves so that one of the waves changes some physical characteristic of the other. Before discussing synthesizer applications of these phenomena and their effects on the sound produced, we shall turn our attention to wave modulation. The particular methods of combining waves to be discussed here are (1) simple addition, (2) gating, (3) amplitude modulation, (4) balanced modulation, (5) frequency modulation, and (6) pulse width modulation.

Simple addition We have covered simple addition of waves extensively in our treatment of Fourier synthesis and beats. Simple addition refers to the addition of two or more waves point by point to obtain a complex wave. In general, the component waves need not be pure waves, as they have been in many previous examples. Recall that, if the two waves are in the harmonic series of a certain fundamental, their sum will have a period equal to that of the fundamental. If the two waves are not harmonically related, the sum wave will be continuously changing in shape because of the changing of the relative phase between the component waves.

Gating The technique of gating involves using one wave to turn another wave on and off. The gating signal switches on and off a higher frequency signal. The higher frequency signal is switched on whenever the value of the gating signal is above a certain preset value, called the *gating level.* Figure 5–1 shows a high-frequency sine wave gated by a lower-frequency square wave. The gating level in this example is 0. If the gating wave is a pulse that is on 10 percent of the time, the wave of Fig. 5–2 results. This is the way the signal was obtained for the experiment to determine the velocity of sound in Sec. 2.1. One can consider a simple on-off switch or a synthesizer key pressed by the performer as producing a gating signal. The gating signal need not be periodic.

Amplitude modulation (AM) In the modulation technique, one signal causes a change in one of the physical characteristics of the other, rather than switching it on and off. The higher-frequency signal is usually called the *carrier,* and the lower-frequency signal is the *modulator.* In amplitude modulation the modulator signal changes the amplitude of the carrier. The value of the amplitude of the modulated signal at any instant is dependent on the value of the modulator at that instant. Figure 5–3 shows a fast sine-wave carrier being amplitude modulated by a slow sine-wave modulator. The envelope of the modulated wave, the curve that defines its overall shape, has the shape of the modulator wave, and its axis is displaced up-

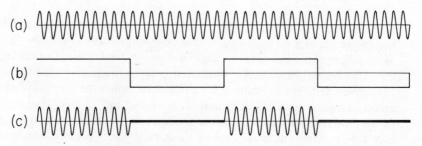

Figure 5–1 Gating of a high-frequency sine wave (a) by a low-frequency square wave (b) producing the gated signal (c).

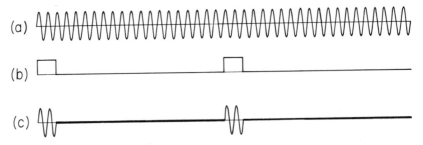

Figure 5-2   Gating of a high-frequency sine wave (a) by a pulse (b) which is on for 10 percent of its period to produce the gated wave (c).

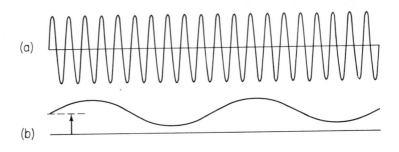

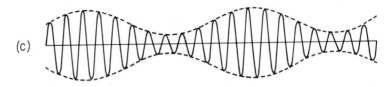

Figure 5-3   Amplitude modulation of a sine wave carrier (a) by a sine wave modulator (b). The envelope (dashed lines) of the modulated wave (c) is obtained from the modulator and its mirror image opposite the equilibrium level of the carrier wave. The offset level is marked by the arrow.

ward from the carrier axis by a certain specified amount, the *offset level*. Because the amplitude of a wave at any instant defines the wave's limit above and below the axis, we can draw the mirror image of the displaced modulator curve to define the envelope below the axis. One hundred percent modulation is said to occur when the amplitude of the modulator is sufficient to reduce the amplitude of the carrier to zero at some time during each period, as shown in Fig. 5-4. That is, the amplitude of the modulator is exactly equal to the offset level. Overmodulation, shown in Fig. 5-5, occurs when the amplitude of the modulator is greater than the offset level.

   If the carrier is an audio-frequency tone and the modulator has a frequency of about 1 to 10 Hz, the effect of amplitude modulation can be heard as a pulsing tone,

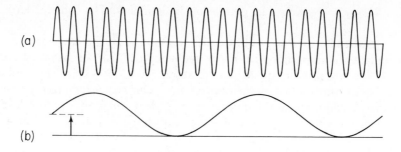

(a)

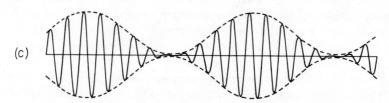

(b)

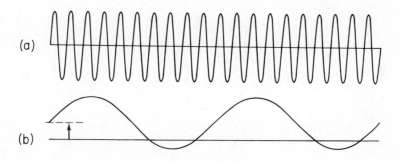

(c)

Figure 5-4 A sine wave carrier of 100 percent amplitude modulated by a sine wave modulator. The carrier (a), the modulator (b), and the modulated wave (c) are shown. The offset level is marked by the arrow.

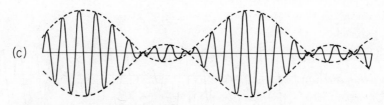

(a)

(b)

(c)

Figure 5-5 AM overmodulation of a sine wave carrier (a) by a sine wave modulator (b) to produce the modulated wave (c). The offset level is marked by the arrow.

continually increasing and decreasing in loudness. The musical term for this is *tremolo.*

AM radio uses an audio-frequency modulator (speech, music, and so on) to modulate the amplitude of a radio-frequency (RF) carrier signal whose frequency lies in the AM radio-frequency band, about 500 to 1800 kHz. The modulated wave is the signal transmitted by the radio station. Our radios then receive the modulated RF carrier and perform the inverse process, demodulation, to reproduce the audio-frequency modulator signal, which we hear on the loudspeaker as music or speech.

Balanced modulation   In the case of amplitude modulation just discussed, it can be observed in Figs. 5–3 through 5–5 that the zero or offset level of the modulator wave is displaced from the zero level of the carrier wave. For this reason, amplitude modulation is sometimes called *unbalanced modulation.* If the offset level of the modulator is 0, *balanced modulation* (BM) or *double-sideband modulation* (DSBM) results. In the world of musical synthesizers, the effect is called *balanced modulation* or *ring modulation* (RM). Figure 5–6 shows a sine wave balance-modulated by a lower-frequency sine wave; the envelope is obtained again by reflecting the modulator wave about the axis of the carrier. Careful inspection of this graph and comparison to the graph of 100 percent amplitude modulation shown in Fig. 5–4 will reveal a slight difference between the shapes of the envelopes, especially for values near zero. In addition, the final DSBM wave undergoes a phase

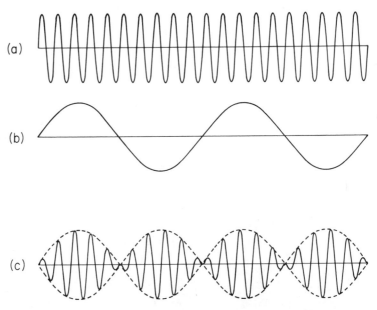

Figure 5-6   A sine wave carrier (a) double side band modulated by a sine wave modulator (b) to produce the modulated wave (c). The offset value of DSBM is 0.

inversion each time the modulator passes through zero. The graph of balanced modulation is identical to the graph of beats shown in Fig. 2–36.

Let us consider more carefully the similarity between beats and balanced modulation. We saw in Sec. 2.4 that, when two equal-amplitude sine waves of frequencies f_1 and f_2 (f_2 larger than f_1) are added, we obtain a wave whose frequency F is the average value of f_1 and f_2: $F = (f_1 + f_2)/2$. The beat rate f_b is the difference between the two frequencies: $f_b = f_2 - f_1$. The balanced-modulation curve demonstrates that this same curve can also be generated by an audio frequency F double sideband modulated by a sine wave of low frequency $f_m = f_b/2$. Each period of the modulating wave produces two beats. Hence balanced modulation can be used to produce a complex wave containing only frequencies f_1 and f_2 from the frequencies F and f_m. The two frequencies thus produced are $f_2 = F + f_m$ and $f_1 = F - f_m$. This is of particular interest when beats are no longer heard because f_m is in the audible range. For example, if $F = 500$ Hz and $f_m = 83$ Hz, the balanced modulator produces the frequencies 417 Hz $= 500$ Hz $- 83$ Hz and 583 Hz $= 500$ Hz $+ 83$ Hz. These two frequencies are not members of a single overtone series; the ratio 583 : 417 is not reducible to a ratio of small integers. The balanced modulator produces nonintegral overtones like those in tuned percussion instruments. If the carrier wave is complex, containing harmonics, each of the harmonics will combine with the modulator to produce two new frequencies; all of these will, in general, be inharmonic.

Frequency modulation (FM)   In frequency modulation, the modulator causes the frequency of the carrier to vary while leaving the amplitude of the carrier wave unchanged. For positive values of the modulator signal, the (modulated) frequency becomes greater, and for negative values of the modulator signal, the frequency becomes less than its normal value; the greater the amplitude of the modulator, the greater is the deviation in the frequency. Figure 5–7 shows a sine wave frequency modulated by a sawtooth wave. Figure 5–8 shows a square wave frequency modulated by a triangular wave. Figure 5–9 shows a triangular wave frequency modulated by square waves of various amplitudes and frequencies. A square-wave modulator has two values; therefore, the modulated wave has two frequencies, one below and one above the carrier frequency. With some thought one can understand that (1) the amplitude of the modulator can be increased, with the result that the frequency variation of the modulated carrier wave becomes greater, or (2) the frequency of the modulator can be increased, resulting in faster changes in the frequency of the carrier. The greater the amplitude of the modulator, the greater the frequency deviation, while an increase in the modulator frequency results only in more rapid alternation of the resulting frequencies.

When the carrier wave is in the audible range, and the modulator has a frequency of a few Hz and an amplitude sufficient to cause a few tenths of one percent change in the carrier frequency, we hear a sound of constant amplitude or loudness whose pitch oscillates slightly (somewhat less than one half-tone). Musicians call this effect *vibrato*.

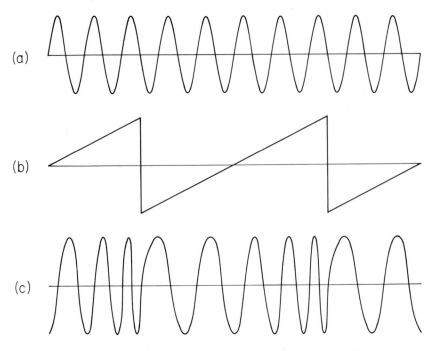

Figure 5-7 A sine wave (a) frequency modulated by a sawtooth wave (b) to produce the modulated wave (c).

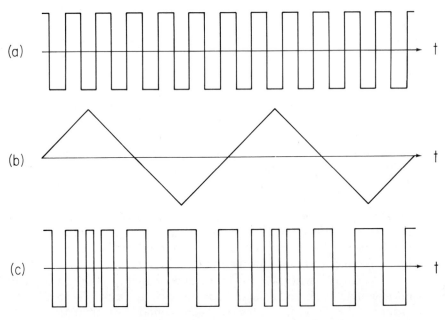

Figure 5-8 A square wave (a) frequency modulated by a triangular wave (b) to produce the modulated wave (c).

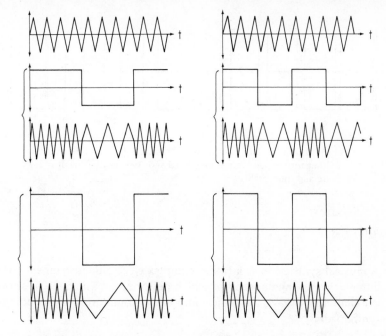

Figure 5-9 Frequency modulation of a triangular wave (top) by various square waves, producing the modulated waves shown below each square wave modulator.

FM radio uses a carrier of about 100-MHz frequency modulated by speech, music, or other sound up to about 75 kHz above and below the carrier frequency. The improvement in the quality of FM radio over AM radio is due to the use of the large frequency range over which the modulation occurs and the nature of the noise sources that affect radio waves. Most radio noise sources affect the amplitude of radio signals, and thus AM signals will be greatly affected. Even though the amplitude of an FM signal is changed if noise is present, the noise has almost no effect, because the demodulation process deals only with the frequency of the signal. An amplitude threshold exists below which AM is less susceptible to noise than is FM, however.

Pulse width modulation (PWM)   Pulse width modulation, which produces a sound unique to electronic instruments, is obtained when the duration of the upper level of a flat-top wave is determined by a modulator wave. For instance, the modulated wave may change form between a square wave (50 percent on time) and a pulse train (very short on time) of the same fundamental frequency, as shown in Fig. 5–10. Pulse width modulation changes the tone quality, as it changes the wave shape and thus the harmonic content of the wave. This unusual effect cannot be obtained on standard musical instruments played their normal way.

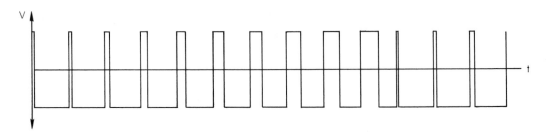

Figure 5-10 Pulse width modulation (PWM). The period of the note remains constant, while the on time for the pulse ranges between 5 and 50 percent. Thus there is a change in harmonic structure and tone quality.

5.2 Synthesizers

A musical synthesizer is a rather complex electronic instrument consisting of many different components, each of which performs a basic task, as shown in Fig. 5–11. We shall now discuss the basic components common to most contemporary musical synthesizers.

Signals Two types of signals are used in a synthesizer: audio signals and control signals. Audio signals are electronic signals with frequencies in the audible range and amplitudes of about 1.5 V. They can be input to a speaker at any stage (after some amplification). Control signals can be either oscillatory or constant in voltage, depending upon their function, and have amplitudes up to about 5 V, in keeping with the requirements of solid-state electronic components. Control signals are used to control the operation of various components in the synthesizer. For instance, a control signal could control the amplifer and thus the loudness of the sound emitted by the speakers.

Keyboard The electronic synthesizer keyboard looks similar to that of a piano except that it contains fewer keys, in general. The synthesizer keyboard provides two functions. First, it provides the control signal that determines the pitch of the note; second, it provides a trigger, or starting signal, that triggers the envelope generator, which can be used to create the attack transient of the tone. Figure 5–12 shows some typical control signals produced by pressing a key.

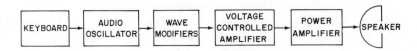

Figure 5-11 Block diagram for flow of audio signals in a typical synthesizer.

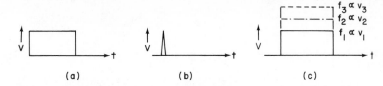

Figure 5-12   Three types of control voltages produced by depressing a synthesizer key. Signals (a) or (b) can be used to trigger the envelope generator or other control modules; signal (c) is used to control the pitch of the note produced by the voltage-controlled oscillator (VCO), and is different in level for each key.

Sample and hold   In general, a keyboard pitch control signal is input to the oscillator, and the audio output from the oscillator is sent to a speaker. Pressing a key will then cause the production of a tone with the pitch associated with that key. The oscillator remembers the pitch and continues to sound at this pitch even when the key is released. This is known as *sample and hold.* The tone will then continue at the same pitch until another key is pressed, at which time the oscillator will remember the new pitch.

Voltage-controlled oscillator (VCO)   The main oscillator for a synthesizer, often called the VCO, takes a control voltage from the keyboard to produce an audio signal whose frequency is determined by the amplitude of the control voltage. If the synthesizer is tuned like a piano (and they almost always are), the keyboard pitch-control voltages will be set so that playing one half-step higher on the keyboard produces a voltage 1.05946 $(= \sqrt[12]{2})$ times that of the preceding note. This is called *equal temperament* and guarantees that after one octave of 12 half-steps the voltage will double. The VCO produces a frequency directly proportional to the voltage; for example, doubling the input voltage doubles the frequency and produces a one-octave change in pitch.

Most VCO units produce several standard waveforms: sine, triangular, sawtooth, square, and pulse train. Often the square wave and pulse train are produced by utilizing a flat pulse whose on time is variable between near zero (pulse train) and 50 percent (square wave). Some wave shapes that can be produced by such a system are shown in Fig. 5-13. By means of an appropriate wave form containing the desired set of harmonics and the appropriate filters (discussed later), a large number of combinations of overtones can be produced.

Voltage-controlled amplifier (VCA)   A VCA simply takes the audio signal input to it and amplifies this signal (increases its amplitude) by an amount proportional to some input control voltage. The keyboard, VCO, and VCA are the minimum components necessary to produce a musical line including rests (moments of silence); a block diagram of the components used is given in Fig. 5-14. The keyboard determines the pitch of the VCO using any desirable wave form. The keyboard trigger control causes the VCA to pass the signal when a key is depressed

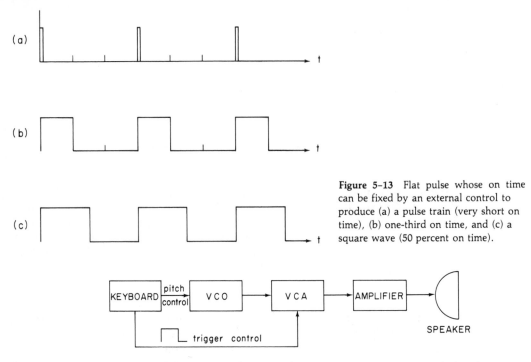

Figure 5–13 (a)

Figure 5–13 (b)

Figure 5–13 Flat pulse whose on time can be fixed by an external control to produce (a) a pulse train (very short on time), (b) one-third on time, and (c) a square wave (50 percent on time).

Figure 5–13 (c)

Figure 5–14 Block diagram indicating a simple type of keyboard control of pitch and duration of notes.

and stop when the key is released. Of course, this simple on-off gated tone is not very interesting, so we need to introduce additional elements: envelope generators, filters, and modulators, which are included in the box labeled "Wave modifiers" in Fig. 5–11.

Envelope generator   Most musical sounds do not merely come on, remain at a constant amplitude, and stop abruptly as does the tone discussed previously. Rather, for a wind instrument, there is an attack period during which the sound may be a bit louder than during the longer period when it is being held. For a plucked string instrument, the attack period is immediately followed by a continual decrease in the amplitude of the sound. An important part of the sound of these instruments is this "envelope," which determines the amplitude of the note at each time throughout its duration.

The envelope generator determines this envelope and has four sequential functions, as illustrated in Fig. 5–15. The *attack* (A) is the time interval after depressing a key during which the initial rise of sound occurs. The *decay* (D) period is the time required for the amplitude to decrease from its attack peak to its steady-state or *sus-*

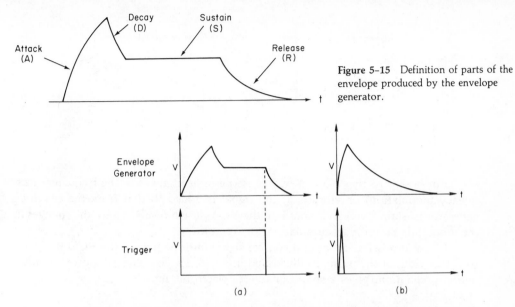

Figure 5-15 Definition of parts of the envelope produced by the envelope generator.

Figure 5-16 Trigger (control) signal from keyboard and envelope generator signal for a normal wind instrument sound (a) and a plucked string sound (b).

tain (S) level, at which it remains until the key is released. Upon release of the key by the performer, the *release* (R) period ensues. The attack, decay, and release periods and the amplitude level for the sustain are all independently controllable by knobs on the envelope generator; the duration of the sustain period is determined by the duration the keyboard key is depressed. Figure 5-16 shows the relationship between the keyboard trigger signal and the envelope generator for two typical cases. A reasonably long attack time followed by a high sustain level is typical for a wind instrument envelope. A very rapid attack time followed by a long release with no sustain period is necessary to achieve a plucked string sound. A longer release time is used to simulate a long reverberation time, such as that found in a cathedral.

The envelope generator can also be used to control harmonic content in the wave by putting its signal into a filter in addition to or in place of its normal input into the VCA, thus allowing production of transients. The normal flow diagram using an envelope generator is shown in Fig. 5-17. For simple cases, and in some elementary envelope generators, the decay control is missing, and the envelope is generated by attack, sustain, and release only.

Low-frequency control oscillator This unit, sometimes referred to as a control oscillator or a low-frequency oscillator, is used solely for control purposes. It will normally be used to produce a sinusoidal control signal with a frequency between 0 and about 25 Hz; this is below the audible range. If the signal from this

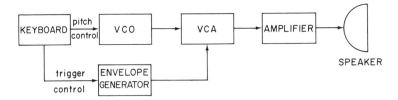

Figure 5-17 Block diagram of normal signal flow for control of the shape of notes by means of an envelope generator.

oscillator is sent to the VCO, it will cause periodic changes in the frequency (frequency modulation) or vibrato. If the signal is added to that from the envelope generator (sustain level) and input into the VCA, it will result in periodic changes in the amplitude (amplitude modulation) or tremolo.

It is also possible to input a control signal into the part of the VCO that controls the width of the pulse, as illustrated in Fig. 5–13. This effect, known as *pulse-width modulation*, gives a sound unique to electronic music.

Filters A filter is an electronic device used to remove some frequency or frequency bands from the frequency spectrum of the signal produced by the VCO. Several types of filters are normally available on a synthesizer.

The low-pass filter allows passage of low-frequency components while removing high frequencies, whereas a high-pass filter removes the low frequencies while allowing high frequencies to pass. Graphs showing the ratio of output to input signal amplitudes for typical low-pass and high-pass filters are shown in Fig. 5–18. The transition frequency f_F of the filter can usually be set by the operator using an external control voltage or a knob on the unit, whereas the frequency range over which the transition occurs is usually an inherent property of the filter.

A bandpass filter passes only a small frequency band around its frequency f_F. The ratio of the frequency of the filter to the frequency range passed by the filter is called its Q: $Q = f_F/f_W$, so a filter with a small frequency width has a high Q. Both f_F and Q can be controlled by separate knobs on the unit or by a control voltage. A notch filter, or band-reject filter, is the inverse of the bandpass filter and rejects all frequency components in a narrow band. The characteristics of bandpass and notch filters are shown in Fig. 5–19.

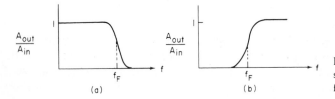

Figure 5-18 Ratio of output to input signal versus frequency for a low-pass filter (a) and a high-pass filter (b).

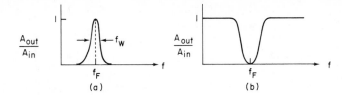

Figure 5-19 Ratio of output to input signal versus frequency for a bandpass filter (a) and a notch filter (b).

For some filters, the frequency f_F is fixed. In more sophisticated filters the frequency of the filter varies with the audio-frequency input so that it always affects the same harmonic components of the complex wave, creating a more uniform Fourier spectrum for all notes. Such a filter, where the frequency of the filter f_F varies with the frequency of the audio tone f_1 (that is, $f_F : f_1$ is a constant), is called a *tracking filter*. For instance, a tracking filter can be set to reject all harmonics above the third for a complex tone of any frequency.

Noise generator   A noise generator, or white-noise generator, produces white noise. It is most generally used for special nonmusical sounds, but is required to produce the sound of a cymbal or a tuned snare drum, for example.

A white-noise generator and a bandpass filter can be used to produce a very striking wind sound. The white noise is input into the bandpass filter and then sent to the VCA. Hand knobs control the frequency and Q of the filter. By adjusting the Q of the filter, either a narrow or a relatively wide frequency band (pitch center) can be obtained; by adjusting the center frequency of the filter, this pitch center can be raised or lowered. Simultaneous adjustment of both controls causes the wind to howl and sound as if it were changing its speed.

Colored noise, consisting of white noise with some frequency ranges missing or reduced in amplitude, is obtained by filtering white noise.

Ring modulator   A ring modulator performs balanced modulation with the two signals that are input into it. These two signals can be an audio signal and a control signal, or they could be two audio signals. Because, in general, ring modulation produces signals with nonintegrally related frequency components, the ring modulator is useful in synthesizing tuned percussion sounds, such as those from a chime or bell. Two features in the tuned percussion sound such as a bell are (a) nonintegral frequency components and (b) a sharp percussive attack with long release time and no sustain. These two features can be achieved using the synthesizer setup shown in Fig. 5-20. Two audio signals from the keyboard are used in ring modulation while a short attack and long decay envelope generator function are used. It is necessary for the two frequencies to track; that is, the two frequencies must go up and down together as different notes are played. The (nonintegral) ratio of the frequency components in the signal produced by the ring modulator will then remain constant, producing similar tone quality for all bell notes.

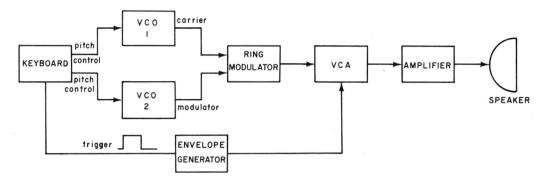

Figure 5-20   Block diagram for production of tuned percussion sounds using a ring modulator.

5.3   *Use of the Synthesizer*

In the previous section we introduced most of the important components of electronic music synthesizers and illustrated their use with diagrams showing how they would be connected to produce certain sounds. We shall now discuss more complex synthesizer programming, which can be used to obtain an infinite variety of sounds.

One of the more interesting recent uses of the synthesizer is for filtering and modulating the sounds of other musical instruments and voices. The sound of the standard instrument is picked up by a microphone or other pickup and amplified to the level of audio signals in the synthesizer. Modules of the synthesizer are then used to modify the sound of the instrument before it is amplified and output to a speaker.

A fascinating type of control available on some of the more recent synthesizers is the "joy stick." This allows mixing of two different control functions in any arbitrary ratio. It is particularly effective in varying the location of the sound smoothly and rapidly between several speakers or adjusting two filters or mixers simultaneously.

Because of the way in which the signals are transferred from the keyboard to the VCO, most synthesizers are capable of producing only one note at a time. However, recent attempts have been made at producing two or more notes with some success. It is relatively easy to use two VCO units; two or more notes can be played simultaneously on some of the larger synthesizers. A problem that has not been completely solved is production of two notes whose wave forms differ, say for use in polyphonic music. The lower line can be of one timbre while the upper line is of another timbre, but if the lines cross, it has not been possible for the synthesizer to retain the proper identification. A method of tricking the synthesizer into preserving identification of crossing lines is to play the top line an octave high on the keyboard and tune its VCO an octave low to compensate.

As requirements for the control of synthesizers and the speed and complexity of the signal response have become more stringent, it has been natural to use computers for the control of large synthesizers. The necessity of very sophisticated controls is particularly evident in electronic voice synthesis. Vowel sounds are relatively simple to produce using standard oscillators and filters, even on the smallest synthesizers. Sibilants (for example "s" and "z" sounds) can be produced readily by addition of some white or pink noise. However, consonant sounds are extremely difficult to produce because of their rapid and very complex transients. This problem has not yet been solved satisfactorily even with use of some of the larger computer-controlled synthesizers available at this time. Analysis of transients in musical instrument tones and attempts to reproduce them with computer-controlled synthesizers are areas of contemporary acoustics research. Solutions of these problems should give us fuller appreciation of what gives various instruments their unique sounds.

Another limitation of the synthesizer in mimicking sounds of musical instruments lies in the production of the original wave. The electronically generated synthesizer waves are composed of exact harmonics, whereas many acoustical instruments contain inharmonicities. Thus a musical sound that depends on inharmonicity of its overtones, as in the piano, cannot be faithfully reproduced by most synthesizers.

The most effective musical use for the synthesizer is in producing new, uniquely electronic sounds, and not in attempting to reproduce sounds of other musical instruments or the voice. Most popular synthesizer recordings use features and sounds unique to the synthesizer. With proper control, sound effects, such as wind, thunder, bombs falling, dogs barking, passing trains, and others, can be obtained.

The synthesizer is now one of the most ubiquitous musical instruments; it is used in recordings, movie scores, rock bands, radio, and television programming. The list of new developments and variations of synthesizer hardware is virtually endless and cannot be covered in our brief survey.

5.4 *Sound-on-Sound Recording*

Because most synthesizers play only one musical line at a time, it is necessary to use the technique of sound-on-sound recording to produce music consisting of many musical lines. We shall discuss the procedure for the production of monaural sound-on-sound recordings, for which only a standard stereo tape recorder, a microphone, and a connecting cable are required, as illustrated in Fig. 5–21. The tape deck must possess the additional feature of a mixer to combine microphone and one other input for each channel or an additional mixer must be provided, as will be seen.

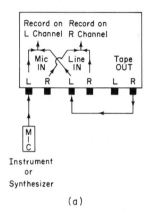

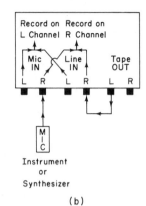

Instrument
or
Synthesizer

(a)

Instrument
or
Synthesizer

(b)

Figure 5-21   Wiring diagram used for production of sound-on-sound recordings.

In a sound-on-sound recording, each successive musical line is added to those previously recorded. To begin, a single instrumental line is played into the microphone and recorded on the left channel of the tape, using the configuration of Fig. 5-21(a), except with no input to the left channel "line in." The microphone and patch cord are then arranged as shown in Fig. 5-21(b). The second line is played into the microphone onto the right channel, while the previous line recorded on the left channel is simultaneously played back and input to the right channel. The levels for each signal must be adjusted with the mixer to give the proper balance. At this point, the two different musical lines are recorded together on the right channel. Arranging the system again as in part (a), the first two musical lines are played back by the right channel and input to the left channel as the third line is simultaneously played directly through the microphone onto the left channel; the original recording on the left channel is erased and written over in this step. After this stage, three musical lines are recorded on the left channel. This process is repeated until all the musical lines have been added to the recording. After this process, the desired music is monaural (one track), while the other track lacks the final line. In actual practice, a microphone is not used for synthesizer productions. For regular instruments, each line must be played independently into the microphone, but with a synthesizer the signal can be input directly into the recorder, thus eliminating one stage of the signal processing and thereby reducing noise.

Use of more sophisticated tape decks will allow stereophonic tape production or even music containing many independent channels. Another production technique is to record all the lines independently of one another and mix them into the desired stereo or quadraphonic format in a single operation.

While in principle this seems rather simple, the complexity of performance practice and the sophistication of the recording equipment make such productions difficult. An attempt at such production gives one a deep appreciation for the technical competence and ingenuity of those who produce high-quality synthesizer recordings.

EXERCISES

1. (a) Describe the wave shape and the effect on the sound of an audio tone undergoing (1) gating, (2) amplitude modulation, (3) frequency modulation, and (4) balanced modulation.
   (b) Define in context the carrier and modulator waves.
   (c) What are the musical terms for amplitude modulation and for frequency modulation?
2. Draw the following waves, indicating for each how you proceeded in setting up the axes for each carrier wave and its modulator wave: (a) sine wave gated by square wave; (b) triangular wave amplitude modulated by a sine wave; (c) square wave frequency modulated by a square wave; (d) sine wave balance modulated by a sine wave.
3. A 440-Hz sine wave is used as a carrier in a ring modulator, with a 75-Hz sawtooth as the modulator wave. List some of the prominent frequencies in the modulated wave.
4. Differentiate between audio and control signals as used in musical synthesizers. For each, give two examples of how they might be used. What are the frequency ranges for the two types of signals?
5. Describe what each of the following synthesizer functions does: (a) sample and hold, (b) voltage-controlled oscillator, (c) voltage-controlled amplifier, (d) envelope generator, (e) control oscillator, (f) low-pass filter, (g) high-pass filter, (h) bandpass filter, (i) notch filter, (j) noise generator. Give an example where each of the functions might be used.
6. Draw a synthesizer block diagram that would be used to produce the sound of a (a) guitar, (b) bell, (c) flute, (d) drum, and (e) toy slide whistle rising in pitch.
7. Explain what sound-on-sound recording is and why it is used with synthesizers. Listen to some synthesizer recordings and describe what functions might have been used to produce what you hear. If the electronic equipment is available, make your own sound-on-sound recording using a synthesizer or other instrument.

REFERENCES

Appleton, J. H., and R. H. Perera, eds. *The Development and Practice of Electronic Music.* Englewood Cliffs, N.J.: Prentice-Hall, Inc., 1975.

A collection of articles dealing with many aspects of electronic music, synthesizer, and recording techniques.

Friend, David, Alan R. Pearlman, and Thomas D. Piggott. *Learning Music with Synthesizers.* Milwaukee, Wisc.: Hal Leonard Publishing Corporation, 1974.

A review of fundamental physics (Chaps. 1 to 4 in this book), an introduction to synthesizers similar to this chapter, and further details, including use of block diagram patch sheets for the ARP Odyssey.

Owner's Manual, The ARP Electronic Music Synthesizer, Series 2600. 8th prtg. Lexington, Mass.: ARP Instruments Inc., 1977.

This book describes the basics of electronic music synthesis, with block diagrams and ideas for producing various sounds, as well as a few patch sheets specifically designed for the ARP 2600.

Recordings

BEAVER, PAUL, AND BERNARD L. KRAUSE. *The Nonesuch Guide to Electronic Music.* Nonesuch Records, HC-73018 Stereo.

A two-record set containing a great number of synthesizer sounds and special effects; it is accompanied by a booklet with explanations of the sounds and how to achieve them.

CARLOS, W. *Switched-on Bach.* Columbia Records, Stereo MS7194.

This is a classic in the field of electronic music, containing selections from the works of Bach performed on the musical synthesizer.

CARLOS, W. *A Clockwork Orange.* Warner Brothers, Stereo Record WB2573.

From the sound track of the movie by the same name, illustrating the use of the synthesizer in combination with other musical instruments. Noteworthy is the voice synthesis in Carlos's version of Beethoven's 9th Symphony.

6

The Human Ear and Voice

In this chapter we shall first study the human ear and the process by which sound waves are received and transmitted to the nerve endings that convert mechanical vibrations into electrical impulses.

Our study will relate the basic anatomy of the peripheral auditory system to the important features of the place theory of hearing, the most widely accepted basic explanation of how the ear functions. We shall cover the basic features of the frequency and amplitude response of the ear and the mathematical concept of logarithms, which is related to the nonlinearity of both the amplitude and frequency response of the ear. A brief discussion of some examples of hearing phenomena conclude this study. Detailed discussion of the central auditory system, such as the auditory cortex, will be generally avoided.

We shall then study the parts of the vocal system and their relationship to the production of speech and musical sounds. We shall analyze speech patterns as we did musical instrument sounds.

6.1   *Peripheral Auditory System*

The human ear is one of the most amazing organs of the body; it possesses an incredible range of sensitivity in both its frequency and amplitude response. For example, the ear responds to vibrations over the range of frequencies from about 20 Hz to 20 kHz, a factor of about 1000 in frequency, whereas the eye is sensitive only

137

to the range of electromagnetic waves having wavelengths between about 400 and 700 nanometers (billionths of a meter), a factor of less than 2 to 1 in frequency. By analogy, we might say that the eye sees less than an octave. The ear adequately responds to a range of pressure variations of about 1,000,000 to 1.

The peripheral auditory system is divided into three parts: the inner ear, the middle ear, and the outer ear, as shown schematically in Fig. 6-1. The outer ear performs the function of focusing the sound waves onto the eardrum. The pinna acts as a collector, gathering sound waves and concentrating them into the auditory canal to a limited extent. It also aids in sound localization. In some animals, such as cats, a movable pinna further aids these ends. The auditory canal transmits the sound waves to the eardrum. In addition, the length of the auditory canal helps to protect the very sensitive eardrum against some shocks and intrusion by external objects.

The eardrum separates the outer and middle ears. Perhaps it is inappropriately called a "drum," because the waves in the eardrum are more in the nature of traveling waves in a flap or a loose drumhead than standing waves with resonances in a tight drumhead. A very important function of the eardrum is to separate the outer and middle ears physically so that the air pressure will not rapidly equalize between the two. This would reduce the amplitude of the vibration of the drum caused by pressure changes of the impinging sound wave. Even a very small hole in the eardrum will reduce the pressure difference, particularly at low frequencies. The slower changes in air pressure caused by changes in the weather are equalized by air flow from the throat to the middle ear through the eustachian tube, which can be seen in Fig. 6-1.

The middle ear consists primarily of the bone chain of three *ossicles*, the hammer, the anvil, and the stirrup. These bones convert the small-amplitude vibrations

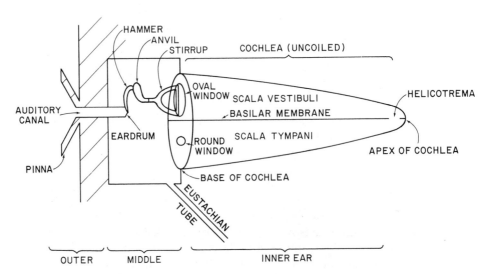

Figure 6-1   Schematic cross section of the peripheral auditory system.

of the eardrum into the larger-amplitude oscillations required to set up waves in the fluid of the inner ear. Attached to the ossicles are muscles that help to limit the vibration of the bones for very large amplitude continuous sounds, and thus prevent damage to the middle ear. Unfortunately, sharp loud noises such as gunshots occur too quickly for the protective mechanism to prevent damage to the middle ear if the noise is sufficiently intense.

The principal hearing organ of the inner ear is the cochlea, which is actually coiled on itself, like a snail, but is shown schematically in Fig. 6–1 uncoiled for simplicity. The cochlea contains the nerves that convert the physical vibrations into electrical signals. The width of the coiled cochlea decreases along its length from base to apex; the typical cochlear cross section is shown schematically in Fig. 6–2.

Vibrations from the stirrup enter the inner ear through the oval window, creating traveling waves in the fluid inside the scala vestibuli. The wave passes through an opening at the end of the cochlea called the helicotrema, and returns through the scala tympani to the round window, a flexible region of the base of the cochlea. The round window provides a point of pressure relief for the impinging traveling wave and damps the wave so that it will not reflect.

The scala vestibuli and the scala tympani are separated by the basilar membrane, which runs the entire length of the cochlea. The region along the basilar membrane containing the nerve endings that convert the waves to electrical impulses and transfer the vibrations to the auditory nerve is called the organ of Corti. The basilar membrane is approximately 3.5 cm long and contains about 30,000 nerve endings, often called *hair cells* because of their physical appearance, distributed fairly uniformly along its length. In actuality, the force that activates the hair cells results from a rather complicated shearing motion between the tectorial membrane and the basilar membrane, which occurs whenever a wave travels down the scala vestibuli.

Electrical impulses from these hair cell nerve endings are transmitted to the brain, which relates the sound heard to those previously experienced and interprets the signals as words, music, noise, and so on.

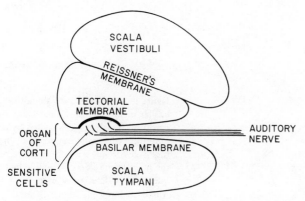

Figure 6–2   Schematic cross section of the cochlea.

In addition to its hearing functions, the inner ear contains the semicircular canals (not shown), which are responsible for detection of gravitational fields and acceleration and play an important role in balance.

6.2   Place Theory of Hearing

The position along the basilar membrane at which the maximum hair cell and nerve response occurs is correlated with frequency. The *place theory of hearing* is based primarily on this observed correlation of frequency with position of response along the basilar membrane.

An important experimental observation is that, over the entire range of audible frequencies, two tones separated by an interval of one octave (a factor of 2 in frequency) excite regions at the same spacing along the basilar membrane, about 3.5 mm; this can be illustrated by the simplified plot of position of response versus frequency shown in Fig. 6–3. Figure 6–3 is not a linear graph, because equal distances along the horizontal axis do not correspond to equal frequency intervals but rather to octaves, that is, equal frequency *ratios.* The full range of human hearing of 20 Hz to 20 kHz encompasses just under ten octaves; the full length of the basilar membrane of 3.5 cm (35 mm) is divided into ten equal intervals of one octave, each of 3.5 mm in length. It is interesting to note that equal musical intervals "sound" the same irrespective of the pitch level at which they are sounded. For identical musical intervals (at any pitch level) the physical spacings along the basilar membrane are approximately equal. It thus appears that the most important factor in determining how a musical interval between two pure tones sounds is the spacing along the basilar membrane of the points that respond to the two tones.

Response of the hair cells along the basilar membrane to a sinusoidal tone is not limited to a single receptor or even to a narrow band of receptors. Rather, the

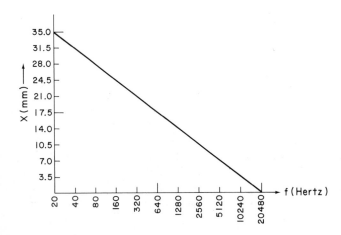

Figure 6-3   The distance along the basilar membrane (measured from the eardrum) which is most excited by a pure tone versus the frequency of the tone. Low frequencies are detected near the helicotrema (large X), while high frequencies are detected near the eardrum (small x).

response is spread out over a relatively wide region along the basilar membrane. The region along the basilar membrane where the nerve endings produce a large response to a sinusoidal audible signal is called the *critical band.* Over most of the audible range the critical band corresponds to about 1.2 mm along the basilar membrane, and includes about 1300 hair cells of the total of 30,000. This corresponds to about 15 percent in frequency, but is larger at lower frequencies. Musically, this is slightly less than a minor third (three half-steps) over most of the frequency range, and almost one octave at frequencies below 200 Hz. The neural mechanism is capable of receiving this broad range of nerve impulses and narrowing this range down so that a single pitch tone is perceived.

Listening to two pure tones sounding simultaneously, we can further investigate the effect of the critical band and experimentally determine its width. Suppose we set one of the frequencies at some value, say 1000 Hz, and vary the other continuously between 20 Hz and 20 kHz, keeping the two amplitudes equal. If the two tones differ greatly in frequency, we will simply hear them as separate. On the other hand, if their frequencies are very close, we cannot hear either tone individually. Instead, we hear a single tone whose frequency is the average value between the two original tones and beats that have a frequency equal to the difference between the two original frequencies. Starting with the two frequencies the same, we hear the single tone with no beats, and, as one frequency is raised or lowered, we hear beats that become faster as the frequency difference increases. About the time that the beats become too rapid to distinguish, both tones will begin to become distinguishable, with a rather low-frequency roughness or coarseness heard in the background. As the frequencies become farther separated, the coarseness effect decreases and finally disappears. The frequency range over which the coarseness is heard is the region where the critical bands of the two pure tones overlap. If we make the simplifying assumption (though not exactly true) that the response of the hairs in the critical band is symmetric in frequency about the center frequency (the frequency of the pure tone), the width of the critical band for a pure tone is just the difference between the two frequencies where the roughness begins to appear as the tones are brought together in frequency.

This effect can also be demonstrated by playing two high notes on recorders or flutes. A dull, low-frequency roughness can be heard when the two tones are a musical interval of a minor third (three half-steps) or less apart, indicating that the critical band is at least the width of a minor third in that frequency range. At low frequencies the critical band is large; a minor third played around G_2 sounds rough. This is a partial explanation of why composers rarely use small musical intervals in the bass range.

What is the smallest frequency change that we can hear, and is it limited in some way by the very large width of the critical band? We can perform another experiment to determine how sensitive our hearing mechanism is to small frequency changes. A sine wave of audible frequency is frequency modulated by another sine wave of frequency of about 1 Hz; this produces a tone that slowly varies up and down in frequency, centered on the frequency of the original sine wave, an effect called vibrato. The greater the amplitude of the modulating wave, the greater the change in the frequency. If the amplitude of the pitch variation is decreased, there will be a point at which the changes in frequency become imperceptible; likewise, if we start with a very small or zero frequency change and increase the frequency limits very slowly, there will be some point at which the pitch change becomes barely perceptible. The frequency difference between the two extreme values at that point is called the frequency *just noticeable difference* or frequency JND. The JND can be measured over the complete range of audible frequencies by changing the original frequency and repeating the experiment. The frequency JND is between about 0.5 percent and 0.6 percent over most of the audible range, but is greater at extreme high or low frequencies. The frequency JND is about one-tenth of the one half-step of musical interval over most of the audible range, increasing to about one quarter-step (half of a half-step) for low bass notes.

The frequency JND is much smaller than the critical band, which is more than a factor of 10 larger at most frequencies. Despite the rather large width of the region excited by a single frequency tone, we can distinguish changes in the position of the peak of that response of less than one-tenth of its width. This neural process of assigning a single frequency to the wide band of excitation along the basilar membrane is called *sharpening*. The sharpening effect is enhanced for complex waves (such as square waves) by the additional frequency information available. As a result of sharpening, two square-wave tones can become much closer in frequency before the occurrence of coarseness due to the overlap of their critical bands; complex tones at musical intervals of less than one half-step are normally easily resolved by the ear.

We have seen that the human ear can generally discriminate two very closely spaced frequencies sounded sequentially, owing to the sharpening of the neural network; however, it is much more difficult to resolve two pure tones whose frequencies are very close when they are sounded simultaneously. The minimum frequency separation between two sinusoidal tones sounded simultaneously, which can be perceived individually, is known as the *limit of frequency discrimination.* For two pure tones, this is between one and two half-steps of the scale (about 7 percent) for low frequencies and rises to about a minor third (three half-steps or about 15 percent) for high frequencies. The limit of discrimination can be demonstrated by using two sine-wave oscillators. They will produce beating between the two tones when their spacing becomes too small for them to be distinguished. However, as most musicians are aware, two notes played only one half-step apart *can* usually be perceived as separate notes. This is due primarily to the additional sharpening effect of the neural network, resulting from the existence of higher harmonics in most musical tones.

The relationship of the JND and the limit of discrimination can be demonstrated by analogy to the sense of touch on the skin of the inner forearm (between the wrist and elbow), using two pencil erasers touching the forearm. Have a friend touch your forearm at either one point or at two points simultaneously to illustrate the following effects. If you are touched once, then again a short distance along the arm, it is easy to feel relatively small changes in the position at which you were touched. If your arm is touched simultaneously in two closely spaced points, you will find it very difficult to resolve the two touches; that is, it is difficult to feel two spatially separate, simultaneous touches. In fact, the spacing between the two touches must be somewhat larger than the minimum discernible change of position of a single touch before the two contact points can be resolved. That is, the sense of touch position along the arm (analog to the JND) is much finer than the sense of resolution of the arm to two separate simultaneous touches (analog to the limit of frequency discrimination). This can be viewed as resulting from the very large region of response along the arm to a touch at a single point, analogous to the width of the critical band. This effect is similar on skin anywhere on the body. The basilar membrane can thus be viewed as a very highly sensitive piece of skin with a particular sensitivity to vibrations in the audible frequency range.

It should be noted that all the experiments discussed here have involved judgments and are basically subjective in nature. In those cases where a more exacting experiment is not available to measure a physical property of the ear, the result is taken to be the average value determined from a number of people with "normal" hearing. Lack of ability to hear some effect as well as others does not necessarily imply a hearing defect, but can be due to a wide variation even among individuals with normal hearing. Nevertheless, despite differences between subjects in musical talent, training, and experience, these experiments should not be considred "unscientific." The important psychoacoustic results are to be related to physical properties of the ear mechanism, and as such remain largely independent of the observer; in that sense, they can be considered "objective" results.

One additional point that should be emphasized is that the concept of the critical band is basically a psychological one, not a neurological one. The experiments described do not *demand* an explanation in terms of the place theory as described. However, while some aspects of this theory have not been verified by direct observation of the hair cell response, most psychophysicists believe that the place correlation exists in humans, and that the place theory provides the most acceptable basic explanation of how the ear functions.

6.3   *Amplitude Response of the Ear*

The ear can sense pressure variations as small as 1 part in 10,000,000,000 of atmospheric pressure, but an oscillation does not become painful to the ear until its amplitude reaches about 1 part in 10,000 times atmospheric pressure (1,000,000 times greater). The *threshold of hearing* is the intensity of the smallest oscillation that can be perceived by the ear, and the *threshold of pain* is that intensity at which a wave becomes painful. These thresholds are plotted in Fig. 6–4 as a function of frequency over the audible range.

The range of pressure fluctuations over which the human ear is sensitive is

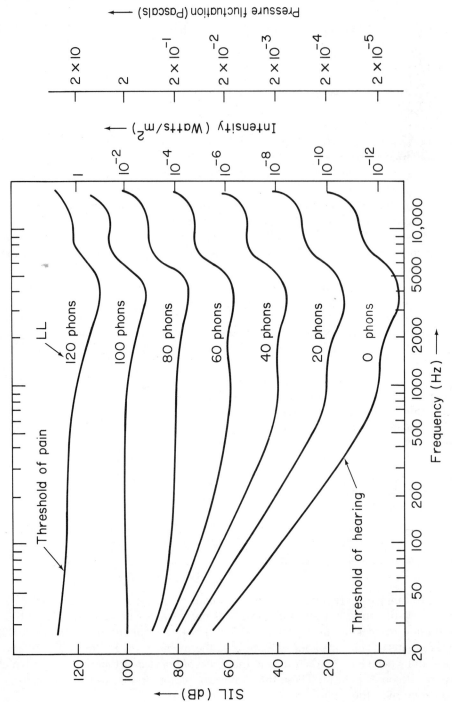

Figure 6-4 Equal loudness curves. The curve with the lowest level marks the threshold of hearing; the curve with the highest level marks the threshold of pain.

significantly greater than the variation in pressure associated with changes in the weather. Our range of pressure sensitivity of over 6 orders of magnitude (factors of 10) is truly amazing. The minimum pressure level is roughly equivalent to placing the wing of a fly on the eardrum and corresponds to an oscillation that causes the eardrum to move about one-tenth of the diameter of an atom. This is the atmospheric pressure difference at sea level between one point and another about $\frac{1}{3}$ inch above or below it.

The extreme sensitivity of the ear is illustrated by the "seashell" resonance effect. When you hold a seashell to your ear, what you hear are the ambient, soft sounds that have been resonated and thereby amplified by the seashell cavity. If the ear were slightly more sensitive to pressure variations, we would be continually bothered by the soft noises created within the body itself. In fact, in an anechoic chamber, where there is no background noise, these sounds are audible. The threshold of hearing is high at frequencies below about 100 Hz; if it were not, we would hear the many low-frequency oscillations in our bodies such as those associated with our heartbeat and blood flow. Another low-frequency oscillation can be observed by listening to muscles of the fingers twitching at frequencies in this range. The low-frequency noise heard when the little finger is inserted into the ear canal is just such a muscle vibration that has been amplified slightly.

The intensity of a sound wave is measured in watts per square meter (W/m²) and is proportional to the square of the pressure fluctuation, which is measured in Pa. For instance, tripling the pressure variation corresponds to increasing the intensity of the wave by a factor of 9. At 1000 Hz, the threshold of pain has a pressure fluctuation 10^6 times as large as that of the threshold of hearing. Thus, the intensity at the threshold of pain is 10^{12} (or 10^6 squared) times greater than that of the threshold of hearing. This can be verified on the two right-hand axes in Fig. 6–4.

Other units shown on the figure, dB and phons, shall be discussed shortly.

6.4   *Logarithms and Sound Intensity Scales*

In our analysis of the frequency response of the ear, we saw that the ear did not respond linearly to frequency stimuli, but rather the same ratio between two frequencies is always perceived to be the same musical interval. A similar type of nonlinearity occurs in the ear's response to signals of varying amplitude; in this case, the basic unit is a factor of 10.

By definition, if $y = 10^x$, then the logarithm of y, written log y, is equal to x, or $x = \log y$. If $y = 100 = 10^2$, then log $y = 2$; if $y = 1 = 10^0$, then log $y = 0$. The logarithm of 1,000,000 or log 10^6 is six, and the logarithm of 0.001 or 10^{-3} is -3. Table 6–1 gives the logarithms of the numbers 1 through 10. These logarithms are fractions between 0 and 1.

Logarithms have two properties that make them particularly useful when dealing with the amplitude response of the ear: (1) numbers with large ranges have

TABLE 6-1   **Approximate logarithms of integers**

log 1 = 0.0
log 2 = 0.3
log 3 = 0.5
log 4 = 0.6
log 5 = 0.7
log 6 = 0.8
log 7 = 0.85
log 8 = 0.9
log 9 = 0.95
log 10 = 1.0

logarithms that do not cover a large range, and (2) steps in equal ratios have logarithms that differ by equal steps.

Because the range of intensity to which the ear is sensitive is so very large (about a factor of 1,000,000,000,000), dealing directly with intensities is cumbersome. It is convenient to define a new scale in which this enormous range can be expressed in a more compact manner. The *sound intensity level* (SIL) scale, with the fundamental unit of the bel, named after Alexander Graham Bell (1847–1922), is such a scale. A decibel (dB) is one-tenth of a bel. The sound intensity level of a tone of intensity I relative to a tone of intensity I_0 is defined as follows:

$$SIL = SIL_0 + 10 \log \frac{I}{I_0}$$

where I_0 is the intensity (in watts per square meter) of the reference tone whose sound intensity level is SIL_0 (in decibels). The choice of this reference tone intensity is equivalent to setting the zero point of the scale. When dealing with human hearing, the threshold of hearing at 1000 Hz is chosen as the reference level. Its intensity has been measured as about 10^{-12} W/m², and its sound intensity level defined as 0 dB.

The logarithm (power of 10) definition of the sound intensity level implies that for every increase in the intensity by a factor of 10 the sound intensity level increases by 10 dB; likewise, for every increase in the intensity by a factor of 100, the sound intensity level increases by 20 dB, a factor of 1000 corresponds to 30 dB, and so on.

The sound intensity level of the threshold of pain at 1000 Hz (intensity of 1W/m²) can be directly calculated using the formula

$$SIL = SIL_0 + 10 \log \frac{I}{I_0} = 0 \text{ db} + 10 \log \frac{1\frac{W}{m^2}}{10^{-12} \frac{W}{m^2}}$$

$$= 10 \log 10^{12} = 10 \times 12 = 120 \text{ dB}.$$

This can be read from Fig. 6-4.

We can go from intensity to SIL, and vice versa. Suppose that we know that the SIL of a given tone is 20 dB. What is its intensity?

$$\text{SIL} = 20 \text{ dB} = \text{SIL}_0 + 10 \log \frac{I}{I_0} = 0 \text{ dB} + 10 \log \frac{I}{I_0}.$$

Therefore,

$$\log \frac{I}{I_0} = 2 \text{ or } \frac{I}{I_0} = 100$$

and thus

$$I = 100 \times 10^{-12} \frac{\text{W}}{\text{m}^2} = 10^{-10} \frac{\text{W}}{\text{m}^2}$$

What is the SIL if two such sources are played simultaneously? The total intensity of the combined sound is just the sum of the two individual intensities, or $2 \times 10^{-10} \frac{\text{W}}{\text{m}^2}$. The SIL is then

$$\text{SIL} = 10 \log \frac{2 \times 10^{-10} \frac{\text{W}}{\text{m}^2}}{10^{-12} \frac{\text{W}}{\text{m}^2}} = 10 \log 200 = 10(\log 100 + \log 2)$$

$$= 10(2 + 0.3) = 23 \text{ dB}.$$

Here we have used the fact that $\log (ab) = \log(a) + \log(b)$, where $a = 100$ and $b = 2$. We can also check this result by applying the formula with $\text{SIL}_0 = 20 \text{ dB}$ and $I = 2I_0$. We are comparing the SIL of two sources to that of the original 20-dB tone. Then

$$\text{SIL} = 20 \text{ dB} + 10 \log \left(\frac{2I_0}{I_0}\right) = 20 \text{ dB} + 10 \times 0.3 \text{ dB} = 23 \text{ dB}.$$

In doubling the intensity, the SIL has only increased by 3 dB. This is true no matter what the original intensity of the sound.

Combination of three identical sources gives

$$\text{SIL} = \text{SIL}_0 + 10 \log\left(\frac{3I_0}{I_0}\right) = 20 \text{ dB} + 10 \times \log 3 = 20 \text{ dB} + 10 \times 0.5 \text{ dB}$$
$$= 25 \text{ dB}$$

and the combined SIL increases 5 dB to give a total of 25 dB.

Combination of ten identical 20-dB sources gives

$$\text{SIL} = 20 \text{ dB} + 10 \log \left(\frac{I}{I_0}\right) = 20 \text{ dB} + 10 \log \left(\frac{10I_0}{I_0}\right)$$

$$= 20 \text{ dB} + 10 \log 10$$
$$= 20 \text{ dB} + 10 \times 1 \text{ db} = 30 \text{ dB}.$$

The combined SIL increases 10 dB to give a total of 30 dB. Again we see this convenient result: An increase in the intensity by a factor of 10 increases the SIL by 10 dB.

Combination of 100 such identical sources gives

$$\mathrm{SIL} = \mathrm{SIL}_0 + 10 \log \frac{100 I_0}{I_0} = 20 \text{ dB} + 10 \times 2 \text{ dB} = 40 \text{ dB}$$

raising the SIL by 20 dB to a total of 40 dB. Thus, if one source produces a SIL of 20 dB, it requires 10 identical sources to produce a SIL of 30 dB (a 10-dB increase), and 100 identical sources to reach a SIL of 40 dB (another 10-dB increase). It would require 1000 such sources to provide a 50-dB SIL, 10,000 for 60 dB, and so on. Similarly, it would require 10^{12} identical incoherent or random sources each of sound intensity level 0 dB to raise the SIL from the threshold of hearing (0 dB) to the threshold of pain (120 dB).

The decibel scale of sound intensity level is important because our ears tend to interpret changes in intensity of a certain number of decibels as being roughly equal changes in loudness, independent of the actual intensity of the signal. That is, increasing the SIL of a 1000-Hz tone in 10-dB steps will be interpreted by the listener as increasing the loudness of the tone by roughly equal increments. This is based on subjective responses of a large number of people with normal hearing.

We can perform an experiment to determine the minimum observable intensity variation of a pure tone as we determined the minimum observable frequency difference; similarly, this quantity is called the intensity JND (just noticeable difference). Such a measurement, made over the entire audible frequency and intensity range, shows that the intensity JND is approximately 1 dB over most of the audible range, varying between about 0.5 and 1.5 dB. The JND is not a certain amount in *intensity*, but is an almost constant *fraction* of the intensity (about 25 percent) of the tone that is being varied, and therefore approximately a constant in the logarithm of the intensity, or a constant in SIL units (dB). This result reinforces our choices of the dB SIL scale as being fundamental in interpreting our response to stimuli of different intensity. Not only is it a more convenient scale with which to work, but decibel units are also more useful and consistent in viewing and interpreting the response of the ear (equal loudness increments and intensity JND) over a wide range of audible stimuli.

One limitation of the decibel sound intensity level scale is that it does not take account of the subjective changes in loudness of different frequency tones whose intensity is the same. Such changes in apparent loudness with frequency arise from the variation of the threshold of hearing and human sensitivity with frequency. Curves of equal loudness are shown in Fig. 6–4. The loudness level or phon scale *does* take this variation into account. The unit of loudness level is the phon. One phon is the loudness level of a pure tone of 1000 Hz whose sound intensity level is 1 dB; 1 phon is equal to 1 dB at 1000 Hz. The loudness level of a pure tone of arbitrary frequency and intensity is equal in phons to the sound intensity level in decibels of a 1000 Hz tone that is judged to be of the same loudness as the original tone. The curves on Fig. 6–4, called Fletcher-Munson curves, represent tones whose loudness is the same, two of which happen to be the threshold of hearing (0 phons) and the threshold of pain (120 phons).

One additional scale of loudness should be mentioned: the scale of subjective loudness, measured in units of sones. The preceding scales can be considered as physical scales, derived from physical measurement of some quantity like intensity or by comparing an arbitrary tone with one whose intensity is known with respect to a physical scale. On the other hand, the subjective scale is set up to take account of judgments as to what intensity sounds twice or three times as loud as a reference intensity. The subjective scale employs a new unit, called the sone, which marks equal steps in subjective loudness. Because of the subjective nature of the sone scale and the objective nature of the decibel scale, the two are not simply related.

This difference can be illustrated in the following way. When two 50-dB tones of arbitrary frequency are added, the sum is 53 dB. It has been shown, however, that the resulting tone will be judged as twice as loud only when the two tones differ greatly in frequency. When the two tones are close in frequency and their critical bands overlap, subjective loudness additivity breaks down; the combined tone is judged to be considerably less than twice as loud as either component tone.

Each scale discussed has features that make it useful in certain specific fields of study. The sound intensity level in dB, the scale that is most closely associated with physically measurable quantities, is very convenient and is therefore used most often in the world of physics and engineering. We shall use this scale regularly in our study of recording and reproduction of sound and in our study of room acoustics.

6.5   *Periodicity Pitch and Fundamental Tracking*

If two or more tones whose frequencies are successive harmonics in some overtone series are sounded simultaneously, an additional frequency will be "heard" by most listeners, particularly trained musicians. The neural mechanism of the ear is able to identify the two frequencies as being members of an overtone series, and assigns to the total stimulus the frequency of the fundmental. This phenomenon is called the *periodicity pitch, missing fundamental,* or *subjective fundamental.* If one plays a sequence of pairs of tones that have a frequency ratio of 3 to 2 but have different fundamental frequencies, the ear will recognize the periodicity of each pair of notes and supply the fundamental tone for each; this is called *fundamental tracking.* For instance, if tones of frequencies of 200 and 300 Hz are played simultaneously, followed by simultaneous 300- and 450-Hz tones, you will track a 100-Hz, tone changing to a 150-Hz tone. Fundamental tracking is used in the production of very low frequency organ tones by sounding the second and third harmonics together and allowing this auditory mechanism to insert the fundamental. By using two pipes in the 16-foot octave, the designer can eliminate the need for bulky and costly 32-foot fundamental pipes. This effect is distinct from the phenomenon of difference tones, which will be discussed shortly.

6.6 Aural Harmonics

Let us consider in more detail some effects of the nonlinearity of the response of the ear to intensity. Figure 6–5 shows a large-amplitude sine wave. Beneath that is a graph related to the logarithm of this sine function. This change of shape might represent the distortion of an incoming sine wave by the nonlinearity of the ear mechanism. That is, the top graph might be the graph of the pressure reaching the ear; due to the logarithmic response of the outer and middle ear mechanism to intensity, the wave in the cochlear fluid might look more like the lower graph. Clearly, a Fourier analysis of the new wave would show other frequencies present in addition to the frequency of the original sine wave. These additional frequencies generated by the nonlinear amplitude response of the ear are harmonics of the original wave, because the period of the distorted wave is the same as that of the original wave. These frequencies are called *aural harmonics* and play an important role in hearing. Aural harmonics of a pure tone become significant when the tone is "loud," that is, when the pressure varies over several orders of magnitude.

Particularly important are the aural harmonics generated when two tones are played simultaneously at about the same intensity level. Such harmonics, called *combination tones*, or *sum and difference tones*, occur when the intensities of the incoming waves again are great enough that the nonlinear response of the ear becomes important. In general, if the original frequencies are f_1 and f_2, combination tones will be generated at the frequencies $f_c = |\,nf_1 \pm mf_2\,|$, where n and m are any integers (1, 2, 3 . . .) and the vertical lines denote the absolute value of the quantity inside (for example, $|\,5.1\,| = 5.1$, $|\,-3.6\,| = 3.6$). If the plus sign is used, the tones $f_+ = nf_1 + mf_2$ are called *sum* tones; if the minus sign is used, the tones

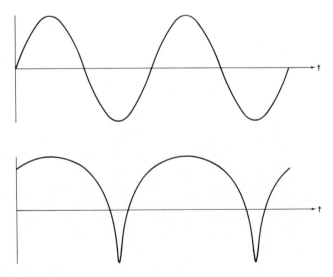

Figure 6–5 A sine curve and a curve logarithmically related to it.

$f_- = |nf_1 - mf_2|$ are called *difference* tones. The larger the numbers n and m, the softer are the combination tone sounds.

The difference tone $f_1 - f_2$ was first pointed out by the violinist and composer Giuseppi Tartini (1692–1770) in the early eighteenth century. He listened for difference tones as an aid in tuning double stops during violin performances.

For example, for two tones having frequencies of $f_1 = 500$ Hz and $f_2 = 700$ Hz, respectively, the combination tones are listed in Table 6–2 along with the numbers n and m. In general, these tones are significantly softer than the original tones; nonetheless, the first few difference tones can be heard clearly if they are well within the audible region and lower in frequency than either of the original tones. On the other hand, the sum tones are very difficult to hear, because in addition to being somewhat softer, they are always higher in frequency than the louder original tones, and are effectively covered up or masked by the louder, lower frequency tones (see Sec. 6.8 on masking).

Combination frequencies exist as components in the complex wave moving through the cochlear fluid and, as such, are physically real, that is, actual physical vibrations. The nonlinearities in wave shape are introduced into the wave by the entire ear mechanism and exist by the time the wave is transmitted through the cochlear fluids. We can therefore verify the existence of any sum or difference tone by producing beats with a pure tone slightly different in frequency from the combination tone. For example, playing a very low amplitude 1201-Hz sinusoidal tone simultaneously with the loud 500-Hz and 700-Hz tones will create beating at 1 Hz between the sine wave and the sum tone $f_+ = 500 + 700 = 1200$ Hz. Combination tones can be contrasted with the periodicity pitch or missing fundamental, which

TABLE 6–2   **Combination tones produced from 500- and 700-Hz pure tones**

n	m	f_+ (Hz)	f_- (Hz)
1	1	1200	200
1	2	1900	900
2	1	1700	300
2	2	2400	400
3	1	2200	800
3	2	2900	100
1	3	2600	1600
2	3	3100	1100
3	3	3600	600

$$f_+ = nf_1 + mf_2 \qquad f_1 = 500 \text{ Hz}$$
$$f_- = |nf_1 - mf_2| \qquad f_2 = 700 \text{ Hz}$$

exists only in the nervous system as an electrical impulse, and not as a real physical wave.

We "hear" bass notes coming from very small loudspeakers, such as those on small transistor radios, particularly when the music is loud, even though the speakers cannot produce such frequencies. Two mechanisms contribute to this phenomenon. Difference tones created by adjacent harmonics in the overtone series of the music reinforce the fundamental, and the fundamental tracking mechanism aids the perception of the sequence of overtones as being associated with the fundamental frequency, and therefore supplies the sound of the fundamental.

In the experiment where we varied one sinusoidal oscillation in frequency about another fixed oscillation, we heard beats that rapidly became higher in frequency until they "felt" like a coarse buzzing effect. The frequency of this buzzing or coarseness, or of the beating, is just the difference frequency between the two oscillations; further separation of the two oscillator frequencies increases the frequency of the coarseness into the audible region, where it is heard as a difference tone.

6.7   Ohm's Law of Hearing

We saw in Chap. 4 that the principal factor in determining the quality of a steady complex wave is the intensity of the various overtones. If the peripheral auditory mechanism introduces nonlinearities into the incoming sound wave, and the central ear mechanism creates and inserts fundamental frequencies into the observed spectrum, what effect has this on the quality of any particular sound? In the late nineteenth century, the two physicists Georg Simon Ohm (1787–1854) and Hermann von Helmholtz first formulated the theory, known as Ohm's law of hearing, which states that the sound quality of a complex tone depends only upon the amplitudes of its harmonics and *not* upon their relative phases. This assumption is basically true, as we have seen in our experiment involving the sounds of various shapes of standard waves. We have also seen that distortion nonlinearities occur when the amplitude of a wave is large, leading to slight changes in quality due to the additional harmonics.

While Ohm's law well describes the relative effects of amplitude and phase of the higher harmonics on the quality of a complex audible tone, we can perform an experiment to demonstrate its limitations. If two pure tones at an interval of nearly, but not exactly, one octave (just about a factor of 2 in frequency ratio) are sounded simultaneously, a type of "beating" somewhat different from that discussed in Chap. 2 will occur, called *second-order* or *quality beats*. Quality beats are particularly strong when the intensity of the higher-frequency tone is reduced somewhat below that of the lower-frequency tone. In the case of regular (first-order) beats, the two tones alternately interfere constructively and destructively, creating a change of amplitude that is perceived as variation of the intensity of the sum

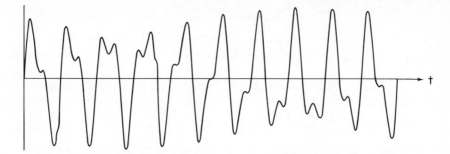

Figure 6–6 Complex wave formed from fundamental and slightly mistuned second harmonic whose amplitude is one-half that of the fundamental. Slight changes in tone quality occur as the pattern changes shape and are called second-order, or quality, beats.

wave. In the case of second-order or quality beats, as shown in Fig. 6–6, the amplitude of the sum wave remains approximately the same, while the wave shape is continually changing at a rate that depends upon the deviation of the two frequencies from the exact ratio of 2. While the sound of the beating is somewhat different from that of first-order beats, it is clear that the mechanism of the ear is quite capable of detecting changes in the wave form.

Second-order beats are not due to nonlinearities in the ear, which might produce a second harmonic of the lower tone that beats, in the conventional sense, with the higher tone. We know this because second-order beats can be heard even when the two original tones are soft, and nonlinearities are insignificant.

The relative phase of the two components changes continually and correlates with a perceived tone quality change. Thus Ohm's law of hearing breaks down in this example. This phenomenon is only audible when the lower frequency is below about 1500 Hz.

Fortunately, Ohm's law is for the most part valid under conditions except in laboratories where attempts are made to disprove it. Tape recorders, except extremely expensive models, produce substantial changes of relative phase between the harmonics of a note (this is called *phase distortion*), and yet for the most part render what we judge to be faithful sound reproduction.

6.8   Masking

The existence of one tone affects the perception of an *identical* tone with less intensity. In fact, because the just noticeable difference of intensity is about 1 dB, the second tone must be raised to about one-quarter of that of the first before the listener can detect its presence:

$$\begin{aligned}
\text{SIL} &= \text{SIL}_0 + 10 \log \frac{I}{I_0} \;= \text{SIL}_0 + 10 \log \frac{1.25 I_0}{I_0} \\
&= \text{SIL}_0 + 10 \log 1.25 = \text{SIL}_0 + 10 \times 0.1 \text{ dB} \\
&= \text{SIL}_0 + 1 \text{ dB}
\end{aligned}$$

This effect, called *masking*, plays an important part in the perception of combinations of tones and in our ability to reject unwanted sounds in favor of those in which we have some interest.

To investigate the phenomenon of masking, we observe a soft pure tone of variable frequency, called the *masked tone*, in the presence of a louder tone that is fixed in frequency, called the *masking tone*. The *masking level* is the sound intensity level to which the masked tone must be raised in order for it to be heard in the presence of the masking tone. In the case of two identical frequencies, the masking tone is very effective, and the masking level is relatively high. In general, masking occurs readily for frequencies above the masking frequency; that is, low-frequency tones readily mask higher-frequency tones. On the other hand, the masking levels for frequencies less than that of the masking tone are relatively low, and even low-intensity tones of low frequency can be easily heard. In general, the masking level drops as the masked tone gets farther in frequency from the masking tone. The first difference tone between two loud sine waves less than an octave apart lies below the louder tones in frequency and can readily be heard. Conversely, the first sum tone has a frequency above those of the louder tones and generally is masked effectively.

It is important for the ear to be able to reject unwanted aural stimuli and noise in favor of the desired information. A standard hearing test includes a test of the ability of the subject to pick out a pure wave signal in the presence of others as well as background noise. Each ear can be checked individually by playing white noise into the other ear. This procedure eliminates the possibility of the second ear hearing the sound being presented to the first ear by bone conduction, or transmission of the vibrations through the skull. Masking helps explain our ear's ability to pick up a single conversation in the presence of others.

Masking is relevant to the perception of the quality of complex tones. We perceive complex tones as a whole, and not as a series of individual pure tones (the harmonics present in its overtone series) for a number of reasons: The difference tones between adjacent harmonics reinforce the fundamental; the high harmonics are spaced closer together than the critical band and cannot be perceived as individuals; the high harmonics tend to be masked by the lower harmonics, including the fundamental; and the fundamental tracking mechanism reinforces the fundamental as the pitch of the note.

Masking is related to the critical band, in that tones mask each other when their critical bands overlap. This effect can be observed in some types of music, such as Bach's fugues. When the melodic lines are closely spaced, we hear a melody with rich harmonic texture. As the lines become more openly spaced, each can be heard more clearly as an individual melody line.

Thus far we have considered effects dealing with both ears together or that apply identically to both ears. We have seen that the paths of the sound from each ear to the brain are similar. In certain binaural effects it is also important to realize that the paths from the two ears are separate; the actual vibrations entering each ear do not mix physically. Rather, the information about these vibrations "mixes" as neural signals in the brain.

If two very soft pure tones of about the same frequency are played simultaneously, one into each ear, a different type of beat, called *binaural beats*, may be heard. In the case of regular first-order beats, there is a physical mixing of the two beating tones. In binaural beats, the two tones must be soft enough that no bone conduction or direct physical transmission of the vibration through the skull occurs. This ensures that there will be no physical mixing of the two waves. In the case of binaural beats, the only mixing of the two waves occurs in the brain, where they are no longer actual physical waves, but rather sequences of neural impulses that will be interpreted by the brain as pure tones. The two tones cannot interfere in the physical sense, and thus no physical beating can occur. When the two tones are presented in this way, some people are able to hear a type of beating that sounds very different from either first-order (regular) or second-order (quality) beats.

Whereas it is possible for most people to match two pitches using first-order beats, it is extremely difficult to match pitches using binaural beats. If the two pitches are heard in separate ears, it is common for even a trained musician to be in error by over one half-step, and less experienced subjects often have larger errors.

The binaural experiment illustrates that the paths of the aural network to the brain are separate, and clearly demonstrates the necessity of actual physical mixing (that is, addition of displacements) of two waves in order to obtain normal (first-order) beats. A significant mismatch in pitch perception between the two aural networks, known as *diplacusis*, occurs to some extent in everyone, and is rarely a serious handicap to a musician. A musician with severe diplacusis, however, may perceive the pitch for the part of the orchestra on his or her left as different from the part on the right, and this makes matching pitches with other members of the ensemble very difficult.

Our ability to localize sound sources in space is another binaural phenomenon. Two different processes are used, one for low frequencies (between 500 and 800 Hz) and another for higher frequencies.

Localization of low frequencies involves the phase or time difference between the arrival of a wave crest at one ear and its arrival at the other. Because of the spatial separation of the ears, a crest of a low-frequency wave will, in general, arrive at the ears at different times. The ear-brain system is sensitive to this small difference and uses it to localize the source. Simultaneous arrivals would indicate that the source is directly in front (or behind) the observer, while an earlier arrival in the right ear would indicate that the source is somewhere to the right. The time difference corresponds to the angle of the source position relative to the observer.

The phase of high-frequency tones *cannot* be used for localization because there is ambiguity as to the relative phase. Imagine a sound source placed in front of and somewhat to the side of an observer. The path difference from the source to the different ears could equal one wavelength of the particular tone emitted. Crests would consequently arrive at the two ears simultaneously, even though one was emitted one period before the other. Based on the phase information alone, the ear-brain system might erroneously judge the source as directly in front of the observer.

Instead of phase information, the system uses intensity information for high-frequency localization. Because the wavelengths of the high-frequency tones are smaller than the diameter of the human head, they do not diffract around the head as well as the low-frequency waves. The head, therefore, has a sort of sonic "shadow" on the side away from the source. Thus, one ear will receive a more intense signal than the other. The ear-brain system is sensitive to this intensity difference and can use it to localize the source. Because the longer-wavelength, low-frequency waves *can* diffract around the head, the intensity difference between low-frequency waves at the two ears is small and unreliable for localization.

Binaural phase sensitivity is also important in sound localization when selectively listening to one of many conversations. Both binaural phase sensitivity and masking effects can be used to select which of many sound sources an observer will attend to. Binaural phase (as well as intensity sensitivity) is fundamental for stereophonic musical perception.

An experiment illustrating the extreme sensitivity of the aural system to differences between time of arrival of a signal at the two ears can be performed using a stethoscope. Tapping on the face of the stethoscope can be used to locate the center of the device. If the center of the stethoscope membrane is tapped, waves strike the two ears simultaneously. If, however, a point to the side is tapped, the waves strike one ear slightly before the other. In this manner, the center of the membrane can be located to an accuracy of about 1 mm.

Another important phenomenon involving sound localization is called the *precedence effect*. In a shopping mall, there may be a large number of overhead loudspeakers, but the music always seems to emanate from the nearest one, regardless of its relative intensity. This demonstrates that sound localization is determined by the earliest arriving wave. In Chap. 8 we shall see that one must consider the precedence effect when designing auditoriums in which speaker systems are used.

6.10 *Hearing Loss*

Hearing loss can be divided into two basic types: temporary threshold shifts and permanent loss. Hearing loss due to damage or aging of the central hearing mechanism is truly permanent, but certain types of outer or middle ear problems can be corrected at least partially through surgery. Permanent hearing loss can be caused by physical damage to the ear mechanism, by disease, by drugs, or by the natural aging process.

Considerable study has been made recently on the effects of loud noises on the functioning of the ear. It is clear that prolonged exposure to loud sounds, particularly around or above 80 or 90 dB, can produce a temporary threshold shift. It is less clear what permanent damage, if any, is done by the multitude of loud noises to which most of us are regularly exposed. Continuous exposure to amplified rock music and other loud sounds has been strongly implicated as a source of hearing damage.

While there is a built-in muscular control in the middle ear to limit damage to the ear by sustained loud sounds, the ear is incapable of reacting rapidly enough to protect itself against very short bursts of sound, such as gunshots. Such sharp noises, if not excessively loud, may cause temporary threshold shifts in certain frequency ranges; if extremely loud, they can cause permanent loss of hearing by forcing some part of the ear mechanism beyond its elastic limit.

Diseases, such as infections of the middle ear, can also result in permanent impairment of hearing. An unfortunate disease-related hearing impairment may occur to a fetus when a woman in the first (or possibly early second) trimester of pregnancy contracts the three-day measles, or rubella. When this occurs, immediate and irreversible hearing loss in the baby may result, apparently from malformation of parts of its inner ear. Rubella often carries only light symptoms and goes away very quickly, so the disease can escape detection but still affect the baby. For this reason, most doctors recommend vaccination against rubella for all women of child-bearing age.

Permanent hearing loss has been produced by certain drugs, such as streptomycin, when consumed by small children. Ringing in the ears while under medication can be an early indication of such hearing loss. Fortunately, these side effects are well known, and the use of such drugs has been appropriately limited.

Normal aging of the skin and hair cells in the organ of Corti results in decreasing flexibility and springiness and therefore in a lessened response, particularly in the high-frequency range; such a normal aging process is called *presbycusis*. In its extreme, it can raise the threshold of hearing even at lower frequencies and necessitate use of a hearing aid. One effect of presbycusis is a widening of the critical band (particularly at high frequencies), and thus a widening of the masking pattern. As a result, background noise more readily masks the incoming desired audible signal in some persons suffering from presbycusis. Because simple amplification by a hearing aid increases the intensity of both the desired signal and the noise, hearing aids do not improve the clarity of signals in such cases. Amplification of the desired source, such as talking louder or using a microphone at the speaker's lips, must therefore be used to obtain significant improvement.

Another common disorder is *tinnitus*, the presence of neurally generated sound sensations, sometimes described as a constant "ringing" in the ear. A common cause is believed to be exposure to intense narrow-band sound, which selectively damages hair cells in some small region on the basilar membrane.

If the hearing loss is due to damage to the outer or middle ear, hearing can often be restored at least in part by surgery. In particular, there are techniques for replacing or bypassing the ossicles with restoration of nearly full hearing. The eardrum can also be repaired or in some cases replaced by a skin transplant. On the other hand, damage to the inner ear mechanism is apparently irreversible, and thus far it has not been possible to restore hearing loss due to the inner ear mechanism or the central auditory system.

A technique for identifying hearing loss is for the person to obtain an *audiogram*, which shows the threshold of hearing for each ear over the entire audible spectrum. Other audiograms can be made later to determine if further hearing loss has occurred. Such tests can only reveal certain types of hearing impairment. It is felt by some that hearing articulation (the ability to distinguish sounds with different transients) may suffer significantly even when the audiogram of a subject has shown only slight degradation.

6.11   *Anatomy of the Vocal Tract*

The sound produced by a musical synthesizer is obtained by filtering or modulating a source of sound such as a standard wave or some type of noise. As in the case of the synthesizer, vocal sounds are produced by a source of sound followed by a filtering and resonating system. The source of sound for the human voice is the vibration of the vocal folds, and the filter consists of the resonant cavities forming the vocal tract: the larynx, the pharynx, the mouth, and the nasal cavity.

A physical principle relevant to the operation of the vocal folds, as well as the lips of a brass instrument player and the reed of a reed instrument, is Bernoulli's principle, named for the Swiss mathematician and physicist Daniel Bernoulli (1700–1782). Bernoulli's principle states that in a fluid flow situation (say water or air) the pressure in the moving fluid is lower at places where the speed of flow is greater than at places where the flow is slower. An example is the "draw" of a chimney. Wind blowing across a chimney results in a low-pressure region just above the chimney, which pulls the smoke up the chimney. The pressure is lower just above the chimney than it is near the fireplace below, and the smoke is pushed out of the chimney. A beach ball or balloon can be supported by a column of blowing air owing to the wind force pushing the ball upward. Even if the ball shifts slightly to one side of the wind column, it will return to near the center of the column. This is because the air pressure is lower on the side of the ball facing the moving air than the normal air pressure farther from the moving air column. The resulting pressure difference results in a Bernoulli force that pushes the ball back into the column.

In the case of air rushing through the vocal folds, there is a low-pressure area in the constricted region between the folds, leading to a Bernoulli force. The Bernoulli force and tension in the vocal folds cause the folds to close. Similarly, the

Bernoulli force tends to close the lips of a player playing a brass instrument or the reed of a reed instrument.

The basic anatomy of the vocal tract is sketched in Fig. 6–7. Air coming from the lungs is pushed through the vocal folds by muscles in the chest and stomach area, including the diaphragm. When the air is started through the vocal folds, the Bernoulli force causes the folds to close. Immediately after the vocal folds close, air pressure builds up in the trachea, rapidly forcing the folds open once again. The burst of air through the vocal folds again creates the Bernoulli force, and the cycle is repeated. This process repeats at a rate determined by the tension in the folds, which in turn is controlled by muscles. The rate of opening and closing determines the frequency of the resulting vocal sounds. High muscle tension can be used by men to force the vocal folds to vibrate at a high frequency, producing a *falsetto* voice.

The vibration of the vocal folds is modified according to the resonance characteristics of the remaining elements of the vocal tract. The most salient factor determining the Fourier spectrum of sustained vocal sounds is the resonant behavior of the air column formed by the larynx, the pharynx, and the oral cavity. This column acts as a closed tube, the closed end at the vocal folds and the open end at the lips. As we saw in Sec. 3.4, the resonant frequencies of such a tube are its odd harmonics, and in this case these frequencies can be manipulated by adjusting the

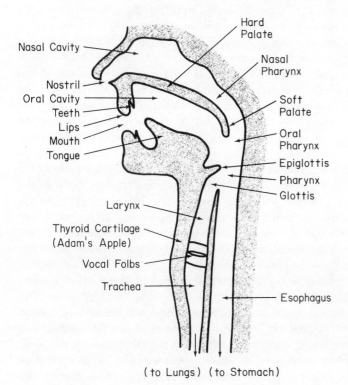

Figure 6–7 Schematic cross section of the human vocal tract. (After Backus, 1977).

shapes of the several cavities. The resonances associated with these frequency regions, called the *vocal formants,* are described in the next section.

The nasal cavity and the oral cavity, which can be adjusted in size and shape with the tongue, lips, and soft palate, lead to additional formant regions, with typical sounds resulting from their use.

The epiglottis protects the larynx against intrusion of food or other matter, and the thyroid cartilage, or Adam's apple, protects the vocal folds and the throat area in general from external damage.

6.12   *Vocal Formants*

As mentioned in the previous section, the sound source for the vocal system is the vibration of the vocal folds; this vibration is brought about by a combination of the force of the air from the lungs causing the folds to open, followed by a Bernoulli force that causes them to close. The frequency of oscillation of the vocal folds is determined primarily by the controlled tension in the vocal folds, whereas the amplitude of the oscillation is affected by increasing or decreasing the rate of the air flow between the folds.

In general, the length of the "tube" forming the primary vocal resonator, between the vocal folds (its closed end) and the lips (its open end), is between 17 and 18 cm. Vocal formants are frequency regions in which harmonics have large amplitudes. The resonant frequencies of such a closed air column are at about 500, 1500, 2500, and 3500 Hz, and so on. The principal vocal formants are then at approximately these frequencies; however, they can be manipulated simply by either constricting or enlarging the diameter of the column at the position of one of the antinodes or nodes in the formant standing wave. Any component frequencies of the vocal fold oscillations near one of these formants will be resonated by the vocal column.

For example, the fundamental or first formant has a velocity antinode at the lips. When the mouth is wide open, the velocity antinode is farther back in the mouth than when the lips are nearly closed. The effective length of the resonant tube, and therefore the lowest formant frequency, can be controlled by the size of the mouth opening. The amplitudes and frequencies of the formants are determined by the complicated shape of the vocal tract. The lips are generally used to adjust the lowest formant frequency, while the higher formants are affected by the tongue and other parts of the vocal tract. The manipulation of the lips need not affect the frequency at which the folds vibrate. By manipulating the tongue and lips, an average adult male can vary the first formant frequency by almost 500 Hz and the second formant frequency by almost 2000 Hz. Fortunately, one need not think about this as one speaks; it comes to us automatically. When the mouth is closed, the nasal cavity determines the formant structure. Because the shape of this cavity cannot be changed significantly, the vocal formant structure cannot be changed as much when the mouth is closed. Consequently, the range of sound qualities obtained with the mouth closed is limited.

Several of the vowel sounds, including ē, e, ā, ah, aw, and ōō, differ primarily by changing the relative frequencies of the first two formants. If you slowly and continuously say ē-e-ah-aw-ōō (as if in terror, or pain, only slowly) you are simultaneously (a) decreasing the second formant amplitude while (b) raising the second formant frequency (until ah), and then lowering it to its original frequency. Typical tongue motion used to adjust the frequencies of the first and second formants can be observed in this way. A good example of a vocal formant occurs in the vowel sound ēē. In this case there is a strong emphasis of harmonics around the frequency of 2500 Hz, the third formant region.

The effect of these formants is significantly different in men's and women's voices. In particular, when a soprano sings in the upper part of her range, the fundamental is generally significantly higher than the normal frequency of the first formant. The best "resonance" in a female voice, however, occurs when the frequency of the first vocal formant is adjusted to be equal to the fundamental frequency of the note sung. A soprano must therefore "drop her jaw" to obtain a very large lip and mouth opening in order to decrease effectively the length of the vocal tract and thereby raise the frequency of the formant. Unfortunately, this makes pronunciation of words while singing in the high range extremely difficult.

An additional feature of the voice, known as the *singing formant*, is relevant in determining the quality of a male singer's voice. One difference between excellent and mediocre male voices is their ability to *project*, or to be heard with a rich tone above the accompaniment, such as an orchestra. An important part of this ability to project is the existence of the singing formant, which lies in the region between about 2500 and 3000 Hz. Emphasis of vocal harmonics in this range increases their amplitudes above that of the average of the orchestral accompaniment and allows the voice to be heard above the orchestra. Those male voices having a dominant singing formant have better projection than singers lacking such a formant. The normal speaking voice generally lacks emphasis of frequencies in the region of the singing formant.

An unusual example of manipulation of the vocal formants is the case of certain Tibetan monks who can produce the illusion of singing two notes simultaneously. Singing the note whose frequency is about 63 Hz, the first formant frequency is adjusted to about 315 Hz and the second to about 630 Hz. These correspond to the fifth and tenth harmonics of 63 Hz. Because every harmonic in the overtone series of the frequency of 63 Hz is present, that note is heard. The emphasis of 315 and 630 Hz creates the illusion that a note of 315-Hz frequency is also being sung simultaneously. The fifth harmonic of any note is up a musical interval of two octaves and one major third, so the combination of the two notes makes a pleasant-sounding, though unusual-sounding, chord.

An amusing experiment involves inhaling a lungful of helium and singing or speaking. The vocal tract remains the same length. Because helium gas is considerably lighter than air, the velocity of sound is greater in helium. Using the relation $f = S/\lambda$, we see that, when the speed of sound increases and the wavelength remains the same, the frequency must increase. The frequency at which the vocal folds oscillate does not change significantly, but the vocal formants are increased in frequency, leading to the typical "Donald Duck" sound.

6.13   *Analysis of Vocal Sounds*

We have now seen the effects of the vocal formants on certain vocal sounds, particularly the vowel sounds. Vowel sounds are the simplest type of vocal sounds to analyze, because they are generally held for a longer time than the consonants, particularly the plosives, speech sounds that require the closure of the oral passage, such as the "p" sound in "cap."

Sounds, such as "n" and "m," can also be easily analyzed because they are long lasting. In both these cases, the mouth end of the vocal tract remains closed, and the sound is therefore dominated by the formants of the nasal cavity. The nasal cavity acts like a Helmholtz resonator, with a formant at the resonant frequency of the resonator. Other continuants can be similarly analyzed, as in the case of ordinary vowel sounds.

Sibilant sounds (for example, "s" and "z") include a component of almost white noise. In the case of voiced sibilants, like the letter "z" or "v," a discrete frequency spectrum is superimposed on the continuous spectrum forming the letter "s," in the case of "z," because the vocal folds are producing sound. Figure 6–8 shows the Fourier spectra for typical "s" and "z" sounds; the voiced effect of the "z"

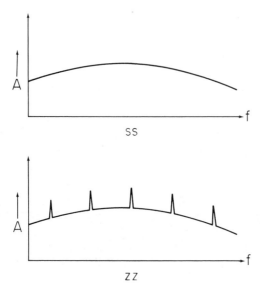

Figure 6–8 Idealized Fourier spectra of "s" and "z" sounds.

sound is the origin of the discrete Fourier components superimposed on the noise spectrum of the "s" sound.

Thus far we have dealt only with steady-state, continuous sounds, which can be characterized by a single Fourier spectrum graph giving their Fourier components at any time. If the sound changes, and we wish to describe a sequence of sounds forming a real word or diphthong, we must increase the information in the graph to account for the change of the spectrum with time. Such a graph is called a *sound spectrogram* or *audio spectrogram.* In this case, we are not so much interested in each harmonic as such, but rather we are interested in groups of harmonics; in particular, we want to look at the formant emphasis versus time. In a sound spectrogram we plot the intensity of the various harmonics or formants on the vertical axis versus time along the horizontal axis. Simplified spectrograms of the vowel sounds ēē and ōō are shown in Fig. 6–9. The vertical line at a given time is more dense for Fourier components of greater intensity. There is no change with time for these sounds, so the graphs remain the same as time passes. Also shown are spectrograms for the consonants "s" and "n." The sibilant "s" is characterized by its noise spectrum, roughly equal amplitudes for all frequencies, whereas the "n" possesses a range of frequency components determined largely by the resonant characteristics of the nasal cavity.

Figure 6–10 shows simplified sound spectrograms for composite sounds, "noo," "soo," "nee," and "see," all sung at a constant pitch. At the point where the sound changes, the spectrogram changes form. Figure 6–11 shows real sound spectrograms of the same four composites shown in simplified form in Fig. 6–10. As the intensity of any formant decreases, the line becomes lighter and finally disappears. In these cases the pitch and intensity of the voice naturally fall off toward the end of

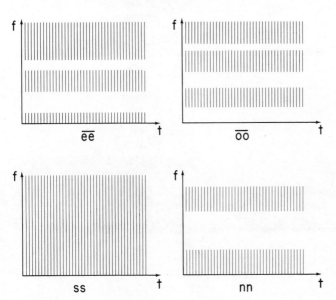

Figure 6–9 Idealized audio spectrograms of the sounds "ee," "oo," "ss," and "nn."

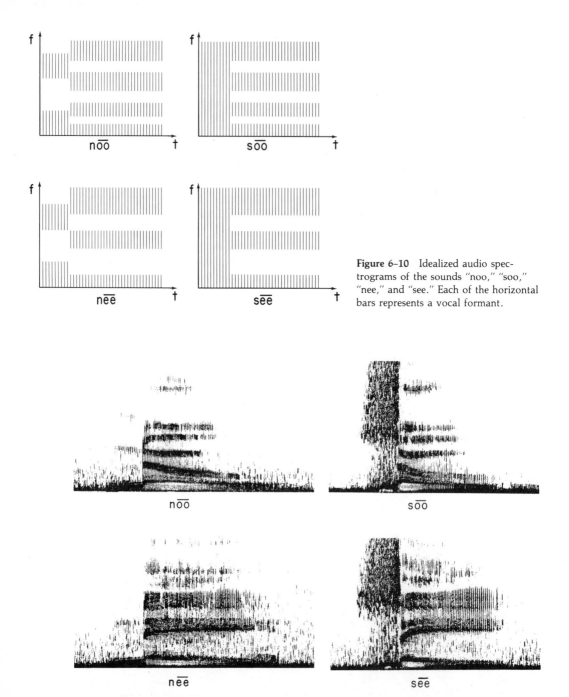

Figure 6–10 Idealized audio spectrograms of the sounds "noo," "soo," "nee," and "see." Each of the horizontal bars represents a vocal formant.

Figure 6–11 Actual audio spectrograms of the sounds "noo," "soo," "nee," and "see," spoken by an adult male.

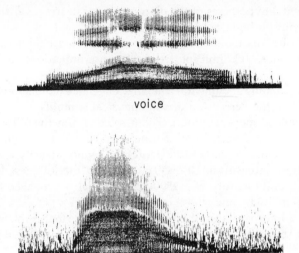

voice

synthesizer

Figure 6–12 Audio spectrogram of the word "wow" spoken by an adult male and produced by a musical synthesizer.

the vowel sound, as can be seen in the spectrograms. The maximum frequency in each spectrogram is 8000 Hz, resulting in a cutoff of the sibilant sound spectrum. Figure 6–12 is a comparison of a real voice and a musical synthesizer producing the word "wow." In this case, as in Fig. 6–11, the continuum of changing vowel sounds, as well as the frequency and intensity contours, can be identified.

Plosive consonants, such as "b," "p," or "d," are very difficult to analyze because they occur quickly. To describe a plosive, a graph of harmonic content as a function of time can be drawn similar to the preceding sound spectrograms, where the entire sound occurs within a period of a few milliseconds. Because this time interval is not significantly longer than the period of the Fourier components involved, it is very difficult to analyze the sound and obtain the amplitudes of the Fourier components.

Electronic voice synthesis has become a very important area of research. With the advent of large computers and methods of analysis of rapidly changing vocal sounds, it is possible to synthesize vocal sounds and even some speech reasonably well. The most difficult sounds to synthesize are the plosives, because of their rapid sequencing. Considerable work is now being done at research facilities, such as Bell Telephone Laboratories, because of the important application of this work to the area of the telecommunication.

EXERCISES

1. Trace a sound wave through the peripheral and central audio systems to the stage where nerves are stimulated. Describe each component part and its individual function.

2. Discuss the fundamental hypotheses of the place theory of hearing. Where are the nerves that respond to high frequencies? Low frequencies?

3. Define the following and discuss the relationship among (a) critical band, (b) just noticeable difference, (c) limit of frequency discrimination, and (d) sharpening.

4. Define periodicity pitch and fundamental tracking. Give an example in which this effect is relevant to the perception of some musical stimulus.

5. From the graph of equal loudness contours (Fig. 6–4) determine the following: (a) the loudness level of the threshold of hearing at 200 Hz, (b) the sound intensity level of the threshold of pain at 10 kHz, (c) the sound intensity level of a tone of 5 kHz whose intensity is 10^{-6} W/m², and (d) the loudness level, sound intensity level, and intensity of a 1500-Hz tone that is perceived to be the same loudness as a 7-kHz tone whose intensity is 10^{-12} W/m².

6. If the SIL of a single 5-kHz oscillator of intensity I_0 is 24 dB, what is the SIL of 100 such oscillators sounding together randomly or incoherently? 500 such oscillators?

7. Tones of 650- and 800-Hz are sounded simultaneously at high intensity levels. Determine the frequencies of several low-order (low m and n value) sum and difference tones, and state whether each would be likely to be audible. (Be sure to consider masking.)

8. How do second-order (quality) beats and how do binaural beats differ from normal (first-order) beats? How can the three types of beats be produced?

9. What is Ohm's law of hearing? Give examples and applications and a counterexample. Define masking. In general, which tone, 800 or 1200 Hz, would more easily mask a 1000-Hz tone?

10. Discuss two processes involved in spatial localization of sound sources. Which depends on high-frequency information? Which depends on low-frequency information?

11. What are audiograms and how are they used? What should a normal audiogram look like? Draw an audiogram for someone with a permanent threshold shift. What might cause such a shift? What is presbycusis?

12. What is Bernoulli's principle? How is it used to explain the operation of the human vocal folds?

13. What are vocal formants and how are they important in determining vocal sounds? What is the singing formant and how does it help a singer?

14. What is a sound or audio spectrogram? Draw possible audio spectrograms for the vocal sounds "oo," "ee," and "ah." Draw simplified audio spectrograms of the sounds "zz" and "ss." How do they differ?

REFERENCES

BEKESY, GEORGE VON. *Experiments in Hearing.* New York: McGraw-Hill, Book Co., Inc., 1960.

The magnum opus of the modern theory of the ear, including the basic experiments validating the place theory.

KRELL, JOHN. *Kincaidiana, A Flute Player's Notebook.* Culver City, Calif.: Trio Associates, 1973.

General book on flute playing and technique, containing the music for the "Trio for Two Flutes."

ROEDERER, JUAN G. *Introduction to the Physics and Psychophysics of Music,* 2nd ed. New York: Springer-Verlag, New York, 1975.

The best modern source of up-to-date information on the ear and psychoacoustics at the undergraduate level.

SUNDBERG, JOHAN. "The Acoustics of the Singing Voice." *Scientific American,* March 1977.

This excellent article describes the fundamental principles relating to voice production and vocal formants.

VOROBA, BARRY. *Experimenting in the Hearing and Speech Sciences.* Eden Prairie, Minn.: Starkey Laboratories, Inc., 1978.

Describes many experiments illustrating various features of the ear and the voice.

Recordings

The Music of Tibet; The Tantric Rituals, An Anthology of the Worlds' Music, No. 6, Society for Ethnomusicology, Anthology Record and Tape Corporation.

This exotic record presents an unusual mode of chanting by Tibetan monks who appear to sing two notes simultaneously, accompanied by reprints of articles explaining the religious significance of the chanting and the acoustical features of the sounds produced.

Voice of the Computer, Recording DL 710180, Decca Records.

This is the original record using the computer to generate sound. It contains Shepard's tones (1) ascending and descending by half-steps in an example and (2) used in glissando form in a suite of electronic music.

VAN LENTEN, D. H. *He Saw the Cat.* Bell Telephone Laboratories.

This fascinating tape uses a computer to synthesize speech and singing. It deals with the basic vocal sounds, including inflections.

7

Sound Recording and Reproduction

In this chapter we shall discuss the principal features of high-fidelity audio systems and their components. Following a brief discussion of basic electrical circuits, Ohm's law of electrical circuits (different from Ohm's law of hearing), and electrical power, we shall apply these concepts to audio reproduction systems. We shall then discuss properties and characteristics of the individual components in high-fidelity sound recording and reproduction systems. The aim of this discussion is to familiarize the reader with the physical principles on which the operation of this equipment is based, to become acquainted with the operational aspects of audio components, and to aid the reader in any future choice of reproduction systems or components.

7.1 *Electrical Circuits and Ohm's Law*

A simple electrical circuit is analogous to a hydraulic system consisting of a water pump, a tube carrying the water around in a complete circuit from the exit of the pump back to its entrance, and a constriction in the tube. The analogous hydraulic and electrical systems are drawn schematically in Fig. 7–1. In the case of a hydraulic system, the pump pushes the water around the tube with a force that is dependent on the operation of the pump. The constriction provides a resistance to the water flow, limiting the rate at which water can circulate through the system. The water flow can be increased by enlarging the diameter of the constricted region of the pipe

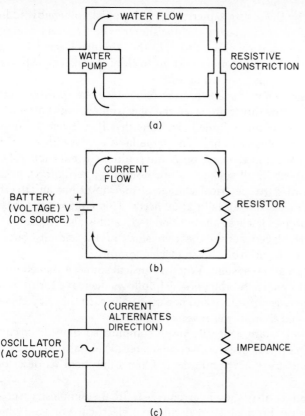

Figure 7–1 (a) Simple hydraulic circuit, and its electrical analog for the cases of (b) direct current, or dc, and (c) alternating current, or ac.

or by increasing the pressure with which the water is pumped. This can be expressed as an equation:

$$\text{water flow} = \frac{\text{pumping pressure}}{\text{resistance due to the constriction}}.$$

An electrical circuit can be thought of as a hydraulic system where a battery or power supply replaces the water pump, an electrical resistor replaces the constriction in the tube, and the flow is the flow of electrons, or electrical current, rather than water. The equation relating these three quantities, known as *Ohm's law* (of electrical circuits), can be written as

$$I = \frac{V}{R} \text{ or } V = IR$$

where I is the electrical current in amperes (symbol A), V is the electromotive force or voltage in volts (symbol V), and R is the electrical resistance in ohms (symbol Ω, capital Greek letter omega). The battery provides the source of electrical current, which is effectively limited by the resistor. Increasing the applied voltage by adding more identical batteries in series or replacing the one shown with a battery of larger voltage will cause more current to flow through the circuit. Decreasing the resistance of the resistor will also allow more current to flow, consistent with both our hydraulic analog and the Ohm's law equation.

This simple system is a direct-current (dc) system, since the (positive) electrical current always flows in one direction, from the plus to the minus battery terminals. In reality, it is the negatively charged electrons that flow from the minus terminal on the battery to the plus terminal. The same laws apply with a slightly different mathematical formalism in the case of an alternating current (ac) system. The electrical power available at wall sockets in our homes is alternating current; 110-volt 60-cycle ac means that the electrical voltage at the socket has an effective value of 110 volts, and oscillates sinusoidally at 60 hertz. This voltage thus causes an electrical current that reverses itself at a rate of 60 hertz and is also sinusoidal in shape when a resistance is placed across the two sides of the circuit. Such a resistance might be a light bulb, an iron, or an electric motor.

Another ac system is an audio system. When a musical sound is picked up by a microphone, the electrical signal it produces (which follows the wave form) varies about the zero in voltage, between positive and negative values. In this way current is forced in one direction, and then in the reverse direction.

Alternating-current circuits are generally more complicated than dc circuits, and the ac analogue to Ohm's law is also more complicated. We shall, however, ignore such complications and treat ac circuits as if Ohm's law held as described earlier.

An ac electrical circuit is also shown in Fig. 7–1. In the simplest case, an oscillator replaces the battery, producing an oscillatory electrical voltage, which causes an oscillating electrical current of the same frequency to exist in the resistor. In the case of alternating current, the resistance is called *impedance*. The current is limited by the impedance of the resistor, and can be increased in amplitude by either increasing the amplitude of the voltage applied by the oscillator or decreasing the impedance of the resistor, or both.

When electrical current exists in an electrical appliance, the electrical power of the battery or oscillator is transformed into another type of energy, such as heat, light, sound, or other mechanical energy. For example, if a light bulb is connected to a battery as shown in Fig. 7–1, the electrical power is converted into light and heat energy.

The mathematical formula for power in an electrical system is

$$P = VI$$

where P is the power delivered to the resistor in watts (symbol W), V is the voltage applied, and I is the current in the resistor, as before. Using Ohm's law, $V = IR$, the formula for electrical power can also be written as

$$P = I^2R = \frac{V^2}{R}$$

Thus we see that the greater the voltage and current applied to the resistor, the greater the power it consumes. Increase of power available to an application can be achieved by increasing the applied voltage or decreasing the resistance of the device.

In an audio system consisting of many individual components, it is necessary to *match the impedance* of an output device with the input of the component into which its signal is fed. This is done to transfer power efficiently from component to component and achieve the proper signal level throughout the system. We shall discuss this in more detail shortly.

7.2 *Reproduction and Amplification Systems*

We shall now look at entire audio systems to study their general properties and functions, and investigate the ways in which the components are matched to achieve optimal sound production. Following this will be detailed discussions of each of the components.

The typical configuration of audio components in a complete sound system is shown in Fig. 7–2. Signal sources include microphone, disc record player, tape recorder, and AM-FM tuner. The signals from these sources are input into a preamplifier that accepts each signal and produces an output of the appropriate amplitude for subsequent input to the amplifier. The power amplifier, or simply amplifier, accepts the low-voltage signal from the preamplifier and increases its amplitude and power level so that it can drive the speaker or speaker system.

We saw earlier that a sound wave will reflect whenever it passes from one

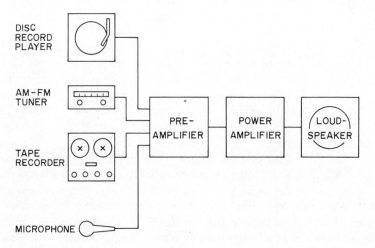

Figure 7–2 Layout of the components in a typical audio amplification and reproduction system.

medium to another of a different sound velocity. Likewise, alternating electrical signals may reflect when passing from one audio component to the next one in the system. In the electrical case, it is not the velocity of signal propagation that creates this effect. Rather, the impedances of the two devices are what determine how much of the signal will be transmitted and how much will be reflected. Signals will be reflected most when the output impedance of the first component does not equal the input impedance of the second component. Such *impedance mismatch* results in loss of signal energy as the signal passes from component to component through the system.

The several components act as links in a chain, accepting signals from one component, processing the signals, and passing them to the next component. The input impedance and the output impedance of a single component can be different, and they should be chosen to match the impedances of the neighboring components. For example, the output impedance of a microphone must match the input impedance of the preamplifier into which its signal is fed; the output impedance of the preamplifier must match the input impedance of the power amplifier; and so on.

If any component does not have the proper output impedance, or the source output impedance does not match the input impedance of the component into which the signal is fed, the efficiency of energy transfer will be decreased. In extreme cases, the signal may be almost entirely lost. If the input impedance is too high, the output current will be low. If the input impedance is too low, the output voltage will be low. Because the power transferred is the product of the current and the voltage, an intermediate input impedance maximizes the power transferred to the device. The maximum power is transferred when the output and input impedances are matched. Because the power requirements of the speakers are generally large, impedance matching is most crucial between the power amplifier and the speakers. Because the power output from a microphone is very low, impedance matching between the microphone and the preamplifier is also crucial.

An additional requirement is that the signal provided by any component be of the proper voltage level and capable of delivering sufficient power (or electrical current) to make the next component function. The range of voltages input to a component is important as well. If too low a voltage is used, the ratio of signal to electrical noise from the component may be too small; if too high a voltage is used, the component may be overdriven and distort the signal. For example, a preamplifier output generally has neither a sufficient voltage level nor adequate output power to drive a loudspeaker.

Several types of microphones are commonly available; they can be subdivided into those with high output impedance and those with low output impedance. If the input to the preamplifier is not of the appropriate impedance, it is necessary to match impedances with an impedance-matching transformer or a low noise audio amplifier, which is inserted into the line between the microphone and the preamplifier input. Microphone signal amplitudes are typically only a few millivolts, and a microphone is capable of delivering only minuscule levels of power, as can be seen from the equation for power discussed previously. The preamplifier into which the signal is fed must therefore have great sensitivity to small signals and the ability to amplify these low-voltage signals. A disc record player is similar to a microphone in that it produces a very low voltage output and is incapable of delivering much power.

Many tape recorders have two separate outputs with different levels. The tape head, which picks up the signal on the tape, produces a relatively low level output. This can be sent to an external amplifier at this stage. Alternatively, the output from the tape head can be amplified inside the tape recorder before being sent to external devices. This is often done to reduce the accumulation of noise that can occur when very low level signals are transported over wires between separate units. This output is then fed into the preamplifier and on to the amplifier and speaker. High-quality tape players and preamplifiers often have the capability of accepting and using signals from either the tape head (low level) or the tape amplifier (high level). The high-level signals from a tape amplifier are of about 1.5-volt amplitude, and are generally input into an impedance of about 50,000 ohms, with a resulting power transfer of a fraction of a milliwatt. Since power levels of the order of at least 10 watts are required to drive most loudspeakers, it is clearly necessary to introduce a power amplifier into the system to drive the speakers.

An AM-FM tuner receives the signals from either an AM or an FM radio station, demodulates the radio frequency (RF) signal to obtain the audio-frequency modulator (the music, announcer's voice, or whatever), and amplifies this signal to the 1.5-volt amplitude level. Again, this signal is fed into a preamplifier whose impedance is about 50,000 ohms; thus the power transferred is only a fraction of a milliwatt.

The preamplifier is the most versatile component in an audio system by virtue of the numerous functions that it performs. It is capable of accepting signals of various impedance and voltage level and providing the appropriate output signal level and impedance for the power amplifier. It also contains the controls (such as volume, loudness, filters, and balance) for adjusting the output signal to various listening conditions and input signals. The preamplifier produces a 1.5-volt output, which is sent into the power amplifier, which has impedance of about 50,000 ohms.

The power amplifier is an extremely simple unit; its sole purpose is to accept a signal of about 1.5 volts at an impedance of about 50,000 ohms and increase its voltage and current such that the power is sufficient to drive a loudspeaker. It usually has no controls; all controls are on the preamplifier. Whereas the power level input into the amplifier is in the milliwatt range, typical power levels from power amplifier's input into speakers are between 10 and 100 watts. It requires typically less than 1 watt of acoustic power to obtain a comfortable listening level in a normal listening room. Speaker systems typically vary in their efficiency to convert electrical power into acoustic power between about 1 and 10 percent. Thus, 1 watt of acoustic power could be obtained, for example, by using a 100-watt power amplifier driving a speaker with an efficiency of 1 percent or a 10-watt amplifier driving a speaker with an efficiency of 10 percent.

The choice of an amplifier and loudspeaker system must reflect the efficiency and impedance of the speaker and the volume of the sound desired. Most loudspeakers have an impedance of 8 ohms, but a large number have 16-ohm impedance, and a few have impedance of 4 ohms. If the impedance match is not proper, efficiency will be lost, and in some cases the speaker will produce a distorted sound and possibly be damaged. Damage can also be caused by overdriving a speaker with too powerful a power amplifier.

Some complete units are available that contain tuner, preamplifier, and power amplifier in a single chassis, eliminating the need for concern about matching between these components. In this case, the prime consideration is matching of amplifier output with speakers.

In general, almost any commercially available disc player with a magnetic cartridge can be used with any preamplifier, and the same is true for tape recorders. There are substantial differences between microphones, and to achieve best efficiency with least noise level, it is necessary to choose properly among various types of microphones.

7.3 Microphones

The two most popular microphones are the dynamic microphone and the condenser or electrostatic microphone. We shall now discuss the physical principles fundamental to the operation of each and complete this section with a discussion of general properties of microphones.

The principle fundamental to the operation of a *dynamic microphone* is the law of magnetic induction, or *Faraday's law*, named after the English physicist Michael Faraday (1791–1867). According to this law, a changing magnetic field within a coil of wire induces an electrical voltage in the coil, as shown in Fig. 7–3. The magnitude of the induced voltage is proportional to the rate of change of the magnetic field; reversing the direction of the field or the sense of field change (increasing to decreasing, or vice versa) changes the sign of the induced voltage. The correlation between induced voltage and the position of the magnet is shown in Fig.

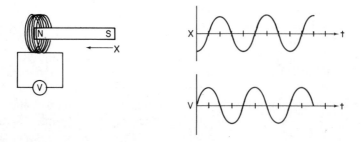

Figure 7–3 Electrical voltage versus time generated by motion of a magnet through a fixed coil of wire.

7-3. If the motion of the magnet corresponds to that of some complex wave, the induced voltage versus time curve will correspond to the wave form of the complex wave. It is important to remember that it is the *change* in magnetic field through the coil, not the magnetic field itself, that induces the voltage, so there is a phase change between the motion of the magnet and the signal this motion creates. This leads to phase distortion for complex waves, as will be discussed in Sec. 7.5.

A dynamic microphone works on this principle, as shown in Fig. 7-4. The sound wave, incident from the left, strikes a movable diaphragm to which a coil is attached, moves the coil about the stationary magnet, and thereby induces a voltage in the coil. A variation of this microphone could have a movable magnet and fixed coil. The result is the same, since the relative motion between the magnet and the coil is identical.

The *condenser* (or *electrostatic*) *microphone* makes use of the electrical capacitance between two parallel plates. A capacitor is an electrical circuit component consisting of two parallel conducting plates on which electrical charge is stored; its purpose is simply to store charge. Capacitance refers to the capability of the two plates to store charge. If equal and opposite electrical charges are placed on the two plates of a parallel plate capacitor, as shown in Fig. 7-5, the voltage between the plates decreases as the distance between the plates decreases, and the voltage increases as the distance between the plates increases. In a condenser, or electrostatic microphone, the sound wave is incident on one of the plates, which is a movable diaphragm. The changing voltage created by the changes in spacing between the two plates follows the wave form of the incident sound wave. The electret condenser microphone is a recent modification of the condenser microphone with excellent frequency and transient response characteristics (see the following). It employs permanently charged capacitor plates; this eliminates the need for external voltage sources. This reduces noise and increases the useful range of signal amplitudes or dynamic range.

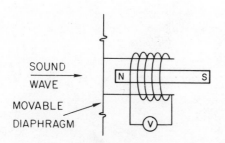

Figure 7-4 Mechanism of a dynamic microphone. Coil moves about fixed magnet, producing voltage measured by meter *V*.

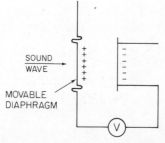

Figure 7-5 Mechanism of a condenser or electrostatic microphone. Fixed charge on plates produces voltage which varies as front plate is moved by sound waves, changing the distance between the two plates.

Several other types of microphones exist that use other physical principles to convert sound into electrical signals, such as the crystal microphone, the carbon microphone, and the ribbon microphone. Crystal microphones are sometimes used in inexpensive systems. Ribbon microphones can be of extremely high quality, but are very expensive and sensitive to damage from shock. Because these microphones are not in general use, we shall not describe their properties. Most telephones use carbon microphones.

Microphones are divided into two principal types by the directional characteristics of their response. A microphone that responds uniformly to sounds coming from any direction is called an *omnidirectional microphone.* Some microphones respond preferentially to signals incident from one direction. The most common of these *directional microphones* is the *cardioid* microphone, so named because the plot of response versus angle looks like a heart shape, as shown in Fig. 7–6, for a high-quality electret condenser microphone. The response of the microphone to a standard sound source at a given angle is plotted as the distance from the center of the graph to the response curve. Response curves are shown at 200, 1000, 6000 Hz, where the 0° position is directly in front of the microphone. An advantage of the cardioid microphone is its ability to pick up a musical performance, for example, while rejecting audience noise. For the microphone shown,

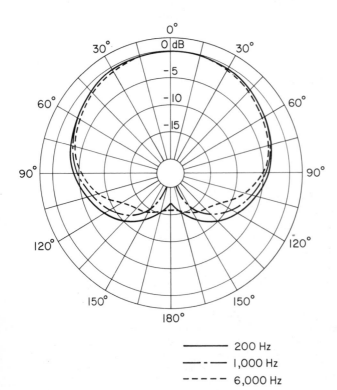

———— 200 Hz
—·—· 1,000 Hz
– – – – 6,000 Hz

Figure 7–6 Directivity characteristics of the Sony ECM-280 Electret condenser microphone. (Used with permission)

sounds from the backward direction are reduced 15 dB below those incident from the forward direction.

A "shotgun" microphone has a response that falls off very rapidly within a very narrow cone about the forward direction. Such microphones are used to pick up a question from one member of an audience, for instance.

Figure 7–7 shows the frequency response characteristics of a high-quality electret condenser microphone over the frequency range from 20 Hz to 20 kHz. Its response in the forward direction (0°) is relatively uniform, or flat, over the entire audio-frequency range from below 30 to above 17,000 Hz. Remember that the ear can just barely hear a change in sound intensity level of about 1 dB. Microphones of lesser quality have response curves that drop off more rapidly at the low- and high-frequency limits of their range or have large wiggles or resonances in their frequency response; that is, their frequency response is not flat. For the microphone illustrated in Figs. 7–6 and 7–7, the frequency response is relatively linear even in the direction (180°) opposite to the direction in which the microphone is pointed.

In addition to the frequency response to steady sound waves, the response to very fast pulses or fast changes in wave shape, called the *transient response,* is also very important. A microphone that is incapable of responding to rapid signals will change the overtone structure of the sound wave, so the electrical signal does not identically match the wave form. Condenser microphones are not only superior in their frequency linearity, but also in their transient response, as illustrated in Fig. 7–8. This is primarily due to the fact that the low mass of the movable diaphragm allows it to respond quickly to the changes in air pressure of the sound wave. The low mass also permits rapid damping of the diaphragm motion, preventing *overshooting.*

A wind screen, which looks like a wire mesh, is often placed around the microphone to eliminate the noise of wind rushing past the microphone diaphragm.

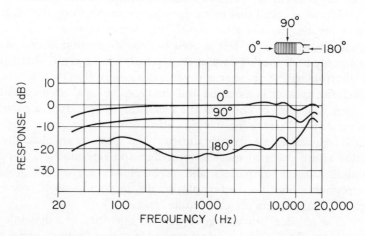

Figure 7–7 Frequency response characteristics of the Sony ECM-280 Electret condenser microphone. (Used with permission)

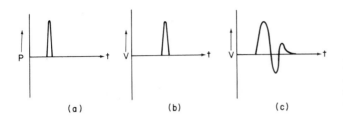

Figure 7-8 Typical transient reponse to a sharp sound pulse (a) by a condenser microphone (b) and a dynamic microphone (c).

The wind screen is often used in recording speech, where very low frequencies are unimportant, and where close proximity of the speaker to the microphone could lead to puffs of air from certain vocal sounds, causing loud extraneous noises. The wind screen also attenuates other unwanted low-frequency noises, such as those from fans, air conditioners, and humming transformers or fluorescent lights. Some wind screens are constructed from foam rubber balls.

Microphones can also be classified as to their type of response. Some, like condenser microphones, respond to the instantaneous pressure. Others, such as ribbon microphones, respond to the pressure gradient, or difference in pressure at two nearby points. The latter type produce a "throaty" sounding recording and are not as well liked, generally.

7.4 Loudspeakers

There are two classes of loudspeakers, electromagnetic and electrostatic. Recall that in a dynamic microphone the motion of the coil through the magnetic field induces the electrical signal in the coil. In an electromagnetic speaker, on the other hand, an ac electrical signal in a coil in a magnetic field is used to produce motion of the coil.

Likewise, recall that in an electrostatic microphone the relative motion of the plates creates the electrical signal. In an electrostatic speaker, an ac electrical voltage between two plates is used to produce motion of one of the plates. Most commonly used speakers are of the electromagnetic class. The names for each type of electromagnetic speaker are generally descriptive of the type of enclosure in which the speaker is mounted and the types of speakers mounted in the box.

A loudspeaker converts the electrical energy from a power amplifier into sound waves. Performing this task efficiently and linearly over the entire range of audio frequencies is extremely difficult, and speakers are generally the weakest link in the entire chain in musical recording and reproduction. Whereas undesirable nonlinearities in amplifiers and preamplifiers are almost nonexistent, and in microphones and disc record players are negligibly small, nonlinearities in even "good" loudspeakers are often quite large.

The physical principle governing the operation of a dynamic loudspeaker is illustrated in Fig. 7-9. If electrical current is passed through a wire in the presence of a magnetic field, a force will be exerted on the wire. This force will be greater for

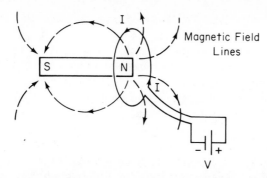

Figure 7-9 Passing electrical current through a wire in the presence of a magnetic field creates a force on the wire. In this case, the force on the current-carrying loop is to the left.

larger currents through the wire. In the case shown in Fig. 7–9, the force on the loop is to the left; if the current in the wire were reversed, the force would be to the right. A simplified diagram of a loudspeaker is shown in Fig. 7–10. When the current is in one direction, the force is to the right, causing the coil and cone to move in that direction, producing a compression. Reversing the current in the coil reverses the force on the coil and the cone will move to the left, producing a rarefaction. Application of a sinusoidal electrical signal causes sinusoidal motion of the speaker cone, resulting in a sinusoidal sound wave; complex waves are produced by applying complex electrical waves to the speaker coil. In the absence of any applied signal, the cone will remain at its equilibrium position.

In general, because of the large volume of air that must be moved to produce an audible wave of very low frequency, a large speaker is often used to adequately reproduce low frequencies. Because of the large mass of material that must be moved in order to move the speaker cone and produce a sound wave, speakers are in general incapable of good response to very fast pulses, and their transient response is similar in character to the transient response of a dynamic microphone.

An unmounted loudspeaker simply set on a shelf and used as is will produce poor sound. As seen in Chap. 2, an enclosure is necessary to prevent low-frequency waves from the rear of the speaker, which are out of phase with those from the front, from diffracting around to the front and interfering destructively with the

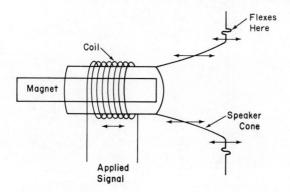

Figure 7-10 Mechanism of an electromagnetic loudspeaker.

waves from the front. Three important types of speaker enclosures that accomplish this are illustrated in Fig. 7–11.

The simplest such "enclosure" is simply a wall with a hole in it into which the speaker is placed. Such a speaker arrangement is called an *infinite baffle,* as the size of the baffle is very large, infinitely large in the ideal case, and there is therefore absolutely no mixing of the waves from the rear and the front of the speaker. Ceiling speakers are often infinite-baffle speakers. Sometimes very large speaker enclosures are called infinite baffles.

Because most people do not have a spare room or an extremely large wall to use as their speaker enclosure, it is necessary to reduce the size of the speaker box to more reasonable limits. The speaker system with the simplest type of smaller enclosure is the *acoustic suspension speaker,* in which the speaker is placed in an airtight box. This prevents interference of the waves from the front and rear of the speaker because the waves from the rear cannot get out. In such a speaker enclosure, the air pressure inside the box creates an additional restoring force to return the speaker to its equilibrium position. As a result of this additional restoring force, the power required to move the speaker cone and produce sound waves is substantially increased; such a speaker system is therefore relatively inefficient in converting electrical energy into sound energy.

Another limitation of this type of enclosure is the relatively rapid fall-off in output level at frequencies below about 60 to 100 Hz. The speaker cone itself can have a resonant frequency of 100 Hz or higher and must be damped. In designing a speaker box, one must be careful to avoid box geometries that yield strong resonances from the waves inside the box; this would lead to distortion of the sound produced. Figure 7–12(a) shows the frequency response of a typical loudspeaker in an acoustic suspension baffle. The graph plots the output of the speaker for a variable-frequency sinusoidal signal of constant amplitude. The region of large response at the low-frequency end shown in graph (a) is due to a resonance in the speaker.

The low-frequency range can be increased and the efficiency of the speaker

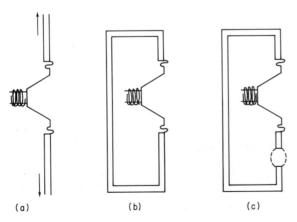

(a)   (b)   (c)

Figure 7–11 Cross-sectional views of (a) an infinite baffle, (b) acoustic suspension, and (c) tuned port loudspeaker systems.

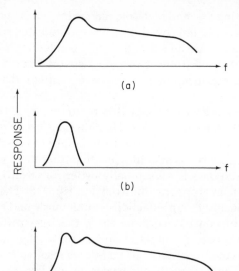

Figure 7–12 Frequency response for (a) an acoustic suspension speaker, (b) speaker box with tuned port alone, and (c) response of speaker and tuned port enclosure combined.

over the entire audio range improved by a simple modification of the cabinet; the result is the *tuned-port* or *bass-reflex* speaker enclosure, shown in Fig. 7–11(c). If a port, or hole, is cut into the speaker box, generally in the front below the main speaker, the box itself will act as a Helmholtz resonator, with a resonant frequency that can be chosen to be slightly below the resonant frequency of the loudspeaker itself. A typical resonance curve for a speaker enclosure with a port is shown in Fig. 7–12(b). If the speaker response is taken with the speaker in the box with the hole, the sum response curve will look like that in Fig. 7–12(c).

The hole is adjusted in size and position so that the Helmholtz-type resonance is at an appropriate frequency and has an appropriate amplitude to make the total response curve look like that in part (c); hence the name "tuned port." In this way, the bass range is effectively extended to a frequency well below the resonant frequency of the speaker alone. Because of the hole, air can move in and out of the speaker enclosure, thus reducing that part of the restoring force and substantially increasing the efficiency of the speaker system; tuned-port speakers are among the most efficient electromagnetic speaker systems.

Some high-quality speaker systems use two speakers: a small speaker, called a *tweeter*, which responds better to high frequencies, and a larger speaker, called a *woofer*, which responds better to low frequencies. Such an arrangement is called a *two-way speaker system*. A two-way system needs a built-in electronic *crossover network* that automatically routes the high-frequency electrical signals to the tweeter and the low-frequency signals to the woofer, and appropriately balances the signals to each speaker at the midrange, where both speakers respond. By using two speakers, the response over the entire frequency range can be improved significantly, and by directing low and high frequencies into their own speakers, the transient response can be greatly improved. In more sophisticated speaker systems, one or more midrange speakers are added to the usual woofer and tweeter, and this requires a more complicated electronic crossover network. For a given speaker system, proper electronic equalization can improve (flatten) the frequency response.

Electrostatic speakers use electrostatic forces to move a thin electrically charged sheet, which creates the sound wave. The thin sheet oscillates between two screens to which the audio output signal from the amplifier is fed. The mass of material moved in the speaker is considerably less than that in an electromagnetic speaker, so the transient response is substantially better for electrostatic speakers, just as with the electrostatic microphone, illustrated in Fig. 7–8. They are therefore capable of much greater fidelity in reproduction of both music and speech. Because electrostatic tweeters can be constructed more easily than electrostatic woofers, a combination of electrostatic tweeter and electromagnetic woofer has been used.

Important limitations of electrostatic speakers have resulted in their remaining less popular at this time. They are relatively expensive and less efficient than most electromagnetic speakers. Being a more recent development, they have yet to gain acceptance among the average audio consumer. As the engineering problems are solved and prices reduced, electrostatic speakers should achieve greater popularity.

Stereo headphones have several advantages over loudspeakers. First, they effectively eliminate most outside sound, and, second, stereo separation is superb. Because the mass of the moving diaphragm is small, the transient response is excellent in high-quality stereo headphones. The frequency response characteristics of good stereo headphones are very good. Although they require less power than most speakers, they have the obvious drawback that only one person can listen at a time. They are also more likely to cause hearing damage when used by unwary persons, since it is easy to produce 130 dB inside a headphone cupped over the ear.

7.5   *Record Players and Records*

The performance of a disc record player involves many principles of mechanics and electromagnetism. As the disc rotates, a number of mechanical forces hold the needle properly in the groove. As the groove passes, rapid oscillations of the stylus result in relative motion of a small magnet and an induction coil, creating an electrical voltage in a way similar to the operation of a dynamic microphone.

Figure 7–13 shows the top view of a disc player. The tone arm is connected to the base at the pivot. The stylus is mounted in the cartridge, which is attached to the other end of the tone arm. The stylus is oriented, as nearly as possible, tangent to the disc groove. This can be achieved by either curving the tone arm itself or mounting the cartridge, and thus the stylus, at an angle to the tone arm.

The *tracking force*, a downward force (into the paper) that tends to keep the

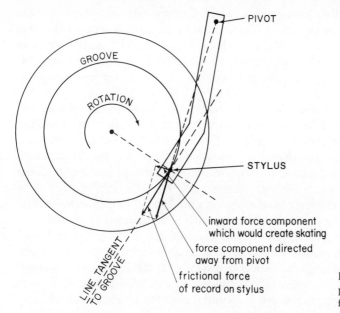

GROOVE

ROTATION

PIVOT

STYLUS

inward force component
which would create skating

force component directed
away from pivot

frictional force
of record on stylus

LINE TANGENT
TO GROOVE

Figure 7-13 Top view of disc record player showing some of the important forces acting in the horizontal plane.

stylus in the groove, is created primarily by the weight of the tone arm. A counterweight is placed on the opposite side of the pivot to limit the tracking force to about 1 to 2 grams in most cases. To be certain that large bumps on the surface of the record do not push the needle out of contact with the groove, a vertical damping force is applied to the arm. This also increases the trackability, or ability of the stylus to follow variations in the groove. This damping is generally accomplished with air or a liquid, as in an automobile shock absorber, and is therefore called *viscous damping*. The inertia of the entire arm, due to its rather large mass, keeps the arm relatively motionless above the groove while the much lighter stylus moves rapidly back and forth in the groove.

The predominant horizontal force is the *frictional force* between the record and the stylus created by the sliding of the groove past the stylus. This frictional force is directed tangent to the groove, as shown in Fig. 7–13. The frictional force can be thought of as having two perpendicular components. The greater of the components is directed away from the pivot and acts to pull the tone arm along the direction of the component. However, the tone arm is held fixed at the pivot and does not move in that direction.

The second, smaller force component acts to rotate the tone arm clockwise about the pivot. This force is called the *skating force*. If the skating force were not canceled, the tone arm would skate, or slide, inward across the disc. To avoid this, an outward, antiskating force is applied to the tone arm. This antiskating force is sometimes applied by a small spring attached to the tone arm near the pivot.

A more recent development in disc players is a straight arm mounted perpendicular to a radius of the disc. The stylus is mounted on the arm so that it is parallel to the groove, and as the record plays, the arm moves inward, keeping the stylus always tangent to the groove. This eliminates the skating force and possible wear to the inner side of the groove. This device, called a *straightarm*, is shown in Fig. 7–14. A straightarm tool is used to cut grooves in the master disc that is used in record manufacturing.

For all turntables, as the record turns, the stylus follows the groove oscillations while the tone arm and cartridge remain relatively fixed. This relative motion between the stylus and the cartridge leads to the induction of a voltage that has a time variation related to the shape of the groove. Just as in the dynamic microphone, the signal produced is proportional to the relative velocity between the stylus and the cartridge.

Two types of magnetic cartridge are now popular: In one type the coil is mounted on the stylus and moves about a magnet fixed with respect to the arm; in the other type, the magnet is mounted on the stylus and moves through a coil fixed with respect to the arm. Some disc players of lesser quality make use of ceramic cartridges. These utilize the piezoelectric effect in a special crystal in which mechanical stress arising from relative motion of the stylus and cartridge produces a small voltage. Because ceramic cartridges are more often used for inexpensive children's record players than in high-fidelity stereo systems, we shall not discuss them further.

It is necessary to control carefully the rotational speed of the disc to obtain

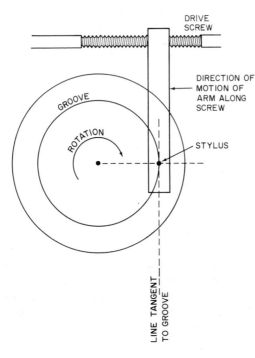

Figure 7-14 Top view of a straight arm disc player.

faithful pitch production. The rotating plate or platter is heavy, to provide a large inertia, and some turntables have an additional flywheel to further increase inertia. The heavier the platter, the less sensitive is its rotation to small perturbations. Some modern turntables have strobe lights to aid in setting and controlling the rotational speed of the platter. If the rotation of the turntable is not uniform, the result can be *wow* (slow changes in the pitch) or *flutter* (rapid quivering of the tone). Wow and flutter combined should be small enough to remain undetected, around 0.1 percent (less than the frequency JND) for a high-quality turntable. Other mechanical vibrations of the motor or drive mechanism lead to *rumble,* low-frequency mechanical vibrations transmitted to the cartridge. Rumble should be at least 50 dB lower than the music on the record in order that the quality of the reproduction not be affected. *Acoustic feedback* results when the sound produced by the record returns from the speaker to the turntable and causes errors in tracking and resultant distortion. Rumble and acoustic feedback can be reduced effectively by using good shock-absorbing mounts for the turntable; these are generally part of a high-quality record player. Total harmonic distortion, defined as the ratio of intensity of distortion introduced by the player to the intensity of the sound being reproduced, is below about 1 percent (slightly higher at high frequencies) from all sources combined, for most good record players. It is also necessary to shield the pickup and signal leads carefully to minimize distortion introduced by pickup of external electrical signals.

The simplest type of record is the monophonic disc, shown in cross section in Fig. 7–15. The stylus sits in the groove, whose walls are at 45° with respect to the vertical. Motion of the stylus is lateral, that is, parallel to the surface of the record and perpendicular to the length of the groove. The stylus must be able to track lateral motions of the groove over the entire frequency range without distortion. *Compliance* refers to the inherent flexibility of a stylus-cartridge system. The higher the compliance, the better the stylus can respond to rapid motions of the record groove. In general, the higher the compliance number, the greater is the trackability of the cartridge. At high frequencies, where tracking is more difficult, the needle tip mass is more important; the smaller the tip mass, the greater is the trackability.

Two important considerations in disc production, preemphasis and equalization, arise from a detail in the application of the law of induction as discussed in

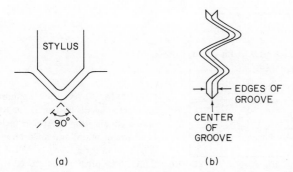

Figure 7–15 Stylus in groove of a monophonic disc record and lateral motion of groove as viewed from above.

Sec. 7.3. Equal amplitude grooves of high and low frequencies will not yield the same amplitude of electrical signal from a magnetic cartridge. Because the motion of the cartridge at low frequencies is slower than the motion at high frequencies, and the voltage produced depends on the relative velocity between the stylus and cartridge, it would be necessary to make the low-frequency oscillations of the groove inordinately large to obtain the proper voltage at low frequency. Such large groove oscillations cannot be tracked by the cartridge, so instead the low-frequency oscillations are attenuated compared to mid-range frequencies. It is necessary upon playback to introduce extra amplification to the low frequencies to compensate for this attenuation; this process is called *equalization.* If you play a record with the volume turned all the way down, you can observe this. Listen carefully to the quiet music coming from the tone arm and cartridge; it lacks bass.

At the high end of the audio spectrum another problem is encountered. The required amplitude of groove oscillation at high frequencies becomes extremely small; in fact, their amplitude becomes so small that the inherent graininess of the record plastic causes oscillations and noise, which become large compared with the amplitude of the signal itself. Therefore, the high-frequency oscillations are increased in amplitude above that required to produce a linear audio signal. This increase in amplitude of the high-frequency components, called *preemphasis,* increases the desired signal level well above the noise level. It requires an appropriate attenuation at those frequencies upon playback; this too is called equalization. The most widely accepted equalization standard is RIAA (Recording Industry Association of America), which specifies exactly how this compensation is to be made. A typical RIAA equalization curve is shown in Fig. 7–16. The phono preamplifier of most stereos assumes use of magnetic pickups and so automatically compensates for RIAA preemphasis. The RIAA curve and preemphasis are designed to make the total system have as linear a frequency response as possible. In the absence of some type of equalization, when playing some old records, for example, there will be high-frequency noise, inadequate bass, distortion of loud musical sounds, and noisy soft passages.

The standard groove configuration for stereophonic disc production has the groove walls at 45° with respect to the vertical, with the left-channel information on the inner groove wall and the right-channel information on the outer groove

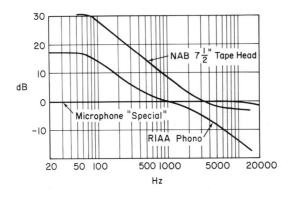

Figure 7–16 Curves showing amplification factor *versus* frequency for RIAA phono equalization, NAB 7½ inch tape head equalization, and microphone equalization for a Dynaco PAT-4 Solid State stereo preamplifier. (Used with permission)

wall. The magnet moves simultaneously through two mutually orthogonal (perpendicular) coils, one of which picks up signals from oscillations of the stylus perpendicular to each wall. The groove configuration in cross section is shown in Fig. 7–17, and the general configuration of the magnetic pickup system is shown in Fig. 7–18 for a moving magnet cartridge.

To allow playback of stereophonic records by monaural systems, it is necessary to reverse the phase of the signals on one of the sides of the groove; that is on one side a positive signal is motion of the stylus away from the groove edge, whereas on the other side a positive signal is motion of the stylus toward the groove edge. The combined motion of the left and right channels both moving in the positive directions then corresponds to lateral (horizontal) motion of the stylus, as in the case of monophonic records. Stereophonic records are then "compatible" with monophonic reproduction systems.

If the skating force is not adequately compensated, the left channel will wear preferentially. *Crosstalk*, when some of the left-channel signal is processed by the right-channel amplifier (or vice versa), can occur if the pickup coil is improperly constructed or if the circuits for the two channels are not adequately separated electrically.

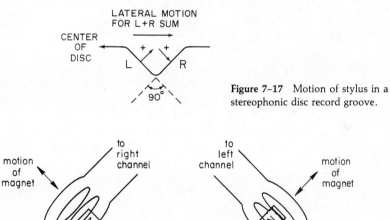

Figure 7-17 Motion of stylus in a stereophonic disc record groove.

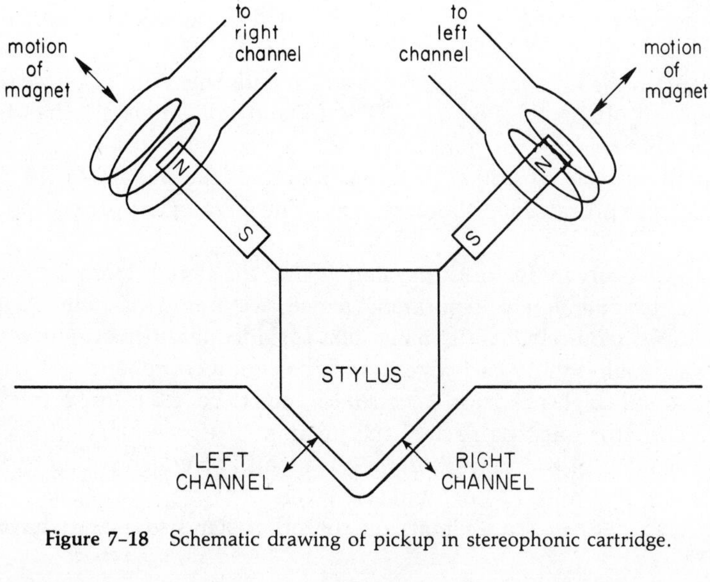

Figure 7-18 Schematic drawing of pickup in stereophonic cartridge.

Quadraphonic sound is so named because it consists of four channels of audio information. Two types of quadraphonic disc production have been used: the quadradisc discrete (CD4) format and the SQ matrix format. Both forms of quadraphonic disc are compatible with stereophonic and monaural reproduction systems, but not with each other, for reasons that will shortly become clear. The four channels of a quadraphonic are labeled left front (LF), right front (RF), left back (LB), and right back (RB). The groove in the case of quadraphonic recording is similar to the configuration used in stereophonic records.

The signals for quadradisc discrete (or CD4) are coded onto the walls of the groove in the following way. The simple sum of the (LF + LB) signals is pressed on the inner wall, and the simple sum (RF + RB) is pressed on the outer wall. At this point, the record would be identical to a stereo disc, with left and right channels. However, superimposed on the inner wall of the groove is the signal (LF - LB), and on the outer wall is superimposed the signal (RF - RB). Whereas the sum signals are simply audio signals, the difference signals are converted to high-frequency FM signals and then pressed onto the groove wall. In particular, the audio signal obtained by adding the left front and left back signals out of phase (the minus sign) is used to frequency modulate a 30-kHz carrier between the limits of 20 and 45 kHz. This frequency-modulated signal is then superimposed on the sum signal as stated.

The signal produced by one pickup is a rather complex sum. For the inner wall, this includes the signal (LF + LB), whose frequency range is 20 Hz to 20,000 Hz, and the signal (LF - LB), the frequency-modulated signal between 20 kHz and 45 kHz. These signals are then input into a preamplifier, where they are processed simultaneously in two ways. Simply amplifying the complex sum signal will occur in one route. If the signal is amplified at this point, as it would be if it were played on a stereo disc player, the signal (LF + LB) would be heard, since the frequency range of the frequency-modulated signal (LF - LB) is too high to be heard. If the complex signal from the pickup is input into a demodulator, the signal (LF -LB) will be reproduced, and the sum (LF + LB) will be outside the audible range. These two (sum and difference) signals are then input into two mixers, one of which adds them to obtain the LF signal and one of which adds them out of phase (in effect, subtracts them) to obtain the LB signal. The LF and LB signals are then fed separately into two channels of the amplifier and on to their respective, separate speakers. Simultaneously, the other pickup coil obtains the signals for the right two speakers, which are processed in the same way. The CD4 signal-processing diagram is shown in Fig. 7–19.

One advantage of this system is that the signals from the four channels can be virtually completely separated, hence the name discrete quad. However, the technical requirements of the system for adequate performance are very stringent. A very high-quality cartridge, which has good compliance for frequencies up to 45 kHz, and high-precision electronics must be used to preserve separation. A demodulator must be part of the preamp. It is necessary to provide special short cables in order to transmit low-level signals at frequencies up to 45 kHz without increasing the noise to unacceptable levels.

The SQ matrix format, on the other hand, does not have the stringent re-

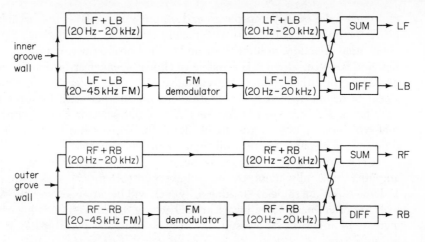

Figure 7-19 Block diagram of signal processing used in production of discrete quadraphonic sound.

quirements placed on its electronic components, but is not able to produce such well-separated channels. In the case of SQ matrix format, the LF channel is cut on the inner side of the groove and the RF channel on the outer side of the groove. However, the LB (RB) channel is cut in a complicated manner to make the stylus move in a clockwise (counterclockwise) corkscrew path as it moves down the record groove. While the two front signals and the two rear signals are well separated, it is necessary to use an SQ matrix decoder in the preamp to separate the front from the rear signals on each side. Unfortunately, this cannot be done nearly as well as in the case of discrete quad. Nevertheless, the greater simplicity of SQ matrix electronics has made this format somewhat popular.

To achieve the trackability required in stereo and quadraphonic reproduction, it has been necessary to develop special stylus shapes. The simplest is the conical, which is adequate only for monoaural records. The higher trackability requirements of the stereophonic record are met with elliptical styluses, where the longer dimension of the ellipse extends across the groove. This shape allows the stylus to respond to the shorter-wavelength, high-frequency oscillations of the record groove. The requirement of compliance up to frequencies of 45 kHz in the case of discrete quadraphonic reproduction dictates the use of even more complicated stylus shapes.

7.6 *Preamplifiers*

The preamplifier serves as the control center for the audio system, accepting the signals from the various sources and processing these signals for input to the power amplifier, which then leads to the speakers. A preamp must have inputs that match the output impedance and signal level of each of the signals required in the system, and must have output impedance and output voltage level that match that of the power amplifier input. Controls typically available on a high-quality preamp include input selection, volume, bass and treble boost or attenuation, loudness, filtering, and balance. Often it is possible to choose speaker arrangement, either stereo, monaural, or either channel individually. On many units it is possible to input a signal from one tape recorder and record on another with the option to play it simultaneously through the speakers. The input selector simply chooses the unit that will serve as the source of the signal for the system; because of the selector (knob), it is possible to leave all signal sources connected at all times.

The volume control varies the level of the signal output to the power amplifier; normally there are no volume controls on the power amplifier itself. Volume is not the same as "loudness," which will be discussed shortly. Volume controls are often made to be logarithmic, because the ear responds logarithmically to increase in intensity. As a result, the volume control is perceived as approximately linear.

Both high-pass (or low-frequency-attenuating) and low-pass (or high-frequency-attentuating) filters are normally available on good preamplifiers. Figure 7–20 shows the transmission of low- and high-frequency filters for a Dynaco PAT-4A solid state stereo preamplifier. In this case there are a single low-frequency filter and three separate high-frequency filters, labeled by the cutoff frequency of the filter in kilohertz. High-frequency filters are useful in removing high-frequency tape hiss or scratch noises on old records, whereas low-frequency filters can eliminate electrical hum or rumble from disc players. In some cases, a special rumble filter is provided specifically for this purpose. Here, as elsewhere, the filter is either in or out (on or off). Limited continuous adjustment of relative signal levels at various frequencies is obtained using the tone controls.

Continuously variable bass and treble tone controls are a fundamental feature of any amplifier. The maximum effect of the tone controls for the PAT-4A preamplifier is shown in Fig. 7–21. Adjusting the bass control changes the amplification factor between the limits shown at the low-frequency end of the graph. Increasing the bass level raises the level at 20 Hz by nearly 20 dB. Similarly, adjustment of the treble control changes the amplification factor at the high frequencies. Setting both controls at their center leaves the levels unchanged for all frequencies.

The loudness control does not refer to volume, but rather deals with production of sounds of constant loudness level (in phons) to the ear at low sound intensity levels. In Chap. 6 we saw that there is a considerable difference between the sound intensity level in decibels for the threshold of hearing at various frequencies. For high sound intensity levels, say above 50 or 60 dB, all frequencies are heard when they are produced at the same level, as can be seen from Fig. 6–4. However, at low overall sound intensity levels, the low-frequency sounds, and to a lesser extent the high-frequency sounds, may be reduced below the threshold of hearing. This can occur when music from ordinary records is played softly by turning down the volume control. In such a case, it is necessary to boost the extreme frequency

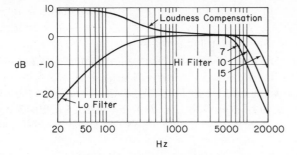

Figure 7-20 Attenuation factors for three high-frequency filters and a low-frequency filter along with amplification factor for loudness control for the Dynaco PAT-4A Solid State stereo preamplifier. Each high-frequency filter is labeled by its cutoff frequency. (Used with permission)

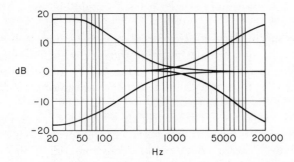

Figure 7-21 Limiting amplification and attenuation factors for bass and treble tone controls (with controls set at maximum and minimum values, respectively) for the Dynaco PAT-4A Solid State stereo preamplifier. (Used with permission)

ranges, particularly the bass, so that the loudness level for all frequencies is approximately the same, and the music will sound the same at the lower level as it did at the higher level. This is accomplished by switching on the loudness control. The amplification factor for the loudness control is shaped to compensate for the increase in the threshold of hearing at low frequencies, as shown in Fig. 7–20. In some preamplifiers, the high-frequency end of the spectrum is also emphasized by the loudness control.

A final control generally available on preamplifiers is the balance, which adjusts the relative output level between the two speakers of a stereo system. In the cases of the controls discussed, either the control applies equally to both channels or separate controls are provided for each channel.

For a high-quality preamplifier, the distortion (either total harmonic distortion or intermodulation distortion) should be below about 0.05 percent. Intermodulation distortion (IMD) is a combination tone (sum or difference) effect in which two tones played together in one channel combine and generate others. This is a nonlinear phenomenon similar to the combination tones produced in the ear. The linearity should be within 1 dB (the intensity JND) over the entire audible range for equalized input signals and slightly better for high-level input signals. Most name-brand stereophonic equipment comes close to these specifications.

The choice of a preamplifier and amplifier system should be made on the features available, the specifications of the unit, and, of course, the price, but most important is the subjective listening response. How good does it sound to you?

7.7 Power Amplifiers

Most power amplifiers are simply "black boxes" with no controls or adjustments, with input impedance of about 50,000 ohms, which accept signal levels of about 1.5 volts and produce an output according to their specifications.

Specifications for a power amplifier include the required input level and impedance, the power output at full-output level, and the impedance of the compatible speaker system. The linearity of a good power amplifier should be within about 0.5 dB over the entire audio-frequency range. The distortion should be less than 0.5 percent (both THD and IMD) for any level between zero and the maximum output specification. Noise caused by the amplifier circuit itself should be nearly 100 dB below the normal signal level.

The linearity is often specified over a frequency range extending well above the audible limit. While this is not particularly useful, the farther the linear region extends, the more likely it is to retain its linearity in the audible frequency range even at high signal output levels; thus many amplifiers have their linearity specified to well above 20 kHz, above the audible frequency range.

If the impedance of the speaker is not appropriately matched to the output impedance of the power amplifier, the power transferred to the speaker will be reduced. For example, the Dynaco Stereo 120-A amplifier, an amplifier of high quality, will deliver up to 60 watts per channel into 8-ohm speakers, the specified impedance. Use of 4-ohm speakers limits the power transfer to 50 watts per channel, while use of 16-ohm speakers further reduces the maximum power output to 40 watts per channel.

Although it takes only a few watts of audio power to fill even a large listening room with sound, most power amplifiers provide considerably more. A greater amplifier power will be required if low-efficiency loudspeakers like acoustic suspension speakers are used, whereas high-efficiency speakers like tuned-port speakers require less power. A 100-watt amplifier providing only 20 watts of power to a speaker will sound better than a 20-watt amplifier working at nearly full power, because it is working more nearly within its linear operating range. Any amplifier operating near peak power output will produce more distortion, especially in transient peaks like percussion sounds, where the level may actually rise instantaneously above the steady-state peak power output level. It is therefore wise to use an amplifier with maximum power output somewhat greater than the desired operating input level of the speaker system.

7.8 AM-FM Tuners

A tuner receives radio-frequency signals from a radio station broadcast, demodulates the signal to obtain the information stored in the modulator signal, and amplifies it to a level appropriate for input into a high-level, high-impedance preamplifier. If the tuner also contains the preamplifier, amplifier, and speaker, it is called a radio.

AM (amplitude modulation) radio stations are assigned carrier frequencies every 10 kHz over the AM radio band, which extends between 540 and 1600 kHz. The audio bandwidth, or (maximum) range of usable frequencies available to any station so that it will not interfere with either adjacent station, is about 5 kHz. Only frequencies up to about 5 kHz can be broadcast, and true high fidelity is therefore impossible for most AM stations. Furthermore, noise caused by the electronic components and static will cause changes in the amplitude of the signal, and lead to noise and distortion of the sound. No provision is available for transmission of stereophonic (or quadraphonic) sound on AM radio as it now exists. While AM radio is adequate for voice transmission, most stations do not transmit music with high fidelity, as do most FM stations.

FM (frequency modulation) radio stations are assigned frequencies every 0.2 MHz (or 200 kHz) over the FM radio band, which extends from 88.1 to 107.9 MHz. This frequency range lies between channels 6 and 7 of the television frequency range. In FM the frequency of the signal changes; in the case of FM radio, the maximum allowable frequency excursion is ±75 kHz, or a total excursion of 150 kHz about the carrier frequency. It is therefore readily possible for an FM station to transmit audio frequencies up to 15 kHz. Since noise and static affect the amplitude of the radio signal received, while the information content of FM signals is stored in frequency variations, FM radio is relatively noise free.

FM radio stations can transmit stereophonically. Two separate audio signals are used to modulate different regions of the frequency band, the second signal being transmitted on a subcarrier 38 kHz above the frequency of the main carrier. The subcarrier is simply a 38 kHz sine wave amplitude modulated by the second signal. An additional 19-kHz audio tone that is transmitted acts as a signal to the FM tuner to find the second signal and use the stereo decoder in the tuner to obtain the two channels of sound. To obtain compatibility between monophonic and stereophonic FM broadcast, the sum signal (L + R) is broadcast in the usual frequency range and can therefore be picked up monaurally. The difference signal (L − R) is transmitted on the subcarrier. The sum and difference of these two signals are used to obtain the left and right speaker signals, respectively. The entire signal, consisting of (L + R), the 19-kHz pilot signal, and the 38-kHz AM subcarrier containing (L − R) is used to frequency modulate the main FM carrier around the frequency of the station.

Quadraphonic sound transmission is possible in the SQ matrix format. The two signals from the magnetic pickup of the disc player in the radio station are used as the two channels of transmitted information. An SQ matrix decoder in the FM tuner then separates these two signals into the four required by a quadraphonic system, just as in the case of SQ matrix quad disc records. Such a system retains the limitation of the SQ matrix format in that it does not provide good separation of the four channels. Proposals have been made for transmission of discrete quad over FM radio, but they are still in the experimental stage.

One important noise problem in FM transmission is high-frequency hiss. This can be reduced significantly by preemphasis of the high-frequency range and equalization by the tuner circuitry, just as is done for disc recordings.

A control labeled FM Muting is often found on good tuners; this attenuates the loud static obtained when tuning between FM stations, but allows the station signal to come through unhindered.

Some specifications on which the choice of a tuner are based are sensitivity to low-level stations (which may be crucial for those living far from the radio stations), signal-to-noise ratio, distortion, frequency linearity of audio output from the tuner, and stereo separation. Existence of an easily readable indicator, to tell that a station is properly tuned, is important. In metropolitan areas, where stations may be assigned to adjacent frequency bands, selectivity, or the ability of the tuner to tune in one station while suppressing signals from adjacent stations, becomes important.

7.9   *Tape Recorders*

A great variety of tape recorders is now available. We shall briefly discuss some of the more popular types, summarize some of the basic features in the operation of tape recorders, and discuss some of their more important properties.

Figure 7–22 shows basic tape configurations in monaural and stereophonic recording. The simplest tape configuration is the two-track (or half-track) monaural, shown in part (a). One-half of the tape is used to record in the forward direction, while the remaining half is used to record in the reverse direction, that is, after the tape has wound onto the take-up reel and the two reels have been reversed. Two-track stereo can be obtained by recording the two channels simultaneously on the two halves of the tape. In this case, as shown in Fig. 7–22 (b), the tape is all used up after going in the forward direction and cannot be recorded in the reverse direction. In the four-track (or quarter-track) configuration, four channels are available and can be used either to record four separate monaural tracks (two in each direction) or two stereo recordings (one forward and one backward), as shown in Fig. 7–22(c) and (d). Notice that this format allows playback of half-track stereo tapes by quarter-track machines. Playing a quarter-track tape on a half-track machine, however, will produce two of the channels properly, but will simultaneously play back the other two channels backward. Eight-track tapes can play eight separate channels of monaural information, four lengths of the tape in stereo, or once in each direction in the discrete quadraphonic mode.

Two types of tape mounting format are in common use. Quarter-inch-wide reel-to-reel tape comes in lengths from about 1200 to 4800 feet; the usual tape speed for this type of tape is $7\frac{1}{2}$ or $3\frac{3}{4}$ inches per second, with $1\frac{7}{8}$ and 15 inch per second speeds available in some tape machines. A typical 7-inch reel of $\frac{1}{4}$-inch magnetic tape holds 1800 feet of tape and runs about 48 minutes in each direction at $7\frac{1}{2}$ inches per second. Reel-to-reel tapes can be edited and spliced if broken and are

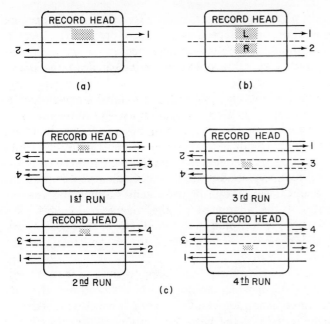

Figure 7-22 Channel configuration on a magnetic tape, showing how recording and playback heads are positioned on tape and location of channels on tape, for the case of (a) half track monaural, (b) half track stereophonic, (c) quarter track, used in monaural mode, and (d) quarter track stereophonic recording.

therefore most often used in situations requiring access to the tape, such as tape editing. Audiocassettes are prepackaged reels of tape about $\frac{1}{8}$ inch wide; the normal cassette lasts from about 30 to 45 minutes per side at a speed of $1\frac{7}{8}$ inches per second, for a total playing time of between 1 and $1\frac{1}{2}$ hours per cassette. Most cassette machines can use either monaural or quarter-track stereo format. Recent engineering improvements have raised the quality of cassette fidelity and reliability to the point where they can be considered for use in high-quality audio systems. An audio-cartridge format, developed primarily for use in automobile sound systems, is also available.

Magnetic tape is constructed from a thin (typically 0.0005 or 0.001 inch thick) plastic base coated with a thin layer of magnetic material. This coating, usually iron oxide, chromium dioxide, or ferrichrome, is in the form of small particles; each of these, in turn, consists of even tinier volumes of the magnetic substance, called *magnetic domains*. For the sake of clarity, in the following discussion we shall assume that the material is iron oxide; the principles are the same for the other coatings. Each of the tiny domains acts like a magnet; that is, each has a north and south pole that respond to the presence of other magnets. These domains are in random orientations on a blank, unrecorded tape. There is, therefore, no net magnetic field on the blank tape, due to the lack of domain alignment.

We can idealize the record head of a tape recorder as a C-shaped piece of iron with a coil of wire around one section, as shown in Fig. 7–23. The electronic signal to be recorded passes through the coil, and by the process called *magnetic induction*, creates a magnetic field in the gap of the magnet. The strength of the magnetic field is proportional to the current in the coil. There are only two possible directions for the resulting magnetic field; either side of the gap can be a north pole or a south pole with the other side of the opposite polarity, depending on the direction of the current in the coil. As the current reverses direction, so does the orientation of the magnetic domains on the tape. Consequently, the periodicity of the audio signal is reflected by the periodicity of the alignment of the magnetic domains. The amplitude of the field in the gap, proportional to the amplitude of the audio signal, determines the degree of alignment of the magnetic domains, and therefore the magnetic field they produce.

Most of the better tape decks have three heads, one to erase, one to record, and one to play back the tape, all of which are located in the head unit through which the tape passes. During recording, the tape is drawn at constant speed past the record head. The magnetic field in the head gap, modulated by the audio-frequency signal in the coil, forces the tiny magnetic domains to reorient. The number of domains that are lined up, and the extent to which they are lined up, that is, the angle that they make with respect to the direction of motion of the tape, is determined by the strength of the magnetic field applied. If the signal current varies sinusoidally with time, the magnetic strength, determined by the number of crystals lined up and the extent to which they are lined up, will also vary sinusoidally along the tape.

To play back the tape, the modulated tape is drawn past the playback head, which is similar to the record head. As the tape passes by the gap in the playback head, the head senses the magnetic field due to the sum of the tiny magnetic domains on the tape. By the reverse of the previous process, a current is generated in the coil around the head. The strength and periodicity of the audio signal are determined by the strength and periodicity of the magnetic field on the tape.

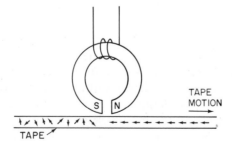

Figure 7-23   Schematic view of the record head of a tape recorder.

In a magnetic system like the tape recorder, the magnetism of the tape is proportional to the electrical current in the coils in the tape head. The current is proportional to the amplitude of the sound being recorded. However, on playback the output of a tape playback head is proportional to the *change* of magnetic field. Since all the Fourier components in a complex wave are changing differently, the playback head will introduce a phase change that is different for each frequency. This leads to a change in the shape of the wave, known as *phase distortion*. Fortunately, according to Ohm's law of hearing, the quality of the sound is not dependent on the relative phase of the various harmonics, so to a large extent phase distortion leaves the sound quality unchanged. Nevertheless, on some of the better and more expensive tape decks, additional electronic circuits are used to eliminate phase distortion.

The magnetic domains in an oxide particle have an inherent resistance to change of their orientations, a sort of inertia, called *hysteresis*. Because of this, very low intensity signals in the record head coil do not produce a corresponding modulation in the tape. An extra signal, called the *bias signal*, or simply bias, is superimposed on the audio signal to overcome this magnetic inertia. The bias signal, a constant-amplitude sine wave with frequency of about 100 kHz, well above the audible range, supplies an extra "push" or "shove" to the magnetic domains to overcome their magnetic inertia, which allows them to then respond to the lower amplitude audio signal modulation. The bias signal is not heard on playback because of its extremely high frequency. Different types of tape respond differently to external magnetic fields and therefore require different bias-signal amplitudes. Many tape decks have switches that allow the user to choose the bias that is most suited to a particular type of tape.

Tapes are erased by an erase head, which is similar to the record and playback heads. In the erase head, however, an extremely high frequency, high amplitude sinusoidal signal is used. The magnetic crystals first feel a strong force tending to orient them in one direction, and then, soon thereafter, a strong force tending to orient them in the opposite direction. The net effect is that the magnetic domain orientations are random or scrambled, leaving the tape with no net magnetization.

As in the case of most phonograph cartridges, which use magnetic phenomena, there are nonlinearities and noise problems in tape recording and playback that require the use of preemphasis and equalization to obtain the highest fidelity and the least noise. Most tape recorders use the National Association of Broadcasters (NAB) standard for equalization, shown in Fig. 7–16. Other recently developed techniques can be employed to remove most of the tape hiss, which is the most serious type of noise generally occurring in good tape systems.

Several systems have been devised to eliminate tape hiss from reproduced sound. One popular system at present is the Dolby noise-reduction system. Inherent tape noise is most prevalent and audible at frequencies above 5 kHz. Most normal music passages are much louder than the tape noise and upon playback effectively mask the noise. When a soft musical passage is reproduced, however, the noise can be heard. A Dolby noise-reduction system takes the input signal components above 5 kHz and boosts them as much as 10 dB, if the musical passage is soft. The boosted signal is then recorded on the tape. During playback, the Dolby device automatically compensates and reduces the intensity of all high-frequency components. This reduces the audio signal to its original level and reduces the tape noise to 10 dB below its original level. A signal-to-noise ratio improvement of 10 dB can thus be achieved. When a non-Dolby tape is played, the Dolby decoder should be switched off; otherwise, the high frequencies will be attenuated.

Another type of noise reduction system that is rapidly growing in popularity is the compander system, for example, the dbx 128, whose operation is shown in Fig. 7–24. Normal live music may vary as much as 100 dB (the *dynamic range*) between the softest and loudest passages. Because of the inherent limitations of the magnetization of tape coatings (saturation) and tape noise, only about a 50-dB dynamic range of signals can be recorded on most tapes. The original concert signal must be *compressed*, that is, the loud signals made softer and the soft signals made louder, for all the signals to be recorded with minimum distortion. Upon playback, the intensity range will be 50 dB, yielding the somewhat unnatural sound to which the modern ear has grown accustomed. If upon playback an electronic *expander*

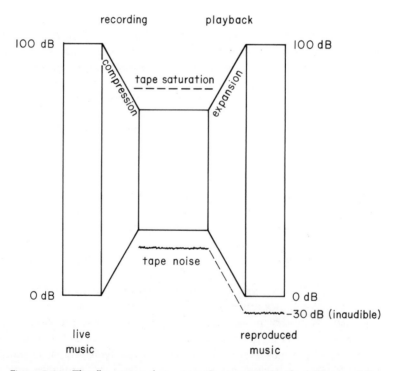

Figure 7-24 The dbx compander system. Compression fits the full range of live sound between the tape noise level and the tape saturation level. Upon playback, expansion restores the full dynamic range and reduces hiss due to tape noise. (Used with permission)

device is introduced, however, the total 100-dB range can be obtained. The expander uses a voltage-controlled amplifier controlled by the overall intensity of the signal to make the loud signals louder and the soft signals softer, returning them to their original live concert levels. Because of limitations in disc recordings similar to those of tapes, expanders can be used to expand a disc player's output with similar beneficial results.

If a perfectly matched compressor-expander system, or *compander*, is used to compress the original signals before tape recording and to expand the signal during playback, not only is the original dynamic range of the music reconstructed, but the noise is reduced as well. This is because the inherent tape noise is below the softest recorded signal on the tape. During playback, the noise is made even softer than the softest signal, typically by 30 dB, as seen in Fig. 7-24.

To achieve good reliability and minimize problems of wow and flutter, most of the better tape units will have three motor drives, one for each type of drive required: one on each reel and one on the tape feed capstan. This has proved significantly more reliable than the use of complicated systems of gears and pulleys.

A typical quarter-track stereo tape deck contains two microphone inputs, two high-level inputs, and mixers in which the relative amplitudes of the signals can be adjusted before recording. Separately adjustable outputs from each channel are normally available. Sound level monitors, called *volume unit* (VU) *meters*, are normally included and can be switched to monitor either the signal being recorded or that being played back. On most of the better machines, there is provision for choice of bias for normal and low-noise tapes, while many units, particularly cassette players, are provided in addition with the Dolby noise-reduction circuit option. Less often, the option is available of output from the tape head or tape amplifier. Tape head output requires equalization when input into the preamplifier, whereas tape amplifier output has been equalized by the preamplifier in the tape deck and can be fed directly into a linear amplifier or a preamp tape amplifier input. Some tape units come with a power amplifier and speaker, but this is intended primarily as a convenience in monitoring the sound when portability is required.

7.10 *Some Recent Innovations*

Several technical innovations have been made possible by recent developments in digital and integrated circuits.

Digital recordings now available use a sequence of digital or exact numerical values to specify the wave form of the recorded music at a sequence of equal time intervals. A small time interval is chosen to obtain a very high frequency response. The digital stage in the recording process occurs between the performer's sound and the master disc used for pressing the plastic discs. As the digital values recorded are essentially impervious to noise, this link can be made virtually noise free. This technique appears to hold considerable promise.

True "digital" record playback has recently been introduced in which a laser beam is used to track digital code numbers etched as tiny pits on the surface of the disc. The numbers represent the wave form of the music, sampled at small time intervals. This technique has dramatically increased the dynamic range, virtually eliminated wow and flutter, and reduced distortion to a small percentage of that in conventional recordings. Because there is no direct contact between the player and the surface being played, the lifetime of a disc is virtually unlimited.

Recent advances in integrated circuits are being used in the impulse noise-reduction system and the subharmonic generator. The impulse noise-reduction system, or click-pop filter, uses integrated circuits to identify clicking and popping sounds due to scratches on disc recordings. The click-pop filter identifies such clicks by responding to their transients, filters out the high-frequency components, and subtracts them (adds them after a phase inversion) from the original signal. The click-pop filter is inserted between the preamplifier and power amplifier.

Live music often has frequencies as low as 27 Hz, frequencies too low to be adequately reproduced by some audio systems. The subharmonic generator is used to increase the amplitude of frequency components in the range below 55 Hz. After identifying components in the 55- to 110-Hz range, their frequencies are divided by 2 and boosted in amplitude. Ideally, this reintroduces or increases the amplitudes of very low frequency components that were lost or attenuated earlier in the recording process.

The passive room equalizer consists of a set of filters inserted between the preamp and amplifier that can be used to equalize the response of the audio system, including the speaker system, in combination with the room in which it operates. A system of about five filters covering most of the audible frequency range is used. By independently adjusting the frequency and bandwidth of each filter, along with its attenuation or emphasis, an audio system can be made to reproduce pink noise to within about 1 dB, the intensity JND, over the entire audible range. We shall return to equalizers in Chap. 8 on room acoustics.

Many new products, like these outlined here, will certainly become popular as increased performance demands are made on audio systems and prices drop owing to technical innovations.

EXERCISES

1. What is Ohm's law of electrical circuits? Explain each symbol and give its unit. If a 100-ohm resistor is connected across the terminals of a 1.5-volt battery, what will be the electrical current? What electrical power is consumed by the resistor? What form does the energy take, and where does it go? If a 100-ohm light bulb is used instead of the resistor, what form does the energy take?

2. Discuss the principal functions of the following components of an audio

system: (a) disc record player, (b) AM-FM tuner, (c) tape recorder, (d) microphone, (e) preamplifier, (f) power amplifier, and (g) loudspeakers. How are they, in general, connected together?

3. What is Faraday's law of magnetic induction? Relate this law to dynamic microphones and magnetic cartridges for disc playback.

4. How does the voltage between two parallel, oppositely charged, electrically conducting plates vary as the distance between them changes? Relate this to the operation of an electrostatic microphone and an electrostatic speaker.

5. List all the forces acting on the tone arm and stylus of a disc record player. Which force opposes gravity? What is the skating force? Which forces are present in a straightarm disc record player?

6. What is bias and why is it needed? How does it relate to the tape recording process?

7. What are the three primary types of microphone with respect to directionality? Which is best suited for the following uses, and why? (a) Recording an orchestra concert with a live audience; (b) recording a large crowd scene for a film; (c) recording a bird call among rustling trees.

8. What is a two-way speaker system? A three-way speaker system? Name the individual speakers that make up such systems and their frequency ranges. What is a crossover network and why is it needed? Sit close and listen to a three-way speaker system when music with low and then high frequency is playing. Try to hear as the different individual speakers present the different frequency sounds. What is a tuned-port speaker system and how does it differ from an acoustic suspension speaker system? Use graphs of the loudspeaker response versus frequency to clarify any points you make.

9. What are preemphasis and equalization, and why are they needed in disc recording?

10. (a) Calculate the power transferred from a microphone to a preamplifier if the voltage is 100 mV and the impedance is 600 Ω.
    (b) Calculate the power transferred from a preamplifier to a power amplifier if the voltage is 1.5 V and the impedance is 50,000 Ω.
    (c) Calculate the power transferred from a power amplifier to a loudspeaker if the voltage is 10 V and the speaker impedance is 8 Ω.

REFERENCES

Backus, John. *The Acoustical Foundations of Music.* 2nd ed. New York: W. W. Norton & Company, Inc., 1977.

Contains a good reference list.

Davis, Don, and Carolyn Davis. *Sound System Engineering.* Indianapolis, Ind.: Howard W. Sams & Co., Inc., 1975.

Good but relatively straightforward book dealing with technical and practical aspects of audio systems.

dbx, Incorporated model 224 pamphlet. Newton, Mass.: dbx, Incorporated, 1980.

Describes the basic principles of the dbx compander system.

JOHNSON, KENNETH W., AND WILLARD C. WALKER. *The Science of High Fidelity.* Dubuque, Iowa: Kendall/Hunt Publishing Company, 1977.

About the same level of this chapter but more complete.

Understanding High Fidelity. Tokyo, Japan: Pioneer Electronic Corporation, 1972.

This booklet contains a collection of articles on various components of the typical high-fidelity sound system that discusses their basic features and operation.

Good sources of technical literature are pamphlets and specification sheets available from audio equipment stores and manufacturers.

8

Room
and Auditorium
Acoustics

In this chapter we shall survey the criteria used to describe the acoustical properties of rooms and auditoriums and relate these properties to physical properties of the room or auditorium. Wave phenomena, such as the inverse square law, Huygens's principle, reflection, refraction, absorption, and diffraction, as well as our knowledge of logarithms, the decibel intensity scale, and noise and spectral characteristics of musical sounds, will be important in our discussion. After a brief look at open-air theaters, we shall see how certain improvements in the acoustical characteristics of this simple structure can be made, resulting in the basic structure of indoor auditoriums. Auditoriums, concert halls, opera houses, and theaters differ from this basic structure in ways that will be discussed.

Attaining proper acoustical characteristics in an auditorium is no accident; good acoustics must be designed into the building, just as are good visibility and adequate ingress and egress. The size, shape, wall and ceiling material, structure, and type of seating must be consistent with the purpose for which the room is designed. Different characteristics are required for optimum listening for speech, chamber music, or large performing groups; furthermore, somewhat different characteristics are required for music from different periods in music history. Home listening rooms differ considerably from performance rooms and will be discussed separately. Suggestions will be made concerning speaker and furniture placement, construction material, and so forth, to aid in obtaining the greatest listening enjoyment from your home audio system.

We shall now survey some of the criteria used in the description of the acoustical characteristics of rooms or auditoriums in which either music or speech is to be presented. In most cases, the descriptive term applied to the room by a performer or critic is related to some physical or measurable property of the sound waves in the room, but in a few cases the characteristic is of a more subjective nature.

Several of the more important acoustical criteria arise from the reverberation time of a room and the characteristics of the reverberant sound. Consider the room shown in Fig. 8–1; the direct sound from the source arrives first at the observer, followed shortly thereafter by the reflected sound. The reflected sound is composed of the original wave reflected from the four walls, the ceiling, and floor, plus the successive reflections of these waves. Because of the decrease in intensity of the wave as it propagates (the inverse square law) and the absorption of some of its energy each time the wave reflects off a surface, the reflected waves a listener receives decrease in intensity as time passes. If we create a sharp sound, such as a snap or a handclap at time $t = 0$, the sound heard by the listener might typically appear as shown in Fig. 8–2. Figure 8–2 is an idealization; the successive pulses would be wider and the entire intensity curve smeared into a more continuous curve because of the continuous range of distances over which the sound wave can pass from the source to the observer.

When a musical tone is attacked and held, the observer will hear something else. If the intensity of the initial sound is I_0 at the observer, the observer will hear the direct sound at this intensity, shortly thereafter reinforced by the reflected waves, with the sum reaching a peak intensity greater than I_0. When the steady tone is released, its intensity will decrease exponentially toward zero, as shown in Fig. 8–3. The time required for the sound to decrease from its maximum intensity to $1/1,000,000$ of its intensity is defined as the *reverberation time*, T_R. Other ways of stating this are that the reverberation time is the time required for the sound intensity to decrease to 10^{-6} of its original value, or for the SIL to decrease by 60 dB. The

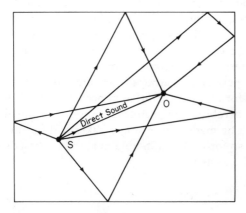

Figure 8–1   Some paths of direct and reflected sound from source to listener in a room. (After Backus, 1977)

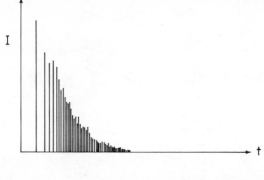

Figure 8-2 Intensity versus time for a sharp sound pulse created at point *S* and observed at point *O* of Fig. 8-1. The direct sound is the first pulse; reflected pulses follow. (After Backus, 1977)

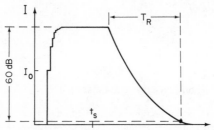

Figure 8-3 Intensity versus time for a steady-state tone in the room shown in Fig. 8-1; t_s denotes the time when the source stops, and T_R is the reverberation time.

reverberation time of a room is determined by its volume (in cubic feet), the total area of the surfaces both within and enclosing the room (including floor and ceiling), and the absorptive characteristics of the surfaces. The reverberation time may be different for different frequencies.

The reverberation time is the most important characteristic for determining the suitability of a room for various applications. The reverberation time that works best for a particular application (speech, chamber music, and so on) in a specific room will depend on the nature of the application. Shown in Fig. 8-4 are lines indicating the approximate ranges of reverberation times appropriate for the various types of rooms and applications. There is no completely accepted "ideal"; this figure is a general guide rather than a rigid rule.

Some of the more important acoustical characteristics, along with the physical properties of sound from which they arise, are discussed next.

Liveness   Liveness is simply a term that is equivalent to the physically measurable reverberation time. The longer the reverberation time, the more "live" the room is. The most acceptable reverberation time depends on the nature of the sound or music and on the size of the room. Furthermore, even subtle differences in the mood or period of the music when played by the same ensemble or orchestra can ideally be suited to rooms with different reverberant qualities, as we shall see. The suitability of a particular hall to certain periods or moods of music can be a significant factor in the type of music that will sound best in that hall, and therefore

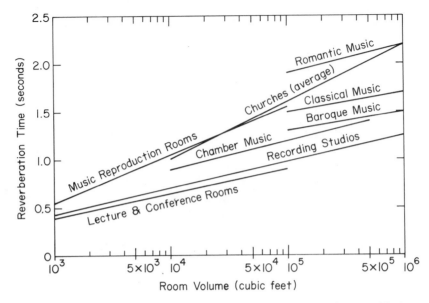

Figure 8-4   Ideal average reverberation time versus room volume for several basic types of rooms. Also shown are the optimum average reverberation times for several important types of music.

in the type of music for which the group performing in that hall becomes re-knowned.

Intimacy   Intimacy refers to how close the performing group sounds to the listener. Intimacy is achieved whenever the first reflected sound reaches the listener less than about 20 ms after the direct sound, a condition easily met in small halls. For large halls this can be achieved by placing a reflecting canopy above the performers, tilted so as to direct the reflected sound toward the audience. Such a canopy is sometimes used in large cathedrals above the pulpit to achieve intimacy of the speaker's voice to the congregation. A canopy also forms the basic part of the structure of a bandshell, helping to increase the sound level by reflecting the upward sound out toward the audience.

Fullness   Fullness refers to the amount of reflected sound intensity relative to the intensity of the direct sound; the more reflected sound, the more fullness the hall will have. For slow, romantic music performed by very large groups, fullness is required, whereas chamber music or music from the classical or baroque periods does not require great fullness. In general, greater fullness implies a longer reverberation time.

Clarity   Clarity, the acoustical opposite of fullness, is obtained when the intensity of the reflected sound is low relative to the intensity of the direct sound.

Great clarity is required for optimum listening to speech and is particularly important when performing early orchestral music, such as the fast movements of the Mozart symphonies. In designing a symphony orchestra hall, it is necessary to strike a compromise between fullness and clarity to allow adequate performance of both early and later music. In general, greater clarity implies a shorter reverberation time.

Warmth   Warmth is obtained when the reverberation time for low-frequency sounds is somewhat greater than the reverberation time for high frequencies. The "ideal" reverberation time, as a function of frequency, required to obtain the feeling of warmth is shown in Fig. 8-5. Above 500 Hz the reverberation time should be roughly constant, but below that frequency it should increase to about 1.5 times the reverberation time for high frequencies. This is accomplished by the proper choice of material for the walls and ceiling, which will be discussed in a later section. If the reverberation time is too long at low frequencies, the sound can become muddy and will lack clarity.

Brilliance   Brilliance is the opposite of warmth and exists if the reverberation time for high frequencies is larger relative to that of the low frequencies. For an average room, the reverberation time for the low frequencies is slightly greater than that for the high frequencies owing to the nature of the absorption from the walls, floor, and ceiling. In a more brilliant room, the difference is less pronounced; the reverberation time for high frequencies is more nearly equal to that of the low frequencies. If the reverberation time for high frequencies is much too long, a continual high-pitched ringing sound may result.

Texture   Texture refers to the time structure of the pattern in which the reflections reach the listener. The first reflection should quickly follow the direct sound to achieve intimacy, with successive reflections following quickly thereafter. To achieve good texture, it is generally necessary to have at least five reflections within about 60 ms after arrival of the direct sound. An important related requirement is that the overall sum of the reflected waves combined should on the average decrease uniformly in intensity. Shown in Fig. 8-6 are two curves of sound intensity versus time, as in Fig. 8-2. Figure 8-6(a) shows good texture, whereas Fig. 8-6(b)

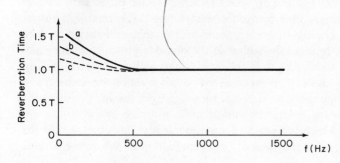

Figure 8-5 Relative reverberation time versus frequency (a) in a "warm" room, (b) in a normal room, and (c) in a "brilliant" room.

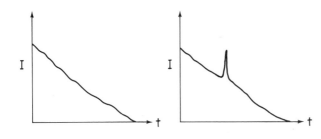

Figure 8-6 Intensity versus time for reverberations in a room with (a) good texture, and (b) poor texture due to a large, late group of reflections.

shows poor texture owing to a late-arriving intense reflection. The structure of the hall should not create any single unusually large echo or have a tendency to focus the sound; otherwise, poor texture may result.

Blend Blend refers to the mixing of the sound from all the instruments of the orchestra or ensemble over all points in the audience. In a concert hall with poor blend, a member of the audience in some specific location may hear one section or player louder than the other players. The simplest technique for achieving proper blend is to mix the sound from the various instruments and voices on the stage before distributing the sound to the audience. This does not imply use of electronic mixing equipment, but rather simply mixing of the sounds of the performers by appropriate reflecting surfaces surrounding the performing area. It is also necessary that there be no surfaces that focus or direct the sound from some part of the performing group to some small section of the audience. All the sound must be diffused over the entire audience by appropriately shaped reflectors.

Ensemble Ensemble refers to the ability of the members of the performing group to hear each other during the performance, allowing an optimum sense of musical ensemble or the ability of the players to play together effectively. To achieve good ensemble there must not be strong reflections that are delayed longer than the duration of the fast notes in the music being performed. This puts a practical limit on the size of the stage for a variety of musical applications. The lateral distance between the nearly parallel sides of the stage should be less than about 65 feet for use with a classical orchestra, one of the larger performing musical ensembles. A more serious problem is the sense of ensemble in the performance of an opera between the members of the orchestra in an orchestra pit and the singers (soloists or choir) on the stage. One method of improving the ensemble is to put reflectors in front of the orchestra, to reflect some of the orchestral sound back to the singers. This, however, reduces the sound of the orchestra to the audience and results in part of the audience hearing only reflected sound with no direct sound, an undesirable situation. Another way to increase the sound level of the orchestra to the singers is to use small loudspeakers aimed at the stage from the wings. Ensemble in an opera remains one of the most problematic areas in opera production.

Another situation in which the sense of ensemble leads to problems is in the performance of a marching band, particularly on a football field, where the band is spread out over a large area. The time delays for the sound of instruments at one end of the band to reach the other end are large in comparison with the rhythmic pattern. This can be compensated only by having a single reference by which every player sets his timing. By watching a director, and playing in time with his or her conducting, and *not* with the sounds of the other instruments, the band can be kept synchronous. The director's hand signals arrive at all the members of the band essentially simultaneously, since the speed of light is extremely great. Someone in the stands may not hear the different instruments as synchronized, however, due to the varying distances.

8.2 *Problems in Acoustical Design*

Several important acoustical problems must be avoided in the design of auditoriums and other rooms. Many of these problems result from improper architectural designs. The following are some of the more serious problems, the architectural features from which they arise, and their basis in the fundamental properties of waves, discussed in Chap. 2.

Focusing of sound Domes, curved walls, or ceilings approximating the shape of spheres, ellipsoids, or paraboloids, or series of small straight panels approximating these shapes should generally be avoided. Such surfaces may produce focal points at which the sound is much louder than that in the surrounding region or can lead to the whispering chamber effect, as shown in Fig. 2–17. In the extreme, if part of the orchestra is at one focus of an ellipse and part of the audience at the other focus, that section of the audience could hear some instruments of an orchestra louder than other instruments. Under these conditions, audience noise could also become inordinately large for the performers sitting near a focal point. Most large, gently curving walls at the rear of an auditorium do not focus the sound unacceptably.

Echoes To obtain good texture it is necessary to avoid any particularly large single echoes; the direct sound followed by a large echo gives the illusion of a performance inside a large can. One way to obtain a very large echo is to make the stage in the shape of a parabola with highly reflective walls, with the performers at the focus. The direct sound will then be followed by a large echo obtained when the sound reflecting off the rear of the stage reaches the audience as a plane wave (see Sec. 2.2).

Another important echo effect is the flutter echo, which may be obtained when the two side walls of the auditorium are parallel to each other and made of highly reflecting materials. In this case the sound reflects back and forth between the two sides, creating a rapid sequence of echoes. These reflections can even lead to destructive interference between the various reflected waves, as in the discussion of standing waves in Chap. 3.

Shadows   Acoustical shadows or quiet regions can be produced when there are long overhanging balconies or other structures jutting into the hall. There should be a large open space extending over a large angle in front of each person in the audience. When any observer's view is blocked or partially blocked by some protruding object, such as a balcony, wall, or column, the sound will undergo diffraction around that structure. As we have seen, the diffraction is different for different frequencies, leading to distortion of the sound passing by or around the obstruction.

Another place where diffraction can become important is in the design of reflecting surfaces, sometimes called clouds, which can be lowered into place, say in a hall with very high ceilings, to create intimate acoustics. Small reflecting surfaces with spaces between them can reflect the higher frequencies adequately, but may allow the lower frequencies to diffract through the spaces and be lost, again resulting in a distorted reflected sound and inadequate warmth. On the other hand, this may be desirable in certain situations to compensate for other defects of the room.

Resonances   Resonances are important primarily in smaller rooms where the dimensions of the room may only be up to a few wavelengths long for the lower audio frequencies. One place where resonances almost always appear is in a bathroom, particularly in a shower or stall. This is because the walls are parallel and of highly reflecting materials such as metal, tile, or glass. A resonance can occur whenever any dimension of the room is any multiple of one half-wavelength long for some particular frequency. There will be a node at each wall, and the wave will resonate as though it were in a tube closed at both ends. There can be a large number of resonances in each of the three directions (up-down, left-right, and backward-forward).

The following equation relates the various resonant frequencies of the room to its physical dimensions:

$$f_{xyz} = \frac{S}{2} \sqrt{ \left(\frac{N_x}{x}\right)^2 + \left(\frac{N_y}{y}\right)^2 + \left(\frac{N_z}{z}\right)^2 }$$

where S is the speed of sound in air, x, y, and z are the three dimensions of the room, which is assumed to be in a regular rectangular shape, and N_x, N_y, and N_z, are any integers. The subscript xyz refers to the values of N_x, N_y, and N_z, respectively. Changing any of these three numbers results in a different resonant frequency. For $N_x = 1$ and $N_y = N_z = 0$, the formula becomes

$$f_{100} = \frac{S}{2x}$$

which is the resonance occurring when the wavelength of the wave of frequency f_{100} is twice the x dimension of the room. The room is one half-wavelength long in this dimension. For a small room, such as a bathroom or a practice room with dimensions of 1.75 by 3.5 m by 2.8 m high, some of the resonant frequencies are shown in Table 8–1.

Not only is there a set of resonant frequencies associated with each of the dimensions of the room (waves reflecting back and forth between two walls to produce standing waves), but there is also a new set of complex standing waves occurring when the wave reflects obliquely between two or more of the pairs of walls. At high frequencies, the resonances get more closely spaced, leading to good reinforcement of higher harmonics, whereas at low frequencies the resonances are more widely spaced, giving a less uniform reinforcement. These high-frequency resonances lead to the nice sounds of singing in the shower or practicing your instrument in the bathroom. Practicing in a bathroom may be all right at times, but it clearly gives one a distorted perspective of tone quality.

You might try to find a few resonances by singing in a shower stall and then measure the dimensions of the stall to verify that the formula works. There should be good agreement between your theoretical and experimental results.

The formula for resonances in rooms applies equally to resonances in smaller boxes like speaker enclosures and must be employed in the design of some types of speaker enclosures. It is necessary in such design to separate the resonances of the speaker and the box enclosure of a tuned-port system to provide the most linear response possible.

One way to avoid some room resonances is to avoid parallel walls. In many cases, small music practice rooms are constructed with oblique opposing walls.

External noise Careful design of the architectural features of an auditorium is of no avail if either external or internal noise is loud compared to the sound of the performers. If there are neighboring airport flight patterns, railroads, or highways,

TABLE 8–1 **Resonances for a room with $x = 3.5$ m, $y = 2.8$ m, and $z = 1.75$ m.**

N_x	N_y	N_z	f_{xyz} (Hz)
1	0	0	50.0
0	1	0	62.5
0	0	1	100.0
1	1	0	80.0
1	0	1	111.8
0	1	1	117.9
1	1	1	128.1
2	0	0	100.0
0	2	0	125.0
0	0	2	200.0

special efforts must be made to isolate the auditorium acoustically from the surrounding area. This can be accomplished by building a box within a box, with acoustical insulation between the boxes, as in the design of the John F. Kennedy Center for the Performing Arts in Washington, D.C. Adjacent rooms must also be insulated from one another, as is the case in the Kennedy Center, which has three major performing halls.

Another important source of noise is air conditioners or blowers, which are often sources of white noise or noise rich in lower frequencies. These low frequencies can, in extreme cases, mask out the higher-frequency sounds of the music.

The maximum tolerable noise level depends on the type of environment and the room's purpose. In general, the noise requirement is most stringent in studios and theaters, where words must be understood, and less critical in homes and offices. A partial list of noise requirements is given in Table 8–2; the sound intensity levels are given in dB as defined in Chap. 6.

One way to control the noise level is to match the size of the room to the size and type of performing group. It is also important to provide for sufficient absorption of sounds created by the audience, which, if not immediately absorbed, may become loud enough to mask out the music.

Shielding outdoor auditoriums from external sound is particularly important. This can be done in part by putting the auditorium in a valley surrounded by high mountains. Such noise shielding is especially important at night owing to the very long distances over which sounds will travel as a result of refraction in the atmosphere during times of temperature inversion. Focusing of sound by air currents must also be considered in outdoor open-air auditoriums.

Double-valued reverberation time The reverbertion time in rooms used for recording playback can be different from the reverberation time of the hall in which the music was recorded. If the reverberation time of the playback room is very long, this can lead to a rather confusing sound, which should be avoided. Decreasing the reverberation time or using headphones in the listening room can combat

TABLE 8–2 **Average normal noise levels for several acoustical environments.**

	dB
Studios (recording, radio, TV)	25
Auditoriums and theaters	30
Classrooms or lecture halls	30
Hospitals	30
Homes	40
Offices	45
Restaurants	50

this problem. In the case of speech reproduction, a double-valued reverberation time can actually render the speech incomprehensible.

Unfortunately, the requirements for use of a room for listening are somewhat different from those when the room is used for performance or even rehearsal, so some compromise must often be accepted.

8.3   *Control of Reverberation Time*

We have seen that the reverberation time is the most important characteristic of an auditorium. We shall now study how to control the reverberation time by use of sound absorptive materials.

Recall that the reverberation time is the time required for the intensity of a sound to drop to 1 part in 1,000,000 (1 in 10^6, or 10^{-6} times) of its original intensity. An approximate formula for the reverberation time in seconds is

$$T_R = 0.050 \, \frac{V}{A}$$

where V is the volume of the room in cubic feet and A is the total absorption of the surface of the room in sabins. The reverberation time is proportional to the volume of the room, since the bigger the room is, the longer the time required for all waves to reach the absorbing surfaces. In this formula we use the volume of the room in cubic feet, as architects and engineers work in these units. The unit of sound absorption, the sabin, was named after Wallace C. Sabine (1868–1919), the physicist generally acknowledged as the founder of the science of architectural acoustics. The sabin is an absorption equivalent to 1 square foot (ft²) of perfectly absorbing surface, say a 1-ft² opening in a wall. A 1-ft² opening, 5 ft² of material with an absorption coefficient of 0.2, or 10 ft² of material with an absorption coefficient of 0.1 will each have a total absorption of 1 sabin. The total absorption of the surfaces of a room is

$$A = a_1 A_1 + a_2 A_2 + a_3 A_3 + a_4 A_4 + \cdots$$

where the A_1, A_2, . . . are the areas of the various types of absorbing surfaces and the a_1, a_2, . . . are the absorption coefficients of the respective surfaces. The formula for reverberation time then becomes

$$T_R = \frac{0.050V}{a_1 A_1 + a_2 A_2 + a_3 A_3 + \cdots}$$

This formula has omitted the absorption of sound by the air. While this effect cannot be neglected in the design of large auditoriums, it is insignificant for small halls, and its effect is only significant at high frequencies. This effect would appear as another term in the series in the denominator of the equation.

As an example of reverberation time calculation, consider a room 13 ft by 20 ft with an 8-ft-high ceiling. The volume $V = 13 \times 20 \times 8 = 2080$ ft^3. If the four walls are of plaster with an absorption coefficient of 0.1, the total absorption of the walls would be $A_w = 0.1 \times 8 \times (13 + 20 + 13 + 20) = 52.8$ sabins. If the floor is covered by carpet with an absorption coefficient of 0.3, it adds an additional absorption of $A_f = 0.3 \times 13 \times 20 = 78$ sabins. An absorptive tile ceiling with absorption coefficient of 0.6 would provide an additional absorption of $A_c = 0.6 \times 13 \times 20 = 156$ sabins. The total absorption of all the surfaces of the room is then the sum

$$A = A_w + A_f + A_c = 286.8 \text{ sabins}.$$

The reverberation time then becomes

$$T_R = \frac{0.050 \times 2080 \text{ ft}^3}{286.8 \text{ sabins}} = 0.36 \text{ sec}$$

A room of the same size with all surfaces, including floor and ceiling, of marble with an absorption coefficient of 0.1, would have a reverberation time of almost 1 sec. If an audience is present, the total absorption of the people plus the seats would have to be added to the denominator.

As we have discussed, it is sometimes desirable to provide warmth by making the reverberation time longer for low frequencies than for high frequencies, and in general the reverberation time should be higher for low frequencies. Fortunately, the reverberation time can be adjusted independently over a broad frequency range by the choice of the proper materials to cover the surfaces of the room. Shown in Table 8–3 are the absorption coefficients for a number of building materials at several frequencies. The absorption coefficients are usually given at frequency intervals of one octave, rather than in equal frequency intervals. Using the formula for reverberation time and Table 8–3, the reverberation time can be calculated at a

TABLE 8–3   **Average absorption coefficients for several types of building materials at octave frequency intervals.**

Material	Frequency (Hz)					
	125	250	500	1000	2000	4000
Concrete, bricks	0.01	0.01	0.02	0.02	0.02	0.03
Glass	0.19	0.08	0.06	0.04	0.03	0.02
Plasterboard	0.20	0.15	0.10	0.08	0.04	0.02
Plywood	0.45	0.25	0.13	0.11	0.10	0.09
Carpet	0.10	0.20	0.30	0.35	0.50	0.60
Curtains	0.05	0.12	0.25	0.35	0.40	0.45
Acoustical board	0.25	0.45	0.80	0.90	0.90	0.90

TABLE 8–4 Average absorption in sabins at octave frequency intervals of two types of auditorium seats, of the average adult person, and of an average adult person sitting in an upholstered seat.

	Frequency (Hz)				
	125	250	500	1000	2000
Unupholstered seat	0.15	0.22	0.25	0.28	0.50
Upholstered seat	3.0	3.1	3.1	3.2	3.4
Adult person	2.5	3.5	4.2	4.6	5.0
Adult in upholstered seat	3.0	3.8	4.5	5.0	5.2

number of frequencies to determine if the required warmth or brilliance will be obtained.

It is also necessary to take into consideration the absorptive effect of the audience. Shown in Table 8–4 is the average total absorption in sabins at a number of frequencies for some types of seats with and without people. This is only an approximation, as can be inferred from the rather large range of variation of absorption coefficients for each case. In "tuning" the acoustics of an auditorium before its general use, it is often possible to use blocks of absorbing material in place of people to obtain their effect, as was done in tuning New York's Avery Fisher Hall with artificial "instant people," who were quiet as well as absorbent.

Ultimately, after the room has been designed and built, the listener will judge the acoustics of the room. A combination of listening in concert situations and actual physical measurement of the reverberation times can be used to investigate any possible inadequacies, which then can be corrected by adding to or changing the surfaces.

Anechoic chambers, often used in acoustics research, are rooms whose walls consist of absorbent wedges that project in toward the center of the room. A sound wave reflects off the side of the wedge and then strikes nearby wedges. The multiple reflections off the extremely absorbent material make the room virtually echo free and the reverberation time extremely short.

8.4 Design of Auditoriums

We shall now "design" an auditorium using the basic acoustical laws and properties discussed previously. In this design, we shall start with an open-air auditorium and end up by covering the sides and top to obtain the design of an indoor hall.

The most elementary type of "auditorium" would be an audience seated in front of the speaker or performer out in the middle of a flat field. This has several problems: no shielding from external noise, the problem of rapid attenuation of the sound as the wave progresses near the ground level, and the total lack of any reflected sound. We correct the first situation by locating our auditorium in a small valley between large hills or by providing walls and a ceiling to keep out outside sounds in the case of an indoor auditorium.

Away from any walls, ceiling, or floor (for instance, up in the sky) a sound signal diminishes 6 dB as it doubles its distance from the source. This is simply the inverse square law; the intensity decreases by a factor of 4 each time the distance from the source doubles. (See the exercises at the end of this chapter.) However, near the surface of the ground, or near any other highly absorbent surface, the calculation is not so simple. In Chap. 2 we discussed how the addition of Huygens's wavelets from all along the surface of a wave front produce the new wave front at a later time. If the surface is absorbent, say a grassy field or an audience, resulting in propagation of the wave over only one hemisphere, one half of the wave that normally acts as a source of wavelets for the succeeding wave will then be missing, having been absorbed. The amplitude of the new wave will then be only about half that of its value with all the point sources included. The intensity will therefore drop off near the ground almost as the inverse *fourth* power of distance, or at a rate of about 12 dB as it doubles its distance from the source. To keep the sound level as high as possible, it is necessary to provide each member of the audience with a large open field of view so that the effect of absorption of the waves by the ground or other members of the audience will be minimized. This can be done by raising the stage, sloping the audience seating area, or a combination of these. In early Greek open-air amphitheaters the slope increased with distance from the stage.

To obtain fullness, with a good strong reflected sound, it is necessary to provide a shell over the performers. The shell must be low enough to give a reasonably rapid reflected sound to assure adequate sense of intimacy. Adding sides to the shell provides additional fullness and texture, acts to diffuse the reflected sound, and assures a better sense of ensemble for the performers. The shell should be made of only a few straight sections, without curved shapes, to avoid any focusing or other undesirable reflections.

Additional large flat reflecting surfaces above, behind, and to the sides of the audience (walls and ceiling) can be added to obtain additional fullness and to preserve the sound level. We now have an indoor auditorium. The two sides of the hall should fan out and be of absorbent material to avoid flutter echo and any standing-wave effects.

If a balcony is desired, it should be strongly sloped and should not extend out far enough to cause any diffraction effect for the area directly underneath the balcony. Pillars or support columns that interfere with the audience either in or underneath the balcony should be avoided.

The rear wall of the auditorium can be lined with an absorbent material to prevent strong reflection of the direct sound, if necessary. Materials for the other walls, the seats, and floor can be chosen to provide the proper reverberation time for the desired application.

An additional possibility is to provide removable or retractable absorbers, which can be used to reduce the reverberation time when the hall is to be used for speech and removed during musical performances. Smaller shells could be provided for small groups. Movable ceiling sections could be lowered and the balcony left unused during chamber music performances.

Shown in Fig. 8-7 are the top and side views of an auditorium that meets these acoustics requirements for general use. The last step in our design process is to listen to musicians on stage from various places in the audience section to test for blend and then measure the reverberation time at several frequencies. Different construction materials could then be substituted or appropriate absorbent panels added, if necessary.

An additional consideration in some large halls is the use of electronic amplification. Two basic loudspeaker configurations are popular. In one, speakers are simply installed in the front of the room; in the other, the speakers are spread out more or less uniformly over the ceiling. Calculations must be made to assure that adequate loudness can be obtained while ringing, feedback squealing, and other distortions are avoided. It is important for some types of musical performance to use the speakers to simply support the direct sound of the music and to avoid having excessive amplification. It is also necessary to delay the sound from the speakers so that it reaches the listener after the direct sound has reached the listener. The audience will then perceive the electronic sound as part of the reverberation. The situation where the amplified sound from above reaches the listener before the direct sound does can create the illusion that the performers are on the ceiling. This is due to the precedence effect, according to which listeners perceive the location of the sound source as the location from which the earliest sound arrives.

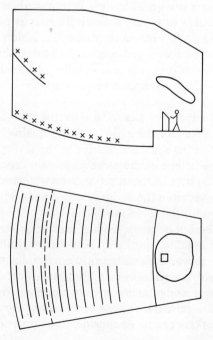

Figure 8-7 Side and top views of an auditorium designed for good acoustics.

As technical improvements are made in the design and installation of electronics and speaker systems, and as more is learned about the use of speakers in large auditoriums, more complex and innovative loudspeaker systems are being used in large concert halls. However, most critics agree that direct unamplified sound in a well-designed auditorium often provides the most enjoyable listening experience.

8.5   *Home Listening Rooms*

While most people will never become involved in the design or construction of auditoriums, many people at some time or another assemble their own audio reproduction systems. Although most of us are not wealthy enough to completely design or rebuild a music room to conform to the optimum specifications for music listening, there are certain things you can do readily and at little cost to ensure maximum enjoyment of your own installation.

Several factors will influence the sound ultimately obtained in your listening room: (1) the type of materials used in the walls, the floor, and the ceiling of the room, (2) the size and shape of the room, (3) types of rugs and furnishings used, (4) loudspeaker placement, and (5) quality of audio components.

In most cases, without undue expense there is not much one can do about the room size and shape or the building materials. If deadening of the room is required, it may be possible to install a suspended ceiling with absorbent tiles or hang rugs from the walls. If it is desirable to retain as much liveness as possible, furnishings can be obtained that offer little absorption of sound. If it is necessary to reduce the reverberation time, overstuffed chairs and couches and other absorbent furnishings can be used. Fortunately, as a usual rule, a normally furnished living room has a reverberation time that is within the range acceptable for use as an audio reproduction and listening room.

The two factors over which one has most effective control, and which probably affect your listening pleasure the most, are the quality of the audio components and the placement of the speakers relative to the sitting areas of the listening room. We have discussed some of the more important criteria in the choice of components for an audio system. One additional consideration should be mentioned here: The output power from the audio system should be matched to the size and quality of reverberation for the room in which it is to be used. Large rooms and rooms with shorter reverberation times require greater power output from the speakers. However, almost any loudspeaker of reasonable quality will produce sufficient power output, when driven by a sufficiently powerful amplifier, to fill a normal home listening room with sound; normally less than 1 watt of acoustic power is required. Use of high-power components is only necessary for unusually large rooms or when it is desired to raise the volume of the music above the normal level.

Placement of loudspeakers can be of considerable importance, particularly in

Figure 8-8 Areas of good listening for stereophonic and quadraphonic music.

listening to stereophonic or quadraphonic programs. Figure 8-8 shows the approximate extent of the listening areas over which the stereo or quadraphonic effects can be reasonably heard with normal positioning of speakers. Quadraphonic music is best heard from the center of a square room, with the speakers positioned at the corners facing inward. For optimum listening to stereo sound, the listening area should be confined to a region between the two speakers, and not much farther from one than from the other, to assure balance of intensity. The listening area should begin back some distance from the speakers; most speakers are somewhat directional, and some space is required for the sound to spread out uniformly owing to diffraction effects.

Figure 8-9 shows possible placement of speakers against one wall of a rectangular room. The speakers should in general be placed symmetrically in the room, and can be placed either against the wall or out from the wall. The closer the speakers are to the wall and corner (positions 1 and 4), the greater will be the support of the bass frequencies by the coherence of reflected waves off the corner or wall. However, if the speakers themselves provide sufficient bass sound, it might be desirable to move them out from the wall (out of the plane of the paper) to position 2 to obtain the greatest uniformity of sound. Raising the speakers to position 3 allows exposure to direct sound from the speakers unhindered by either reflections from the corners or by absorption from the floor. Placing high-quality speakers at position 3 and slightly out from the rear wall is possibly optimal.

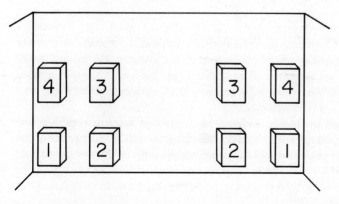

Figure 8-9 Good positions for stereo loudspeakers along a wall of a listening room.

One additional recent innovation should be mentioned at this point: The development of reasonably inexpensive equalizer circuits has provided us with a way of "tuning" our listening rooms, in combination with the audio system, to remove any strong resonances that may tend to cause emphasis of certain frequencies. The average power per unit frequency of orchestral music, and to a good approximation of other types of music, falls off at about 3 dB per octave. In the audio engineering field, noise that falls off in power per unit frequency at the rate of 3 dB per octave is called pink noise, as discussed in Sec. 4.4. In tuning a room, pink noise is played through the audio system and picked up by a high-quality microphone that is connected to a sound-level meter to measure the intensity of the pink noise at the position of the microphone. The microphone is placed at the optimum position for listening to the speakers. If room resonances or resonances in the speakers exist, the sound level will be higher than that of the pink noise at those frequencies.

On the other hand, there may be frequencies at which the room absorbs sound particularly strongly or at which the speaker responds less well. In either case, *equalizers* consisting of special filters can be used to modify the power amplifier signals sent to the speaker to provide pink noise in the listening area, which falls off with frequency at exactly the right rate. In most cases, it is only necessary to use about four to six separate frequency bands spanning the frequency region up to a few thousand hertz to provide a response that is within about 1 dB of the ideal pink noise. This procedure is limited; in suppressing a very stong resonance it may also suppress frequencies on either side of the resonance to below the desired intensity level.

Other techniques can be used to remove or limit the most serious room resonances. Because these resonances are sometimes the result of reflections between parallel surfaces forming standing waves, it might be possible to install a wall or an extension of the existing wall at an angle. Alternatively, it might be possible to cover adjacent walls with a rug or some other absorbent wall hanging to reduce reflections.

EXERCISES

1. Define each of the following acoustical characteristics of rooms and relate them to the features of the reverberation time: (a) liveness, (b) intimacy, (c) fullness, (d) clarity, (e) warmth, (f) brilliance, (g) texture, (h) blend, and (i) ensemble. How might the presence or absence of an audience affect these properties and the musical performance?
2. Describe the following problems arising from improper acoustical design and discuss how these problems could be avoided or corrected: (a) flutter echo, (b) excessive liveness, (c) poor texture, (d) problems relating to the precedence effect, and (e) focusing.

3. A room is 25 ft × 40 ft × 30 ft. The front (25 ft) is covered by glass, the back (25 ft) by curtains, and both sides (40 ft) by plasterboard. The ceiling is of acoustical board, and the floor is made of concrete. What is the reverberation time of this room at 125 Hz? at 1000 Hz? How could these reverberation times be increased or decreased? Discuss the acoustical properties of this room. For which purposes might this room be suited? For which purpose might it not be suited?

4. Design a lecture hall with emphasis on clear visibility and good acoustics. Compute the reverberation time for the room you design for 125 and 1000 Hz, including absorption effects of a full audience. What changes might you make if the room were to be used for a chamber music concert?

5. Design a home listening room suitable for stereo system listening and piano and small ensemble rehearsal. Compute its reverberation time. What might you do to increase intimacy? warmth? liveness?

6. With the appropriate calculation, show that an inverse square decrease in intensity is equivalent to a fall-off of 6 dB for each factor of 2 in distance from the source.

7. Intimacy is achieved when the first reflected sound arrives less than 20 ms after the direct sound. What extra path length must the reflected wave travel to achieve this delay? Could a cathedral have intimate acoustics? A lecture hall? A recording studio?

8. What can be done to improve listening in your dormitory, apartment, or home music listening area? How can transmission of sound to neighboring areas be reduced?

REFERENCES

BACKUS, JOHN. *The Acoustical Foundation of Music.* 2nd ed. New York: W. W. Norton & Company, Inc., 1977.

BENADE, ARTHUR H. *Fundamentals of Musical Acoustics.* New York: Oxford University Press, Inc., 1976.

See their reference lists.

BERANEK, LEO L. *Music, Acoustics, and Architecture.* New York: John Wiley & Sons, Inc., 1962.

Perhaps the most complete modern work on architectural acoustics. Fairly advanced.

GEERDES, HAROLD P. *Planning and Equipping Educational Music Facilities.* Reston, Va: Music Education National Conference, 1975.

A summary booklet describing some of the concerns when designing a practice or rehearsal facility, as well as auditoriums. Includes an excellent recording, *Acoustics for the Music Educator,* illustrating reverberation and other effects.

KNUDSEN, VERN O. "Architectural Acoustics." *Scientific American,* November 1963.

A good survey article for the nontechnical student.

SABINE, WALLACE C. *Collected Papers on Acoustics.* New York: Dover Publications, Inc., 1964.

Contains reprints of the papers on architectural acoustics by the founding father in the field. Some are advanced.

SCHROEDER, MANFRED R. "Toward Better Acoustics for Concert Halls." *Physics Today,* 33, no. 10 (October 1980): 24.

The article discusses recent innovations, with some math, and gives references dealing with improvements in concert hall acoustics.

9

Musical Temperament and Pitch

Unless the reader has listened attentively to musical passages in different temperaments, he or she is unlikely to appreciate the concern regarding these differences. To experience them, one should hear an organ tuned in a temperament other than the usual equal-tempered scale or a recording (see References) that illustrates these differences.

After a brief historical background, we continue with a discussion of the historical change in pitch standard and the effects of this change. Finally, we shall discuss several types of musical temperaments along with their strengths and weaknesses, following roughly historical order. We shall give procedures by which each temperament can be set.

9.1 Background and Historical Perspective

In tuning a keyboard instrument, such as the piano, organ, or harpsichord, two major decisions must be made. First, one must decide the pitch level to be set, say A = 440 Hz (U.S. standard), A = 415 Hz (the approximate value used during the baroque era), C = 256 Hz (the "physics standard," which is now obsolete), and so on. Second, one must decide on a *temperament*, that is, the exact frequencies of all 12 notes in the chromatic scale. After the pitches of the 12 notes within one octave have been set, the remaining notes on the instrument can be tuned by octaves.

Before we discuss temperaments, we shall review the basics of keys, scales, and harmonic relationships. Figure 9–1 is a graphic illustration of the *circle of fifths.* The 12 letters around the perimeter represent the 12 major keys. Notes in adjacent positions are separated by a musical interval of a fifth, hence the name "circle of fifths." The inner wedge may be imagined to be rotated to show the notes that make up the major scale in any key. For instance, to find the seven notes that comprise the C major scale, the inner wedge is rotated such that I is placed in the C position. The note C is called the *tonic* in the key of C major. The Roman numerals on the inner wedge indicate the sequence of notes in the major scale for that tonic. Thus the Roman numerals also indicate the musical interval between the corresponding notes on the perimeter and the tonic. For example, F is the fourth note in the C scale and an interval of a fourth above C, G is a fifth above C, and so on.

You can see that the key of G, adjacent around the circle, shares six of its seven notes with the key of C. The key of G is said to be closely related harmonically to the key of C. Likewise, F is closely related to C. The key of D♭ shares only two notes (F and C) with the key of C and is therefore not closely related to C. Keep in mind the relationships summarized in Fig. 9–1 as the discussion of temperament proceeds.

There is no such thing as an "ideal" or "best" temperament; the particular choice of temperament should reflect the requirements of the music to be performed and the instruments to be used. While the variations between different temperaments may seem minute or insignificant to the untrained listener, to the trained musician they can be very substantial, creating vast differences in the sound of the music. For instance, one can choose a temperament that will have most intervals of

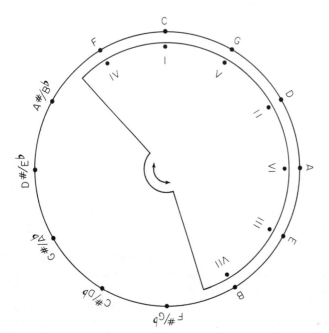

Figure 9-1 The "circle of fifths." The notes along the outer ring are separated by a musical interval of a fifth, and therefore repeat after twelve steps. Numbers on the inner wedge show steps in the major scale. Inner wedge can be rotated to show the notes of a major scale with chosen note as its tonic (I).

a fifth well in tune; that is, notes a fifth apart will have a ratio of frequencies of exactly 3 : 2, but this demands that the intervals of a third will not have the exact frequency ratio of 5 : 4 and thus they will sound somewhat unpleasant. On the other hand, one can choose a temperament based on many exact thirds, but then the fifths will deviate from the frequency ratio of 3 : 2. These two temperaments will sound different. Other temperaments strike a compromise and have some intervals slightly out of tune. In each temperament that we shall study the octave is tuned to the exact frequency ratio of 2 : 1. The single exception to this rule is the tuning of the piano, which will be discussed in Chap. 13.

While the problem of temperament dates back to antiquity, it only emerged as a practical problem with the advent of well-tuned keyboard instruments and string and wind instruments whose pitch could be controlled readily enough that a temperament could be discerned. This occurred in the late medieval period, when music in the church began to use two different parts, rather than the traditional monophonic structure. The Pythagorean temperament, based on exact intervals of fifths, was used throughout antiquity because of its simplicity and basic symmetry, which was considered a philosophical matter of beauty and order. The very earliest two-part music used paralled fifths predominantly and was perfectly in tune in the Pythagorean system.

A more sophisticated temperament was needed in the early Renaissance when music with several parts became commonplace and the interval of a third became more popular. A serious deficiency in the Pythagorean system became obvious: Major thirds were grossly out of tune. (Other deficiencies of that temperament, such as the "wolf fifth," became important several centuries later as musical tastes and styles changed further.) The *just scale*, or temperament, improved the sound of the thirds and had exact tuning for important chords (I-IV-V) in only one key, and acceptable tuning for nearby, harmonically related keys.

The limited number of properly tuned chords in the just scale was not a serious drawback in the Renaissance (about 1430–1650), as the music of that period used primarily these few chords. As the music deviates further harmonically from its good key, the intonation becomes worse, and some keys have abominable intonation in just temperament. For instance, a harpsichord tuned justly in the key of C would sound good in the key of C major but bad in the key of F♯, which is harmonically distant from C.

Just temperament is an example of an *open* temperament. In open temperaments, the intonation problems are worse as the instrument is played in keys harmonically distant from its good key, and there are one or more intervals in which all or most of the tuning error is concentrated. A *closed* temperament, on the other hand, involves tuning compromises so that the instrument is playable in all keys, even though they need not all be equivalent or sound equally good.

Later in the Renaissance it became obvious that the just scale was inadequate to handle the increasingly complex harmonies. Around 1500, there was considerable experimentation with a set of temperaments called *mean-tone* tuning, the most popular of which was the quarter-comma mean-tone temperament. (We shall discuss tuning errors and commas shortly). This scale was based on improved tuning of thirds at the expense of well-tuned fifths, which did not seem as harsh as mistuned thirds. As the quarter-comma mean-tone temperament is an open system, there are certain intervals that remain uncorrected in which a disproportionate amount of the tuning error is concentrated.

Closed temperaments, in which all keys could be played with some success, were first seriously used during the seventeenth century, although they had been discussed for some time. Harmonic developments of the late Renaissance and baroque (about 1600–1750) required keyboard instruments to be played in many keys and to be easily tuned. The term "well-tempered," made famous by J. S. Bach, originated with the music theorist Andreas Werckmeister (1645–1706) in his treatise, *Musicalische Temperatur*, published when Bach was six years old, and was applied to any closed temperament. At that time, keys containing many sharps and flats were not used often, and the tuning of these keys was compromised in favor of the keys more often used. Nevertheless, as Bach demonstrated in his Well-Tempered Clavier, all keys could be used within a single temperament. It appears now that Bach was not necessarily advocating the use of equal temperament, and although the temperament that he preferred is unclear, some historians feel that he favored a closed temperament developed by Werckmeister.

Many closed temperaments were developed during the seventeenth and eighteenth centuries and a few became popular. While the use of the equal-tempered scale was first documented in the early seventeenth century, it was not widely used at first. Most organs were tuned to unequal closed temperaments during the eighteenth century, and regular use of such tuning systems continued into the middle of the nineteenth century.

Since the nineteenth century, virtually all instruments have been tuned in equal temperament. In this closed temperament there is a single frequency ratio between any two notes separated by the same musical interval. In equal temperament, corresponding intervals in all keys have the same frequency ratio and therefore sound the same, except for absolute pitch level. An instrument tuned in this temperament can be played in any key with equal success, but because the tuning error is spread over all keys there are no keys that have the pure sound of the better keys in unequal closed temperaments. This "equality" was considered a disadvantage in the baroque period; it took nearly two centuries for the equal temperament to achieve regular use. Romantic and modern music use a wide variety of keys and modulations, and equal temperament is the only practical temperament for such music. However, the persistent beating that occurs in any equal-tempered organ or piano can be bothersome.

There is a movement now among some organists to return to unequal closed temperaments. This has been done most effectively where eighteenth-century churches have been restored and baroque style organs installed or renovated. The baroque organ literature sounds very good in these temperaments, especially in the acoustical environment designed for such music. Many scholars and historians believe scales like that of Werckmeister are considerably better than the equal-tempered scale in many keys. Although Werckmeister's scale is slightly worse in a few keys, this is more than compensated for by the beauty of the sound in its best keys.

Other closed, unequal temperaments were developed by Kirnberger, Valotti, and Neidhardt. Each temperament has good and bad features and may be best suited to slightly different circumstances. These scales are easier to tune than the equal-tempered scale, which requires some subtle judgments or modern electronic equipment for proper tuning.

Several points in music history and tradition are related to problems associated with temperaments. For instance, some types of musical ornamentation arose in part from the desire to avoid inherent intonation problems by using a trill, passing tone, or some other simple ornament. This permitted the performer or composer to avoid lengthy use of an ill-tuned note. Such practice was recommended by music theorists of the Renaissance. Organist and theorist Arnolt Schlick, in his book *Spiegel der Orgelmacher und Organisten* (1511), as quoted by Franssen and van der Peet (1970), stated, "the descant may be disguised or hidden with a little pause at its start or a sequence of short notes, a grace, run, shake or ornament—name it as thou wilt—so that the harshness of the close shall not be noticed . . . ."

Finally, the idea that each key possessed its own harmonic flavor or emotion arose from the use of the unequal closed temperaments. Because of the nature of the compromises, each key is slightly different from the others in an unequal temperament. Perhaps this variety is one of the subtleties and beauties of earlier musical tradition that has been lost in the contemporary era of "progress."

Table 9–1 shows the ratio of frequencies between notes within an octave tuned

TABLE 9–1　**Frequency ratios for notes of the Pythagorean, just major, quarter comma mean-tone, Werckmeister no. 1, and equal-tempered scales.**

Pythagorean scale		Just major scale		Mean-tone scale		Werckmeister scale		Equal-tempered scale	
C	2.0000	C	2.0000	C	2.0000	C	2.0000	C	2.0000
B	1.8984	B	1.8750	B	1.8692	B	1.8793	B	1.8877
B^b	1.7778	B^b	1.8000	B^b	1.7889	B^b/A$^\sharp$	1.7778	B^b/A$^\sharp$	1.7818
A	1.6875	A	1.6667	A	1.6719	A	1.6705	A	1.6818
A^b	1.5802	A^b	1.6000	A^b	1.6000	A^b/G$^\sharp$	1.5802	A^b/G$^\sharp$	1.5874
G	1.5000	G	1.5000	G	1.4953	G	1.4951	G	1.4983
F$^\sharp$	1.4238	F$^\sharp$	1.4063	F$^\sharp$	1.3975	G^b/F$^\sharp$	1.4047	G^b/F$^\sharp$	1.4142
F	1.3333	F	1.3333	F	1.3375	F	1.3333	F	1.3348
E	1.2656	E	1.2500	E	1.2500	E	1.2528	E	1.2599
E^b	1.1852	E^b	1.2000	E^b	1.1963	E^b/D$^\sharp$	1.1852	E^b/D$^\sharp$	1.1892
D	1.1250	D	1.1250	D	1.1180	D	1.1175	D	1.1225
C$^\sharp$	1.0679	C$^\sharp$	1.0417	C$^\sharp$	1.0449	D^b/C$^\sharp$	1.0535	D^b/C$^\sharp$	1.0595
C	1.0000	C	1.0000	C	1.0000	C	1.0000	C	1.0000

TABLE 9–2   Frequencies of notes between C_4 and C_5 in the Pythagorean, just major, quarter comma mean-tone, Werckmeister no. 1, and equal-tempered scales, based on $C_4 = 261.626$ Hz.

Pythagorean scale	Just major scale	Mean-tone scale	Werckmeister scale	Equal-tempered scale
C 523.25	C 523.25	C 523.25	C 523.25	C 523.25
B 496.67	B 490.55	B 489.03	B 491.67	B 493.88
B$^\flat$ 465.12	B$^\flat$ 470.93	B$^\flat$ 468.02	B$^\flat$/A$^\sharp$ 465.12	B$^\flat$/A$^\sharp$ 466.16
A 441.49	A 436.05	A 437.41	A 437.05	A 440.00
A$^\flat$ 413.42	A$^\flat$ 418.60	A$^\flat$ 418.60	A$^\flat$/G$^\sharp$ 413.42	A$^\flat$/G$^\sharp$ 415.30
G 392.44	G 392.44	G 391.21	G 391.16	G 392.00
F$^\sharp$ 372.50	F$^\sharp$ 367.92	F$^\sharp$ 365.62	G$^\flat$/F$^\sharp$ 367.51	G$^\flat$/F$^\sharp$ 369.99
F 348.83	F 348.83	F 349.92	F 348.83	F 349.23
E 331.11	E 327.03	E 327.03	E 327.76	E 329.63
E$^\flat$ 310.08	E$^\flat$ 313.96	E$^\flat$ 312.98	E$^\flat$/D$^\sharp$ 310.08	E$^\flat$/D$^\sharp$ 311.13
D 294.33	D 294.33	D 292.50	D 292.37	D 293.66
C$^\sharp$ 279.39	C$^\sharp$ 272.54	C$^\sharp$ 273.37	D$^\flat$/C$^\sharp$ 275.62	D$^\flat$/C$^\sharp$ 277.18
C 261.63	C 261.63	C 261.63	C 261.63	C 261.63

in the Pythagorean, just, mean-tone, and Werckmeister temperaments (all based in the key of C), as well as equal temperament. The exact frequencies of these notes based on the frequency standard $C_4 = 261.63$ Hz are shown in Table 9–2. We shall refer to these tables regularly.

9.2   Historical Development of Pitch Level

The first thing that is done when a piano is tuned or an orchestra tunes up is to sound a tuning fork (or perhaps an electronic oscillator) at A = 440 Hz, tune the A to that frequency, and then tune all the other notes relative to that A. This has not always been the standard for pitch level; in fact, a strong case could be made for the statement that the pitch level used over the course of the last several centuries has been in a state of almost total anarchy. In general though, over the last 500 years the pitch level has risen approximately a minor third, but during this time frame there have been many variations. There are important implications of this substantial rise in pitch, both for instruments and voices. We shall survey many of these problems here and discuss some in more detail in later chapters.

While there was no systematic choice of pitch level in the medieval and early Renaissance periods, it was generally much lower than the level today, particularly in the church. During the fifteenth century, the pitch level was about a minor third below the level of today, particularly in church music. The implication of this for modern performance practice is clear: If the editor has not transposed the music down about a minor third from its original notation, the performer must do it or suffer the consequences of voice strain and/or uncontrolled sounds due to the un-naturally high tessitura.

Some editors of medieval music transpose the notes down about a fourth as a standard part of the publication preparation. The music of Palestrina and his contemporaries should probably be performed about a minor third lower than written in the original. This transposition can lead to problems if instruments are used. In some cases the instruments will go out of their ranges, while in many situations transposition will result in a key with many chromatic notes, which will be difficult to play on historical instruments, especially wind instruments. This problem is one of the factors motivating construction of early wind and string instruments at their original pitch level, rather than the A = 440 Hz now used.

Although during the Renaissance and early baroque most churches used a pitch level that was at least one semitone below A = 440 Hz, there were some churches and areas that tuned considerably higher, and the level of pitch common in much secular music was often higher than that common in the church. Isolated areas in northern Europe used a pitch level almost one fourth higher than that used today, and some of the music of early baroque or late Renaissance German composers should therefore be performed about a fourth higher than its original notation. Most editions of this music are transposed by the editors to make the ranges appropriate to the standard voice ranges. Bach composed music for churches and performers where the pitch level varied from about a semitone below today's level to almost a whole step higher, depending on where he was at the time or where he intended the music to be performed. Mersenne, in the early seventeenth century, documented pitch levels regularly used in his region that included a range of almost a musical fifth.

The rise and variation in pitch level has created some problems for musicians performing the music at contemporary pitch level, particularly when the composer used the entire range of instruments in combination with voices. Lowering the pitch to make the music comfortable for the singers makes transposition necessary for the instruments, again putting them in awkward keys or extending the pitch level below the range of the instrument. While the larger corrections of earlier Renaissance music could often be accomplished by changing to another instrument in the family, more subtle changes were involved in the later baroque music. By this time, composers were writing for specific instruments and combinations, and it is necessary to use these instruments at the correct pitch in the intended combinations to achieve the tonal color desired by the composer.

Although there remained considerable fluctuation in pitch, by the late baroque the general level of pitch had generally settled down to about a half-step below that of today. Bach wrote much of his music at a frequency of about 415 Hz for A, and Handel used 422.5 Hz as his standard. The invention of the tuning fork by John Shore (ca. 1662–1752) aided the documentation and standardization of pitch level. The standard baroque low pitch of A = 415 Hz is often used by contemporary instrument makers seeking to return to more authentic baroque performance.

Within a limited range, adjustments in pitch level can be obtained by increasing or decreasing the instrument length using a tuning slide on wind instruments and string tension in string instruments. On modern wind instruments one can simply pull out the slide between the mouthpiece and finger holes or valves to elongate the instrument, thus lowering the pitch level, or, conversely, push in the slide to raise the pitch of the instrument. Baroque flute players often carry several barrel sections of different lengths, and choose the one that produces the desired pitch level. This can even be done to change the pitch between baroque low pitch and contemporary pitch, about a half-step, but the relative intonation between the notes suffers considerably in such a large change.

By the end of the nineteenth century the pitch level commonly used increased to about A = 435 Hz, although there was considerable variation documented between major music centers and instrument builders. The pitch used by such prominent composers as Mozart and Beethoven in the late eighteenth and early nineteenth centuries was still almost one half-step below contemporary pitch. During this time, much of the motivation for the rising pitch level came from the strings, which tended to sound more brilliant when tuned higher. While this created severe problems for wind instruments, generally necessitating major redesign of the finger hole positions on woodwinds and tube lengths on brasses, the strings were also affected by the changes they themselves wrought. The increase in tension of the strings and the concomitant increase in forces on the back and belly of the instruments made additional reinforcement of the wood necessary, and demanded use of a steel string rather than the traditional gut for the highest string of the violin. A violin built in the seventeenth or early eighteenth century, such as those by Stradivarius or Amati, and still in use today, is in many respects not the same instrument as it was when it was built owing to the modifications required by the rise in pitch, as well as other modifications. A Stradivarius violin performed at today's pitch level differs in tone quality from one performed at the pitch level of Stradivarius's day. This is also a factor in the return to baroque pitch and performance practice.

By the beginning of the twentieth century, as the greater exchange of musicians between orchestras all over Europe and the United States became commonplace, a pitch standard became necessary. The response to this pressure was to establish A = 440 Hz as the international standard in about 1920. Most modern instrument makers now use this as their standard.

However, there remains pressure to raise this level even further among some of the major symphony orchestras, with pitch levels at 442 to 444 Hz being used occasionally, and even a level as high as 448 Hz being used by one major American orchestra. Some European orchestras use an even higher standard of pitch level. These changes are usually resisted by the wind players. Increases of 2 to 3 Hz require some modification of blowing technique to obtain proper intonation, while changes of 4 to 5 Hz or more require major modifications of the woodwind finger hole design or tube lengths for brasses to make the instrument comfortable to play in tune. A pitch level of about A = 444 Hz is often used in Europe, and some European instruments, including many recorders, must be pulled out at their barrels in order to play in tune properly at A = 440 Hz.

Since the late baroque and the invention of the tuning fork, the data on pitch level have been documented as an exact science. Previous to this pitch level had to be determined by less direct techniques. Some of the sources are existing historical wind instruments, detailed designs of wind instruments, and notes of scientists and artists containing certain details about the instruments and their design. Bell-tuned organ pipes are also an important source of historical pitch data and will be discussed in Chap. 10. Other details concerning the effects of the changes in pitch level on instruments will be discussed in the sections describing the instruments.

9.3   *Pythagorean Temperament*

In all the temperaments we shall discuss, octaves are required to be tuned exact; that is, they must have a frequency ratio of 2 : 1. The Pythagorean scale is based on the musical interval of a fifth, which is tuned exact for every fifth within an octave except one.

It is impossible to tune *all* octaves and fifths perfectly; there will necessarily be notes separated by the interval of a fifth that will not be true. To see this, consider the note C_4. Starting with that note, we can tune up and down the keyboard by fifths, as indicated in Fig. 9–2. Going up, we can continue to $C_8^\sharp$, and going down we can tune $D_1^\flat$, which is the enharmonic equivalent of $C_1^\sharp$, that is, the same note on the keyboard. We can then tune the other notes on the keyboard by setting octaves properly. Now we can compare the frequency ratio between $C_8^\sharp$ ($D_8^\flat$) and $C_1^\sharp$ ($D_1^\flat$) using two different methods: one based on octaves (ratio of frequencies 2 : 1) and the other based on fifths (ratio of frequencies 3 : 2). Since there are 12 jumps of a fifth between these two notes, their ratio of frequencies is $3/2 \times 3/2 \ldots$ (12 times) $= (3/2)^{12} = 129.75$. On the other hand, there are 7 jumps of an octave between the notes, so their ratio of frequencies should be $2 \times 2 \times \ldots$ (7 times) $= 2^7 = 128$. This discrepancy ($129.75/128 = 1.0136$) corresponds to a significant part of a half-step.

Thus not all the octaves and fifths can be tuned exactly. As the octaves must remain perfect, there muse be some mistuned fifths. In the Pythagorean temperament, all fifths except one are tuned true or beatless. Starting with C_4, successive fifths are tuned true, going up to $C_8^\sharp$ and down to $A_1^\flat$; then the octaves are properly set. Every interval $C^\sharp$ to $A^\flat$ will be less than a true fifth by the amount determined previously, called the *comma of Pythagoras*. This final fifth is very badly out of tune and is called the *wolf fifth* by musicians owing to its growling character. Its use in music is considered unacceptable.

The Pythagorean scale is the most easily tuned scale among those we shall discuss. After choosing the frequency of the note C_4, true fifths are obtained by eliminating beats between the third harmonic of the lower note and the second harmonic of the upper note. This is indicated in Fig. 9–3 for the particular fifth between the notes G_2 and D_3. When the fifth is exactly in tune, a pure or true fifth, no beats will be heard between the third harmonic of the lower note and the second har-

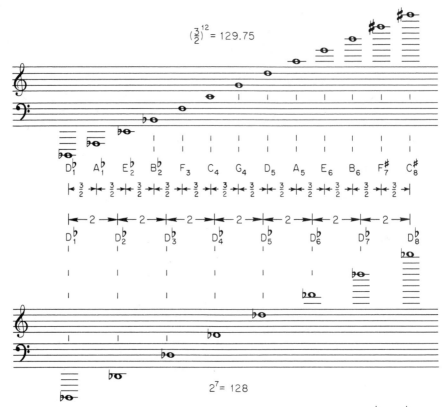

Figure 9-2 Origin of the comma of Pythagoras. Tuning upward from $D_1^\flat$ to $D_8^\flat$ by octaves (factors of two in frequency) produces a note which is different in frequency from its enharmonic equivalent $C_8^\sharp$, obtained by tuning upward from $D_1^\flat$ by fifths (factors of 3/2 in frequency).

monic of the upper note. The frequency ratio between the notes will be exactly 3 : 2.

Fifths are used in succession to set one note of each letter name of the chromatic scale; after that, octaves are tuned beatless to set all the notes on the keyboard to their proper frequencies. Setting the correct value for the first note of each letter name is called *setting the temperament* and is much harder than fixing the frequencies of the other notes by tuning octaves.

One need not go up and down the keyboard to set the temperament, as outlined; one can set the temperament within slightly more than a one-octave range. In practice, the temperament is generally set in the octave around C_4 to C_5, perhaps including a few notes lower. Figure 9-4 outlines a procedure for setting the Pythagorean temperament that involves correct tuning of a sequence of true fifths and octaves, starting from C_4 and proceeding by tuning the intervals shown. The blackened notes indicate the first tuning of the note of that letter name. The remaining notes on the keyboard are tuned in octaves. The wolf fifth between all $C^\sharp$ to $A^\flat$

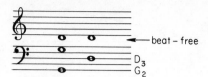

Figure 9-3 G_2 and D_3 are tuned to an interval of a pure fifth by eliminating beats between the third harmonic of G_2 and the second harmonic of D_2 when these two notes are played simultaneously.

Figure 9-4 Sequence for setting the Pythagorean temperament. Blackened notes are the first tuning of a note with that letter name.

intervals results from tuning by this procedure. Frequencies and ratios for a Pythagorean scale are given in Tables 9-1 and 9-2, and Table 9-3 summarizes the derivation of these ratios.

The wolf fifth became problematic when keys using many chromatic notes (sharps and flats) came into common use. A more serious problem historically was

TABLE 9-3 Derivation of frequency ratios for the notes of the Pythagorean scale.

$$C_4 = 1.000 \text{ (start)}$$

$$C_4^\sharp = \left(\tfrac{1}{2}\right)^4 C_8^\sharp = \left(\tfrac{1}{2}\right)^4 \times \left(\tfrac{3}{2}\right)^7 C_4 = \left(\tfrac{1}{2}\right)^4 \times \left(\tfrac{3}{2}\right)^7 = \tfrac{3^7}{2^{11}} = 1.0679$$

$$D_4 = \left(\tfrac{1}{2}\right) D_5 = \tfrac{1}{2} \times \left(\tfrac{3}{2}\right)^2 C_4 = \tfrac{3^2}{2^3} C_4 = 1.1250$$

$$E_4^\flat = 2^2 E_2^\flat = 2^2 \times \left(\tfrac{2}{3}\right)^3 C_4 = \tfrac{2^5}{3^3} C_4 = 1.1852$$

$$E_4 = \left(\tfrac{1}{2}\right)^2 E_6 = \left(\tfrac{1}{2}\right)^2 \times \left(\tfrac{3}{2}\right)^4 C_4 = \tfrac{3^4}{2^6} C_4 = 1.2656$$

$$F_4 = 2F_3 = 2 \times \tfrac{2}{3} C_4 = \tfrac{4}{3} C_4 = 1.3333$$

$$F_4^\sharp = \left(\tfrac{1}{2}\right)^3 F_7^\sharp = \left(\tfrac{1}{2}\right)^3 \times \left(\tfrac{3}{2}\right)^6 C_4 = \tfrac{3^6}{2^9} C_4 = 1.4238$$

$$G_4 = \tfrac{3}{2} C_4 = 1.5000$$

$$A_4^\flat = 2^3 A_1^\flat = 2^3 \times \left(\tfrac{2}{3}\right)^4 C_4 = \tfrac{2^7}{3^4} C_4 = 1.5802$$

$$A_4 = \tfrac{1}{2} A_5 = \tfrac{1}{2} \times \left(\tfrac{3}{2}\right)^3 C_4 = \tfrac{3^3}{2^4} C_4 = 1.6875$$

$$B_4^\flat = 2^2 B_2^\flat = 2^2 \times \left(\tfrac{2}{3}\right)^2 C_4 = \tfrac{2^4}{3^2} C_4 = 1.7778$$

$$B_4 = \left(\tfrac{1}{2}\right)^2 B_6 = \left(\tfrac{1}{2}\right)^2 \times \left(\tfrac{3}{2}\right)^5 C_4 = 1.8984$$

$$C_5 = 2C_4 = 2.0000$$

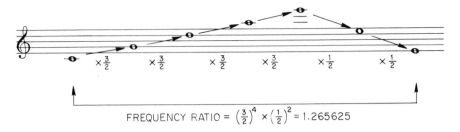

$$\text{FREQUENCY RATIO} = \left(\tfrac{3}{2}\right)^4 \times \left(\tfrac{1}{2}\right)^2 = 1.265625$$

Figure 9–5   The Pythagorean major third.

the use of the interval of the major third. A true beatless major third, formed as the fourth and fifth harmonics in an overtone series, has a ratio of frequencies of 5 : 4, or 1.250. Thirds do not have this ratio in the Pythagorean temperament, as can be seen by considering E_4 and C_4 as diagrammed in Fig. 9–5. Going up four intervals of a true fifth from C_4 to E_6 gives a frequency of $(3/2)^4$ times the frequency of C_4. Going down two octaves from E_6 to E_4 gives a frequency of $(1/2)^2$ times the frequency of E_6. Thus, the frequency of E_4 in the Pythagorean temperament is $(3/2)^4 \times (1/2)^2$ $= 81/64 = 1.265625$ times the frequency of C_4. This ratio differs considerably from the true major third ratio of 1.25 obtained from the notes in an overtone series. The difference between these two thirds, known as the *comma of Didymus*, is a significant fraction of a half-step, causing the Pythagorean thirds to sound out of tune, just as the wolf fifth is out of tune.

9.4   *Just Temperament*

If some of the thirds are to be tuned true, that is without beats, then all the fifths cannot be perfectly in tune. With the historical development of more emphasis on thirds, the *just* scale was devised in which the thirds and fifths of three major chords, the tonic (I), the dominant (V), and the subdominant (IV), of some key were tuned true. The tonic chosen was usually C major. As music of that period used these chords primarily, the deficiencies of the other keys were little noticed. In addition to those three chords, two more chords could be tuned beatless before all 12 notes of the temperament were determined, as shown in Fig. 9–6. The twelfth note in the chromatic scale could be chosen as $B^\flat$, in tune with D (as a major third). Note that the interval of a fifth, $F^\sharp$ to $D^\flat$ ($C^\sharp$) is not in tune, as those notes are obtained via two different routes. We saw earlier that following two different routes of successive intervals results in a discrepancy between the frequencies to be assigned to a single note. In the Pythagorean temperament, one route consisted of sequential octaves, the other route consisted of sequential true fifths, and the discrepancy was called the comma of Pythagoras. In other temperaments, the same principle applies, although the routes are more complicated than in setting the Pythagorean temperament. In this case the fifth from $F^\sharp$ to $D^\flat$ is in the ratio of 1.517, considerably off from the ideal ratio of 1.500.

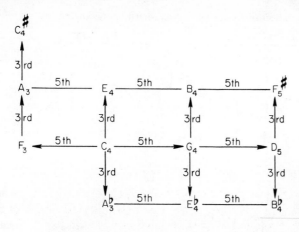

Figure 9-6 Setting the temperament for the just major scale. Arrows indicate the sequence in which the notes can be tuned.

The derivation of the frequencies in the just diatonic major scale in terms of the three major chords is shown in Table 9-4. The ratios of the frequency of any note to the tonic C can be obtained by referring to the ratios in the three chords shown. Notes with ratios of 4 : 5 : 6 form a major chord, which contains a major third (ratio 4 : 5) and a minor third (ratio 5 : 6) in that order.

If the scale is to be tuned for a minor key, a different set of ratios must be used to make the I, IV, and V chords in tune in the minor key. This just diatonic natural minor scale is shown in Table 9-5. The frequency ratios between the notes in a pure minor chord are 10 : 12 : 15, formed from a minor third (5 : 6 or 10 : 12) and a major third (4 : 5 or 12 : 15) in that order. The fifth is perfect, as it is in the major key.

The frequency ratios for the notes in the just major scale can be calculated using the procedure shown in Table 9-6; for notes in a minor scale (such as E♭, A♭,

TABLE 9-4 **Frequency ratios for notes of the just major scale obtained from the three major chords C, F, and G.**

C	D	E	F	G	A	B	C	D
4		5		6				
			4		5		6	
				4		5		6
1.000	1.125	1.250	1.333	1.500	1.667	1.875	2.000	2.250

TABLE 9-5 **Frequency ratios for notes of the just minor scale obtained from the three minor chords C, F, and G.**

C	D	E♭	F	G	A♭	B♭	C	D
10		12		15				
			10		12		15	
				10		12		15
1.000	1.125	1.200	1.333	1.500	1.600	1.800	2.000	2.250

TABLE 9–6   Derivation of frequency ratios for the notes of the just major scale.

$$C_4 \; = \; 1.000 \text{ (start)}$$

$$C_4^\sharp \; = \; \tfrac{5}{4} A_3 = \tfrac{5}{4} \times \left(\tfrac{5}{4} F_3 \right) = \tfrac{5}{4} \times \tfrac{5}{4} \times \left(\tfrac{2}{3} C_4 \right) = \tfrac{25}{24} C_4 = 1.0417$$

$$D_4 \; = \; \tfrac{1}{2} D_5 = \tfrac{1}{2} \times \left(\tfrac{3}{2} G_4 \right) = \tfrac{1}{2} \times \tfrac{3}{2} \times \left(\tfrac{3}{2} C_4 \right) = \tfrac{9}{8} C_4 = 1.1250$$

$$E_4^\flat \; = \; \tfrac{4}{5} G_4 = \tfrac{4}{5} \times \left(\tfrac{3}{2} C_4 \right) = \tfrac{6}{5} C_4 = 1.2000$$

$$E_4 \; = \; \tfrac{5}{4} C_4 = 1.2500$$

$$F_4 \; = \; 2 F_3 = 2 \times \left(\tfrac{2}{3} C_4 \right) = \tfrac{4}{3} C_4 = 1.3333$$

$$F_4^\sharp \; = \; \tfrac{5}{4} D_4 = \tfrac{5}{4} \times \left(\tfrac{9}{8} C_4 \right) = \tfrac{45}{32} C_4 = 1.4063$$

$$G_4 \; = \; \tfrac{3}{2} C_4 = 1.5000$$

$$A_4^\flat \; = \; \tfrac{4}{5} C_5 = \tfrac{4}{5} \times \left(2\, C_4 \right) = \tfrac{8}{5} C_4 = 1.6000$$

$$A_4 \; = \; \tfrac{5}{4} F_4 = \tfrac{5}{4} \times \left(\tfrac{4}{3}\, C_4 \right) = \tfrac{5}{3} C_4 = 1.6667$$

$$B_4^\flat \; = \; \tfrac{4}{5} D_5 = \tfrac{4}{5} \times \left(\tfrac{3}{2} G_4 \right) = \tfrac{4}{5} \times \tfrac{3}{2} \times \left(\tfrac{3}{2} C_4 \right) = \tfrac{9}{5} C_4 = 1.8000$$

$$B_4 \; = \; \tfrac{5}{4} G_4 = \tfrac{5}{4} \times \left(\tfrac{3}{2} C_4 \right) = \tfrac{15}{8} C_4 = 1.8750$$

$$C_5 \; = \; 2 C_4 = 2.0000$$

and B♭), the minor chord ratios can be used. C♯ and F♯ do not appear in any of the three chords used to set the temperament, and these two notes can be set to fit in best with any desired chord. The procedure for setting the temperament is outlined in Fig. 9–6, starting from the note C_4 and proceeding by tuning the intervals indicated by the arrows.

9.5   *Mean-Tone Temperament*

An important observation made during the early Renaissance was that compromising the intonation of the thirds creates more harshness in chords than does compromising the fifths. A number of temperaments were therefore developed that are based on using as many true thirds as possible, at the expense of the fifths. The most widely accepted of these was the *quarter-comma mean-tone temperament*. Referring again to Fig. 9–5, we see that the Pythagorean third, obtained by sequential tuning of fifths, is too large by the amount of the comma of Didymus. The frequency of the note E can be brought down to a true third (ratio of 1.250) by reducing the four intervals of a fifth (C to G, G to D, D to A, and A to E) by one-quarter of that amount, dividing the comma into four parts: hence the name quarter-comma mean-tone temperament. The scheme for setting the quarter-comma mean-tone temperament is shown in Fig. 9–7. Each interval marked 5th* is shortened by a quarter-comma. For the interval of a fifth between B_2^b and F_3, this results in a beat rate of 1.10 Hz. Similarly, the interval of a fifth between F_3 and C_4 has a beat rate of 1.64 Hz, and the fifth between C_4 and G_4 has a beat rate of 2.45 Hz.

All the fifths except one can then be tuned one quarter-comma flat, just as they were tuned true in the Pythagorean scale. Referring to Fig. 9–2, the ratios from D_1^b to F_3, from F_3 to A_5, and from A_5 to $C_8^\sharp$ will each be exactly 5.000, giving a ratio of 125.000 between the frequencies of D^b and $C_8^\sharp$. This factor should be exactly 128 for true octaves, and is 129.75 in the Pythagorean scale, so there will be an even greater wolf tone where the last fifth fails to close. Thus the quarter-comma mean-tone temperament is also an open temperament.

The Pythagorean scale produced chords with thirds that were badly out of tune, and the just scale produced five well-tuned chords with all the others unsatisfactory. The quarter-comma mean-tone scale as presented here yields five chords, which, while they are not true as in the just scale, are nevertheless still quite adequate. The main advantage of this mean-tone scale is that it has an additional chord that is good, and several of the other chords are more usable than with the just scale.

One of the proponents of the mean-tone scale around the beginning of the sixteenth century, Arnolt Schlick, suggested that the frequency of the note $G^\sharp/A^b$ be set halfway between the good $G^\sharp$ and the good A^b. However, if A^b were tuned a true third below C, the intonation defect resulting when $G^\sharp$ was used as the third of the E major triad could be minimized by ornamentation rather than holding the note, while the other notes in the chord could be held since they are in tune.

Despite these advantages, because the mean-tone scale is an open temperament, there are some unsatisfactory keys that simply cannot be used. Therefore, ad-

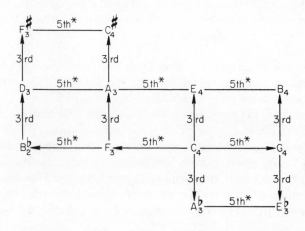

Figure 9–7 Setting the temperament of the quarter comma mean-tone scale. Arrows indicate the sequence in which notes can be tuned. Intervals marked 5th* are narrowed from true fifths by one-quarter of the comma of Pythagoras.

ditional compromises were required in order that more keys could be successfully used, leading eventually to the development of closed temperaments.

The frequency ratios and frequencies for the quarter-comma mean-tone temperament are given in Tables 9–1 and 9–2.

9.6   Closed Unequal Temperaments

The temperaments previously described are all open temperaments, characterized by certain chords and keys that are unplayable, because the tuning error is concentrated in a few chords. The necessity for greater harmonic modulation increased steadily through the seventeenth century, and temperaments that allow unlimited modulation came into regular use by the late seventeenth century. Because the tuning error of these temperaments was distributed over more or all of the triads, all keys were usable, and the temperaments were known as closed temperaments. There are two types of closed temperaments: equal temperament, in which the tuning error is distributed equally over all possible triads, making all triads identical; and unequal temperaments, in which the greater proportion of the tuning error was put into the less frequently used keys, while the more frequently used keys were tuned more nearly to true intervals.

Many closed temperaments were popular at various times and places throughout the eighteenth century. Among the more popular were those suggested by Andreas Werckmeister, Philipp Kirnberger (1721–1783), Johann George Neidhardt (1685–1739), and Francesco Antonio Valotti (1697–1780). The criteria for the "good" temperaments which they developed were that the circle of fifths be "closed," that is, wolf-free, that the thirds and sixths, both major and minor, vary in size systematically so that they could be readily tuned, and that the temperament be playable in any key, favoring, however, the more common keys.

In most closed temperaments the choice of which intervals deviate from the true intervals, and by what proportion, is relatively unrestricted, determined by which errors the artist deems most tolerable. In all cases, tempered fifths are tuned smaller than true fifths, while tempered fourths, major thirds, and minor sixths are larger than the true intervals.

In the Valotti temperament, the thirds in keys farther removed from C major become sharper symmetrically, with the thirds on F, C, and G the best, and those on D♭, G♭, and B tuned farther from true. In addition, the fifths on F, C, G, D, A, and E are slightly tempered, with the remaining fifths being true. This temperament was considered excellent both because of its pleasing variation of tonality between keys and its basic symmetry. Most unequal temperaments are not symmetric and do not vary uniformly in tuning error, as does the Valotti temperament.

Probably the most famous of the closed unequal temperaments was that of Werckmeister, the Werckmeister Correct Temperament No. 1 (sometimes erroneously called Werckmeister III). In the Werckmeister scale, the thirds on F and C are very close to true, those on B♭, G, and D are somewhat farther off, and those on A♭, D♭, and G♭ are off by almost one-quarter of a semitone. The fifths on F, B♭, E♭, A♭, D♭, and G♭ are tuned true, as are the fifths on A and E. The fifths on D, G, B, and C are tempered to differing degrees. In addition to being one of the earliest proposed closed systems, the Werckmeister system is one of the simplest to tune.

The tuning procedure, proceeding in steps from $A_4 = 440$ Hz, is shown in Fig. 9–8. A simplifying feature of this temperament, common to most unequal closed temperaments, is that the notes are largely obtained by either tuning to true intervals or comparing beat rates between two different intervals. Thus it is not necessary to listen to complex beat patterns, as is the case in the equal-tempered scale, as we shall see. Another simplifying feature is that a list of intervals is given that can be compared for beat rate to indicate tuning errors in previous steps, as indicated in Fig. 9–8.

In the Werckmeister temperament, eight of the fifths are true, as in the Pythagorean temperament, and the entire Pythagorean comma is distributed among the four remaining fifths, C to G, G to D, D to A, and B to F♯. The character of the scale therefore changes with key, being close to mean tone tuning in the key of C and more nearly Pythagorean in the key of F♯. Thus, while all keys are usable, the character of the tuning varies from key to key. This type of temperament is one source of the "difference" between the overall character of various keys that still persists to this day in the minds of some musicians.

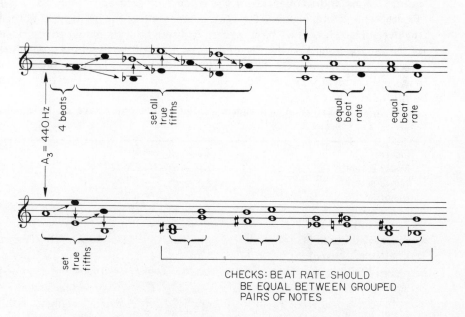

Figure 9–8 Setting the temperament for the Werckmeister correct temperament no. 1. Blackened notes are the first tuning of a note with that letter name.

9.7   Equal Temperament

The natural progression of scale development has taken us through a sequence of open temperaments and unequal closed temperaments. The final step in this progression is an equal closed temperament, the *equal-tempered scale*. Equal temperament is based upon equal intervals between all notes one semitone apart on the keyboard. The only true interval in the equal-tempered scale is the octave, exactly a ratio of 2 : 1 in frequency. Noting that it requires 12 half-steps of equal frequency ratio R to go from one note to the note one octave higher, we can write the frequency relationships shown in Table 9–7. Applying the ratio 12 times to the original frequency raises the frequency to twice its original value:

$$f_{12} = Rf_{11} = R(Rf_{10}) = R(R(Rf_9)) = \cdots = R^{12}f_0 = 2f_0$$

So

$$R^{12} = 2 \quad \text{or} \quad R = \sqrt[12]{2} \doteq 1.0595 \ldots .$$

an irrational number whose first four decimal places are given. Thus the frequency of any note of the equal-tempered chromatic scale is equal to 1.0595 times the frequency of the note a half-tone lower.

The frequencies of the notes of the equal-tempered scale are given in Appendix A, Fig. A–1, for all the notes of the piano keyboard. A comparison of the frequency ratios and frequencies of the equal tempered and other scales is given in Tables 9–1 and 9–2. Figure 9–9 gives a comparison of the frequencies of the notes of the overtone series, as we have written it throughout the book, to the closest notes of the equal-tempered scale. Recall that the notes of the overtone series shown form the true or beatless chords. The fifths (D) are off a slight amount, but not enough to

TABLE 9–7   **Derivation of frequency ratios within one octave of the equal-tempered scale.**

Octave	$f_{12} = Rf_{11} = R^{12} f_0$	2.00000
Major seventh	$f_{11} = Rf_{10} = R^{11} f_0$	1.88775
Minor seventh	$f_{10} = Rf_9 = R^{10} f_0$	1.78180
Major sixth	$f_9 = Rf_8 = R^9 f_0$	1.68179
Minor sixth	$f_8 = Rf_7 = R^8 f_0$	1.58740
Perfect fifth	$f_7 = Rf_6 = R^7 f_0$	1.49831
Augmented fourth ⎫ Diminished fifth ⎬	$f_6 = Rf_5 = R^6 f_0$	1.41421
Perfect fourth	$f_5 = Rf_4 = R^5 f_0$	1.33483
Major third	$f_4 = Rf_3 = R^4 f_0$	1.25992
Minor third	$f_3 = Rf_2 = R^3 f_0$	1.18921
Major second	$f_2 = Rf_1 = R^2 f_0$	1.12246
Minor second	$f_1 = Rf_0$	1.05946
Unison (start)	f_0	1.00000

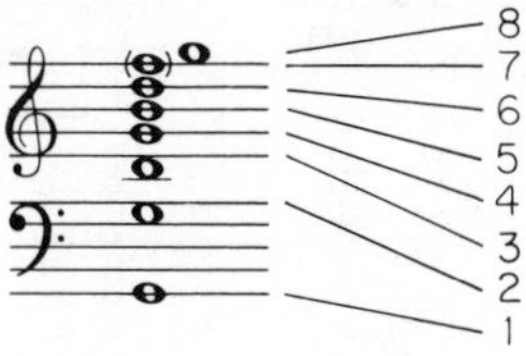

HARMONIC NUMBER	EXACT HARMONIC FREQUENCY	EQUAL-TEMPERED FREQUENCY
8	784.0	784.0
7	686.0	698.5
6	588.0	587.3
5	490.0	493.9
4	392.0	392.0
3	294.0	293.7
2	196.0	196.0
1	98.0	98.0

Figure 9–9   Comparison of frequencies of notes of the overtone series with nearest notes of the equal-tempered scale.

be upsetting to most listeners. However, the major thirds (B) are off a considerable amount from the true major third, and the minor seventh (F) is grossly out of tune, which is why it was put in parentheses in Fig. 3–12. In each case, simultaneously playing the exact harmonic and nearest equal-tempered note will produce beats at a rate equal to the difference between their frequencies.

This system has one important basic feature: All the scales and triads in any key will sound the same, as the frequency ratios from which they are formed are all identical. However, this compromise simultaneously improves the worst keys of earlier temperaments while degrading their better keys. The positive result of this temperament is that all keys can be used, with equal success, being exactly the same in sound except for pitch level.

Some musicians, especially historians whose emphasis is in the baroque period or earlier, feel that the equal-tempered scale is too much of a good thing, and prefer to suffer a little every now and then by listening to a few less true chords to obtain the generally better intonation available in closed unequal temperaments for the often-used keys. Apparently, the musicians of that era also had considerable reservations about the equal-tempered scale, for while it came into universal use in the latter half of the eighteenth century, it had been seriously proposed and thoroughly documented almost 200 years earlier and had been known for several centuries before that. The reason for this lack of acceptance was the considerable beating of all chords in the equal-tempered scale, particularly the thirds. Most people with even a moderate ear for music can hear the difference between the beating of the equal-tempered scale and the purer sound of the Werckmeister scale in its good keys, and many musicians feel that the beating occurring even in a properly tuned piano is very distracting or even annoying.

The prescription for setting the temperament of the equal-tempered scale involves listening to beats between notes a fifth apart, as shown in Fig. 9–3. Applying this prescription to a sequence of fifths, one begins with a beatless interval and narrows the fifth by raising the lower note or by lowering the upper note until the proper beat rate is reached, as shown in Fig. 9–10. Table 9–8 lists the frequencies of both notes of the fifths used in Fig. 9–10, along with the third harmonics of the lower notes and the second harmonics of the upper notes. The beat rate, the frequency difference between these frequencies, is also shown. To set the temperament

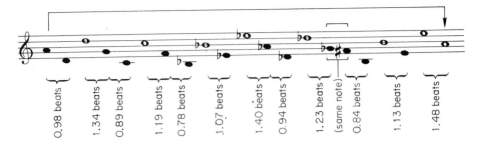

Figure 9-10 Setting the temperament for the equal-tempered scale. Blackened note at right of each bracketed pair is tuned to make the tempered fifth small by the indicated beat rate.

of the equal-tempered scale properly, these beat rates must be set very close to the given values. Notice that the beat rate is proportional to the frequencies of the notes of the particular interval of a fifth involved. There are few exact integer beat rates, and no intervals that can be used as a comparison or check, as in the case of the Werckmeister and other unequal temperaments. This feature makes the equal-tempered scale the most difficult temperament to set properly. However, the systematic nature of the changing beat rate makes it somewhat easier to cope with than implied here. Also, modern electronic equipment can be used to set this temperament, although many piano and organ tuners do not rely on it.

TABLE 9-8 **Calculation of beat rates used in setting the temperament by tuning perfect fifth intervals of the equal-tempered scale.**

Interval	f_{lower}	$3f_{lower}$	f_{upper}	$2f_{upper}$	Beat rate
B_3^b–F_4	233.08	699.24	349.23	698.46	0.78
B_3–$F_4^\#$	246.94	740.82	369.99	739.98	0.84
C_4–G_4	261.63	784.89	392.00	784.00	0.89
D_4^b–A_4^b	277.18	831.54	415.30	830.60	0.94
D_4–A_4	293.66	880.98	440.00	880.00	0.98
E_4^b–B_4^b	311.13	933.39	466.16	932.32	1.07
E_4–B_4	329.63	988.89	493.88	987.76	1.13
F_4–C_4	349.23	1047.69	523.25	1046.50	1.19
G_4^b–D_5^b	369.99	1109.97	554.37	1108.74	1.23
G_4–D_5	392.00	1176.00	587.33	1174.66	1.34
A_4^b–E_5^b	415.30	1245.90	622.25	1244.50	1.40
A_4–E_5	440.00	1320.00	659.26	1318.52	1.48

Despite the problems in tuning and listening to music in the equal-tempered scale, it is destined to remain the most important temperament, as it has for almost two centuries. The demands of unlimited modulation in most romantic and modern music render any other temperament inadequate; equal temperament is probably the only practical choice.

EXERCISES

1. Why are temperaments necessary?
2. List and discuss the important features of each temperament discussed in the text (Pythagorean, just, mean-tone, Werckmeister, and equal temperament). Contrast open and closed temperaments. Which temperaments are open and which are closed?
3. For each of the intervals listed, draw on the staff two overtone series, one for each of the notes that make up the interval. Use the notes you have drawn to explain how the following intervals would be tuned true: octave, fifth, fourth, major third, minor third, major sixth, and minor sixth.
4. Look at the score for a Palestrina mass or other major piece of the same period. Examine the tessitura of each voice and describe the effect of lowering the pitch level by about a minor third.
5. If you have access to such recordings, listen to a recording of the Bach Brandenburg concerti played on modern instruments at contemporary pitch, and then listen to a recording of the concerti played on historically authentic instruments at low pitch. Try to hear differences in the tone, blend, overall sound, and performance styles.
6. Try tuning a harpsichord in the Werckmeister temperament using the procedure given in Sec. 9.6. Listen to the basic chords in various keys to get a feel for the differences from equal temperament. Play, or have someone play, pieces in different keys, such as selections from Bach's Well-Tempered Clavier. Is this temperament usable in all keys? How would you characterize the sound? Play a simple melody transposed to another key and compare the sounds.
7. Listen to an organ (or a recording of one) in an unequal closed temperament. How does it differ from an organ in equal temperament? Can you hear the differences between the keys?

REFERENCES

Barbour, James Murray. *Tuning and Temperament—A Historical Survey.* East Lansing, Mich.: Michigan State College Press, 1951.

The early magnum opus of temperament, covering a large number of temperaments and variants.

BLOOD, WILLIAM. " 'Well-Tempering' the Clavier: Five Methods." *Early Music,* October 1979, pp. 491–495.

A concise article detailing tuning techniques for five closed temperaments popular in the baroque.

FRANSSEN, N. V., AND C. J. VAN DER PEET. "Digital Tone Generation for a Transposing Keyboard Instrument." *Philips Technical Review* 31, no. 11/12 (1970): 354–365.

A summary article dealing with the main concepts in temperament. Included with the article is an excellent record, showing some of the intonation defects occurring in certain temperaments and a comparison of Pythagorean, just, mean-tone, Werckmeister, and equal temperaments.

JORGENSEN, OWEN. *Tuning the Historical Temperaments by Ear.* Marquette, Mich.: Northern Michigan University Press, 1977.

By far the most modern and complete book on temperament; all you ever wanted to know about literally any temperament.

Recording

BACH, JOHANN SEBASTIAN, *Brandenburgische Konzerte 1–6,* Erstaufnahme mit Original Instrumenten in Originalbesetzung, Concentus Musicus Wien. Lechtung: Nikolaus Harnoncourt, Telefunken SAWT 9459/60–A.

This two-record set is perhaps the premiere example of baroque music played at low pitch on replica instruments.

10

Woodwind Instruments

Next to the percussion instruments, the woodwind family, which includes the pipe organ, has the richest variety in both sound and acoustical properties. In this chapter we shall trace a brief history of the development of the contemporary woodwind instruments, summarize the important elements in woodwind tone production, and discuss some of the acoustical details that are relevant to performance technique.

10.1 *History of Woodwind Instruments*

Although the earliest woodwind instruments were primitive bamboo and pipe flutes, the earliest organs can be considered the forerunners of the modern reed woodwind instruments. Wind pressure for these organs was supplied first by water columns and later by hand-driven bellows. The keys were linked directly to valves to allow air flow to the pipes. Each key would control one pipe, which was usually a reed pipe. The *tracker* action organ, common in the baroque era, evolved from these early reed organs.

The original hydraulic organs were bulky, difficult to control, and usually played quite out of tune, problems that were largely responsible for the exclusion of the organ from the church service. Several people were necessary to operate the organ, often including two keyboard players; the keys were large and difficult to depress. In the Middle Ages, bellows and other technical improvements in the organ

245

mechanism helped it to become useful in church settings. Bellows were also used in smaller portable or *portative* organs, popular in the Renaissance. Electrical blowers and switches today replace the bellows and direct mechanical linkage of earlier organs, and progress in materials science and acoustics has enabled modern organ builders to make organs that stay in tune for a longer time and behave more reliably.

A direct precursor to the reed instruments was the bagpipe, which was developed in medieval Spain as a modification of a Middle Eastern ancestor. The later Scottish bagpipe was developed from this original Spanish model. The drones, pipes that produced the same note all the time, were simply reed organ pipes attached to the air sack, a modified bellows. A typical medieval bagpipe might consist of two drone pipes and a chanter, a reed pipe with finger holes providing a scale consisting of up to about an octave.

The *capped-reed* instruments were direct descendants of the early bagpipe; they achieved their pinnacle of use during the high Renaissance and gave way ultimately to the oboe and bassoon. In capped reed instruments like the krumm-horn, kortholt, rauschpfeife, and cornemuse families, the double reed is enclosed in a box or cylindrical enclosure that is connected to the resonant pipe, which in turn is fingered like a recorder or similar woodwind instrument. This allowed the player to blow the instrument directly through a hole in the box in which the reed was enclosed, but without touching the reed with the lips, a technique that permitted considerably greater control of the tone, articulation, and intonation than in the bagpipe.

An even greater degree of tone control was obtained by removing the cap and blowing the instrument directly with the lips on the reed, as in the Renaissance shawm and rackett families, and later in the dolcian, a predecessor to the baroque bassoon. Modifications of the shawm in the late Renaissance and early baroque led to the baroque oboe. Few acoustical innovations over the baroque oboe and bassoon are found in their contemporary counterparts other than the addition of a good bit of machinery and additional holes to provide more adequate intonation of the chromatic scale over the entire range of the instrument.

The recorder and transverse flute families developed in parallel fashion to the double-reed instruments, beginning many centuries ago with primitive bone, bamboo, or other flutes. By the late medieval era, the flute had developed into the basic form used in the Renaissance, and during the Renaissance the flute developed into a family of straight cylindrical instruments.

The continuing acoustical evolution of the flute is very interesting with respect to bore shape, that is, the change of size of the tube along its length. In the Renaissance the flute was a simple cylindrical tube with embouchure hole and finger holes, stopped at the end above the embouchure hole. To achieve a greater range, the bore of the baroque flute was modified to a conical shape with the larger radius at the embouchure hole and the smaller radius at the bell end. As in the cylindrical Renaissance flute, the short end adjacent to the embouchure hole was stopped, and the bore shape in this end section was modified to give the best intonation over the entire range. The modern flute reverted back to a cylindrical bore and achieved the desired range and acceptable intonation by elongating the end section above the embouchure hole and modifying the sizes and positions of the finger holes. The addition of the considerable machinery used in fingering allowed the flute to accommodate the demands of modern music, particularly the range, intonation, expression, and volume of sound that can be obtained. A very important improvement in woodwind design was a system of fingering introduced by Theobald Böhm (1794–1881), a German watchmaker and goldsmith and an amateur flutist. The Böhm fingering system, introduced in the middle of the nineteenth century, was a method of organizing the fingering machinery for the flute. It was later adapted to the clarinet and saxophone families, but has not yet achieved acceptance on the oboe and bassoon families, although significant acoustical improvement could be obtained by its adoption.

Recorders are similar to the flutes acoustically, differing primarily in the way in which the initial noise source is generated. Recorders are *fipple flutes* in which the air from the mouth of the player is blown through a windway and impinges on a sharp edge, producing an edge tone. One of the earliest known fipple flutes is the medieval tabor flute, or one-handed flute, which has only two finger holes positioned near the end of the tube so that both the holes and the end of the flute can be covered. The performer could play the tabor flute with one hand and the tabor, a type of drum, with the other, perhaps dancing at the same time. Such minstrels were very popular during the Middle Ages, and although virtually no written music of this type remains (it probably was never written down, but only passed on informally from performer to performer), paintings and descriptions of such performances abound in medieval history. The bore of the tabor flute is so narrow that it is impossible or very difficult to play the fundamental; in any fingering configuration the octave and the fifth above that could be easily obtained. Because the higher harmonics are produced, the notes playable with any fingering are closely spaced. With three or four sets of closely spaced overtones, one for each fingering, a full scale of notes over more than one octave range could be easily obtained.

The earliest Renaissance recorders were cylindrical instruments with eight finger holes. As in the case of the flute, the baroque recorder developed with a conical bore having the larger diameter at the fipple and the smaller diameter at the bell end. The recorder was the most advanced of the baroque woodwinds, possessing the best intonation and voicing, the result of an incredible number of man-years of development during the seventeenth and eighteenth centuries. This was summarized in extremely detailed formulas relating the bore diameter, hole sizes and positions, and other physical dimensions. While the recorder required centuries to develop into its final form, it was supplanted relatively quickly by the flute in the classical orchestra because of the superiority of the flute in volume and expression.

The clarinet is the youngest of the standard modern orchestral woodwinds; it was developed during the middle of the seventeenth century from the chalumeau, a very soft, single-reed Renaissance woodwind with a narrow cylindrical bore. The saxophone, a single-reed instrument with a very large conical bore, is an even more recent development, having been invented in the middle of the nineteenth century. Although the saxophone is seldom used in the standard symphony orchestra, it has achieved immense popularity in jazz and dance music and is a member in good standing of the contemporary concert band.

Figure 10–1 shows some of the woodwind instruments that acted as important stepping stones in the development of contemporary woodwinds. As in the contemporary instruments, they can be classified by their bore shape (cylindrical or conical), the type of reed (single or double), or the type of noise source (fipple or embouchure hole).

Virtually all Renaissance instruments came in families of different sizes and pitch ranges, the basic sizes being the soprano, alto, tenor, and bass. When appropriate, this basic family was extended in either direction, and when size became a critical factor, the larger members were simply left out or fell rapidly into a state of benign neglect.

Many modern woodwinds come in a family of sizes, as did their Renaissance ancestors. Which members of the family survive as members of the modern symphony orchestra is now a matter of tradition and usefulness in the orchestra.

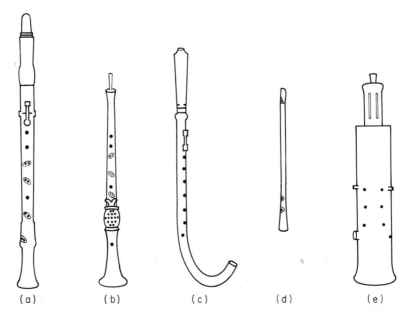

(a)　　　(b)　　　(c)　　　(d)　　　(e)

Figure 10–1 Some ancestors of the contemporary woodwind instruments: (a) the chalumeau, a single-reed cylindrical Renaissance precursor to the clarinet; (b) the shawm, a double-reed, direct blow conical precursor to the oboe; (c) the krummhorn, a cylindrical Renaissance instrument with a capped or enclosed double reed; (d) the tabor flute, a medieval narrow-bore fipple flute played with one hand; and (e) the rackett, a direct blow, double-reed Renaissance instrument with a long cylindrical bore, which has been folded back on itself nine times, a precursor to the bassoon.

Some instruments, such as the piano, guitar, and voice, can be played in most keys with equal ease. There is nothing inherent in the human voice, for instance, that "prefers" the key of C major to the key of $A^\flat$ or $F^\sharp$. This is not true of some other instruments, such as the clarinet family. One easily fingered scale on a $B^\flat$ clarinet sounds $B^\flat$ major. It is convenient, then, to write music for this instrument in which the key that is simplest notationally, C major (no sharps or flats), corresponds to the key most easily played, sounding $B^\flat$. When the clarinet player sees the note C written in music, he or she fingers that note on the clarinet, but $B^\flat$ sounds. Composers must allow for this change between written note and sounded note and write accordingly, or *transpose*. For instance, if the composer wants the $B^\flat$ clarinet to sound a C, a D must be written.

In practice, this system has another benefit. Once a clarinetist learns to play one instrument, say the $B^\flat$ clarinet, it is a simple matter to play a different member of the clarinet family, for instance, the $E^\flat$ alto clarinet. The musician learns one fingering for each written note. When reading the note C, the same finger combination is used on each instrument. Of course, a $B^\flat$ will sound on the $B^\flat$ clarinet, while an $E^\flat$ will sound on the $E^\flat$ clarinet. The musician need not worry: The written music for each instrument has been transposed appropriately so that the desired pitch will sound.

10.2   *Woodwind Tone Quality*

For woodwind instruments, there are three primary elements that determine the tone quality and overall sound: (1) the source of noise or vibration, (2) the size and shape of the bore, and (3) the sizes and positions of the finger holes. In addition, details in construction such as the inside finish of the bore, reed and body materials and construction, and others, contribute to the tone quality. While some of these effects are well understood and their importance generally agreed upon, others are controversial and appear to have no basis in physics. We shall try to discuss the elements that affect the quality of woodwind tones and put some of the more controversial items in perspective using a combination of physics and musical experience. Unfortunately, as we shall see, there is much emotionalism and extremely strong tradition in the choice of musical instruments and the details of their construction.

The most obvious factor determining the tone quality of an instrument is the source of the vibration. Perhaps the simplest in appearance, but not the simplest acoustically, is the *edge tone*, which is used in the flute and recorder families. In the fipple flute shown in Fig. 10–2, a jet of air is blown through the fipple, impinges on a sharp edge, the lip, and creates acoustical noise. Components of this noise that are at resonant frequencies of the air column will be strengthened, and after a short time the oscillations of the air column begin to strongly affect the motion of the air stream about the lip. This feedback of the motion of the standing wave in the tube

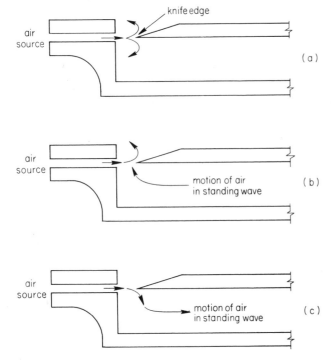

Figure 10-2 Action of sound source of a fipple flute. After initial attack transient period, (a), air jet from source is forced to oscillate back and forth across the knife edge or lip by the motion of the standing wave, as illustrated in (b) and (c).

strengthens the frequency components in the noise spectrum at the resonant frequencies of the tube, and thereby strengthens the tone of the instrument. The changing harmonic structure at the beginning of the tone is described by the general term of *attack transient*. After the initial transient period, the oscillating standing wave in the recorder drives the air stream back and forth across the lip at the sounding frequency of the instrument.

Because there is rapid motion of the air in and out of the opening by the lip, as well as at the other end (at the first open hole or at the bell, if all the holes are covered), both ends are velocity (displacement) antinodes or pressure nodes; the instrument will be one half-wavelength long and will behave acoustically like an open tube. Of course, the antinodes are not exactly at the end of the bell and the edge of the fipple opening, but will be strongly affected by details of the geometry. As all recorder players know, blowing too hard will raise the pitch of the tone. This effectively pushes the antinode farther into the bore, while leaving the bell end almost unaffected, thus shortening the effective length of the tube and raising the pitch. Alternatively, reducing breath pressure allows the antinode to move farther out the fipple opening, thus increasing the effective length of the air column and lowering the pitch.

The sound production mechanism in a flute is similar to that in the recorder, with a few significant differences. The lips must be carefully shaped when blowing onto the edge of the embouchure hole to produce the initial noise source. While the feedback mechanism between the standing wave in the tube and the air stream oscillating across the edge of the embouchure hole is similar to that of the recorder, two additional adjustments can be made on the flute.

First, the flute has a section above the embouchure hole whose bore can be modified and in which a cork end is inserted and adjusted in position to correct some tuning problems. This was very important in the development of the modern cylindrical flute. The second and possibly the most important element in the flute, which allowed it to supplant the recorder and become a member of the modern symphony orchestra, was the greater dynamic expression made possible by the control the player has over the air stream. The player can compensate for intonation defects in the baroque flute by "rolling" the mouthpiece slightly about the lip. This allows a greater range of dynamics with good pitch compensation, leading to considerably greater "expression."

In a reed instrument, the reed is driven by the oscillating air column at the resonant frequency of the standing wave in the tube. While each reed has a resonant frequency at which it vibrates freely, this resonance is usually very broad and does not play an important role in determining the sounding frequency of the instrument. On the contrary, undesirable resonances in the reed are responsible for the assorted high-pitched squeaks and squawks typical of poorly played reed instruments. Such effects are increased when the reed is dry, shaped improperly, has cuts or cracks, or is otherwise incapable of providing uniform vibrations. In most woodwind instruments the natural vibration frequency of the reed is considerably higher than the frequency range of the instrument as it is normally played, even with the damping obtained when the lips are in contact with the reed. One exception to this general rule is that of reed organ pipes. Since each pipe and reed sounds only at one frequency, the reed is generally "tuned," by a technique that we shall discuss later, to sound at a frequency very close to the resonant frequency of the pipe.

A reed end acts as a closed end and thus is the acoustical opposite of a fipple or embouchure hole. Consider what happens in a cylindrical tube with a reed on one end. A large burst of air flowing rapidly through the reed opening will cause it to close by the Bernoulli principle. The compressional air burst travels down the length of the tube, reflects off the open end, and returns to the reed end as a rarefaction. The rarefaction in turn reflects off the reed and travels down the length of the tube, reflecting off the bell end back to the reed end as a compression, forcing the reed open, and allowing another air burst to enter the reed from the player's mouth. Since a burst of air entering the reed from the instrument causes the reed to open, allowing another burst of air to enter from the player's mouth moving in the opposite direction, there is a large pressure variation at the end of the reed. There is therefore a pressure antinode or a velocity node at the reed end. In the clarinet, a cylindrical reed instrument, the reed end is a velocity node and the bell end a velocity antinode, so the clarinet behaves acoustically like a closed tube. This behavior of the reed is modified considerably if the bore shape is conical rather than cylindrical, as we shall discuss later. In reality, for soft tones the reed does not close entirely, but closes only when the volume of sound is increased to a relatively high level. This means that there is some change in tone quality as the volume increases, when the reed closes.

Double reeds function acoustically in the same manner as single reeds; the tone quality of most instruments depends much more strongly on the size and shape of the bore. There are some differences in detail that will be discussed as each of the instruments is surveyed.

While the intensity of tone is largely determined by the bore diameter, the bore shape affects the tone quality. Two features of the sound produced by a woodwind instrument can be controlled by shaping the bore of the instrument. First, the sequence of overtones that resonate in the instrument is adjusted to be exact harmonics. Second, the bore is shaped to preserve this harmonic relation as the finger holes are opened; in other words, the harmonicity of the overtones must remain as the length of the instrument is changed.

There is a large number of shapes for bores; it is convenient to limit the discussion to the family of exponential horns, in which the area of the bore is proportional to some power of the distance from one end. Of all the possible bore shapes, only two satisfy the harmonicity requirements and are physically possible to produce and play. There are the cylinder, in which the area is constant, and the cone, in which the area is proportional to the square of the distance from the pointed end. A complete cone need not be used; any section of a cone will resonate in the same basic manner as an open tube of approximately the same length, even if one end of the cone is a reed end. These shapes can be determined theoretically by computing the details of standing waves in these instruments. They were discovered experimentally by trial and error during the Renaissance; the task of Renaissance wind instrument makers was to produce playable families of instruments with good sounds. The instruments developed during that era evolved into the orchestral woodwinds, which have either cylindrical or conical bores. Considerable modification of the bore shape is used to correct intonation problems and improve the voicing or ease with which some instruments can be played, but the basic bore shape must be either cylindrical or a straight conical section. Figure 10–3 shows examples of cylindrical and straight conical tubes that might be useful shapes for woodwind instrument bores.

A great deal of study and care is involved in designing of the finger holes. Basically, the finger hole opens to allow the air to come in and out, effectively moving the antinode from near the bell end to near the position of the first open hole.

Figure 10–4 shows a hypothetical cylindrical bore fipple flute with a row of finger holes designed to produce a chromatic scale. The instrument behaves acoustically like an open tube, with one end at the fipple opening and one end at either the bell or the open hole closest to the fipple. Each hole is opened in order to play the chromatic scale, with no fork fingerings or other complexities. The length of the instrument is such that one half-wavelength of the note C is the distance L between the fipple opening and the bell. When all holes are open, the effective length is $L/2$, one-half the original wavelength, producing twice the original fre-

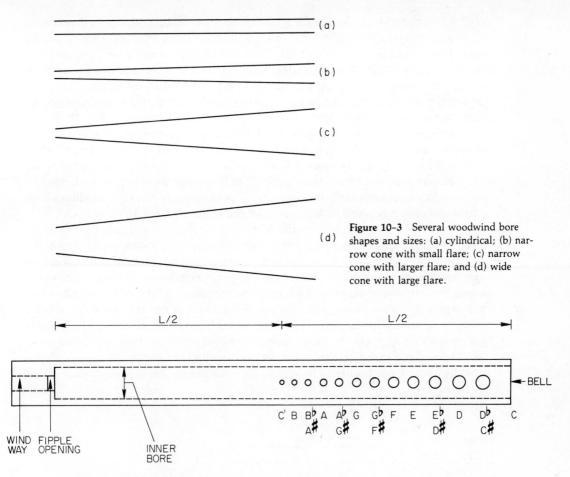

Figure 10-3 Several woodwind bore shapes and sizes: (a) cylindrical; (b) narrow cone with small flare; (c) narrow cone with larger flare; and (d) wide cone with large flare.

Figure 10-4 A hypothetical cylindrical fipple flute with "finger" holes which, if covered successively, would provide a chromatic scale of one octave.

quency or a musical interval of one octave above the lowest note. Notice that the hole spacing increases for holes farther down the tube from the fipple. Closing each hole successively, beginning from the top, must increase the length of the standing wave by a factor of 1.0595 times the previous length to lower the frequency by that factor, or one half-step.

Another way to obtain the note C' is to keep all the holes closed except hole C', allowing air to move in and out through hole C' and thereby forcing a velocity antinode at that position. The instrument will then be two loops or one full wavelength long, and produce the note C'. If both hole C' and hole D♭ are opened, the standing wave will adjust itself so that the antinode in the middle will move slightly upward toward the fipple, while the antinode at the end will be pushed slightly from the D♭ hole toward the bell. This produces two loops of a standing wave whose wavelength is slightly greater than that of the note one octave above D^b; this D^b is therefore slightly flat. By opening the remaining holes successively, notes about one octave above D, E^b, E, F, and so on, can be produced, each becoming successively flatter in pitch relative to the notes one octave above the original chromatic scale. By moving the position of hole C' up toward the fipple, additional notes in the second octave can be played more nearly in tune, since that hole will be nearer to the position of an antinode for more notes, as illustrated in Fig. 10–5. The antinode is not exactly at the position of the hole for all notes, so there remains an intonation problem of varying degree for notes farther from the average. Also, moving the C' hole up the tube causes C' to be out of tune when played with all holes open and renders this note unusable. The hole must therefore be made smaller if it is to play C' more nearly in tune. This repositioning of the C' hole allows it to function as an octave hole, a feature of all the contemporary woodwinds.

One other minor problem for this instrument is that it could only be played by someone with 12 fingers and no thumbs. Additional modifications required to make the instrument playable by a normal human are shown in Fig. 10–6. First, some of the holes must be eliminated. The normal choice is to eliminate the chromatic notes, leaving an instrument capable of playing one complete octave of a diatonic scale. Even with this modification, the spacing of the holes is not comfortable, not the normal positioning for fingers. Therefore, in addition to placement of the thumb hole on the side opposite the finger holes, the positions of the holes are adjusted so that the hand feels comfortable. However, changing the hole positions introduces additional acoustical problems having to do with intonation and voicing.

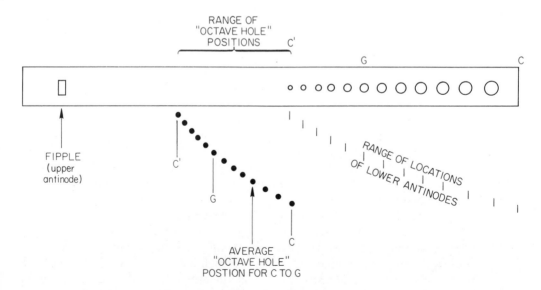

Figure 10-5 Location of average octave hole position for a single hole servicing the notes C through G on the hypothetical fipple flute.

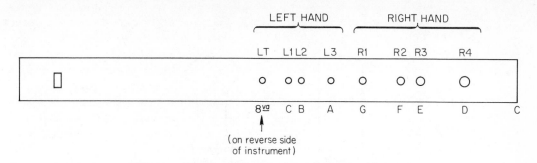

Figure 10-6   Modifications to finger holes of hypothetical fipple flute to make it playable by a human.

Moving a hole up the tube toward the mouthpiece decreases the length of the standing wave and increases the pitch of the note controlled by that hole; moving the hole toward the bell lowers the pitch of that note. Making a hole larger allows the air to flow in and out more easily, moving the antinode up, and thereby shortening the wavelength and raising the pitch; making a hole smaller lowers the pitch of the note. In the Renaissance these factors were used of necessity; positions of the holes were chosen so that the instruments were playable without mechanical keys, and the sizes were adjusted to provide the best intonation. Without this constraint, it is best acoustically for the holes to become progressively larger as the notes go lower, perhaps requiring keys. This not only allows proper intonation, but also creates better *voicing*, that is, more uniform tone quality of all the notes. Listening to chromatic notes played on early instruments gives one an appreciation for the improvements in intonation and voicing that have been achieved by the adjustment of hole size and the addition of rings and keys to modern instruments. The added machinery permits large holes at large spacings to be used comfortably.

Modification of hole positions for comfort (and out of necessity for some of the bigger instruments) and appropriate adjustment of their sizes to retain the proper intonation for the diatonic scale results in a fipple flute very closely resembling the Renaissance recorder, shown in Fig. 10-7. To play the note with the right-hand little finger (R4), that hole is usually placed toward the right side of the line in which the other holes are positioned. Certain other modifications in the hole positions and sizes are then made to provide for best intonation and voicing of the fork fingerings used for some of the more common chromatic notes. The basic baroque recorder finger-hole layout is shown in Fig. 10-8, and the typical fingering chart for a baroque alto recorder is given in Fig. 10-9.

Perhaps the most important and exotic secondary effect of the tone holes is related to the *tone-hole cutoff frequency*. If a resonance curve is obtained experimentally for a woodwind instrument like the hypothetical fipple flute, say for a note with only about half of the holes covered, it will be observed that above a certain frequency the resonances diminish rapidly in amplitude and then disappear. The frequency at which the resonances disappear is called the tone-hole cutoff fre-

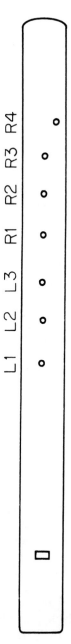

L1 L2 L3 R1 R2 R3 R4

Figure 10-7 Basic design of a typical cylindrical Renaissance recorder.

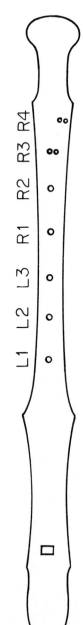

L1 L2 L3 R1 R2 R3 R4

Figure 10-8 Basic design of a typical conical baroque recorder.

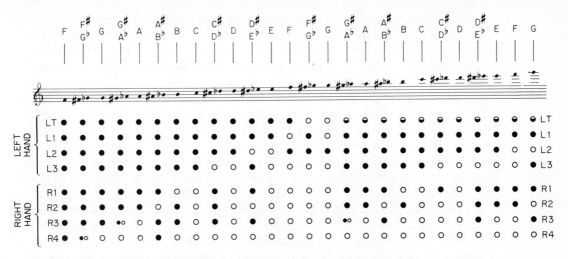

Figure 10–9 Fingering chart for a typical baroque alto recorder. The symbol ○ denotes an open finger hole; ● denotes a closed finger hole; ◒ denotes a half-closed hole (left thumb only for most baroque recorders); and ●○ denotes a half-closed double hole (third and fourth fingers of right hand only for most baroque recorders). This fingering chart will work acceptably for most F recorders, but only alto recorders sound at the pitch written here.

quency. A higher-frequency standing wave is capable of extending its oscillation much of the way along the tube, despite the existence of the array of open holes. At lower frequencies the first open hole will act as a strong cutoff point for the standing wave and effectively preclude the existence of strong oscillations beyond that point. Well below the cutoff frequency the first open hole has by far the strongest influence on the frequency of the note. However, toward the higher ranges of some instruments, particularly the smaller woodwinds, the standing wave extends well into the array of open tone holes, and closure of holes at some distance down the tube can have a dramatic effect. This feature of woodwind instruments makes possible the use of *fork fingerings*. In a fork fingering, one or more holes are left open while holes both above and below this open hole are closed. A very dramatic and surprising fork fingering involves the note D at the top of the range of some Renaissance alto recorders. The note C is played with L1, L2, and L3 covered and LT half covered, as an octave hole. The note D is obtained by *closing* holes R2 and R3 while leaving R1 open and the left hand as it was! Fork fingering is the only way to obtain any of the chromatic notes, the notes other than the original seven notes in the basic scale of the hypothetical fipple flute that we designed.

Another important effect of the open hole array is to increase effectively the size of the first open hole; the holes nearer the upper end of the instrument must therefore be slightly smaller than if they were opened individually to provide the proper tuning.

The holes have other, more subtle effects. Particularly in instruments with thick walls, the volume of the holes is not negligibly small compared to the volume of the tube itself, and the frequencies of the notes will be affected by the existence of the covered holes. Futhermore, as higher notes are played, fewer holes remain along the resonating column. The standing wave is partially broken up by the irregularity of the holes along the tube. This leads to a change in tone quality between various notes of the scale and is thought to be partially responsible for the unique quality of the "throat" tones of the clarinet (the highest notes in the lower register, just below the register break). Another feature that changes the effect of the holes is chamfering of the inside of the holes, that is, tapering the inside edge so that there is not a right-angle corner where the finger hole enters the bore. Chamfering is sometimes used in more expensive instruments as a final very subtle step in adjusting the tone and voicing.

Important improvements can be obtained in woodwind intonation and voicing by modification of the bore diameter from the exact cylinder or straight cone geometry. This is analogous to the adjustment of frequencies of vocal formants by changing the size and shape of the vocal tract. If the size of the bore is increased at the position of a velocity node (pressure antinode), the frequency of that mode of oscillation will be lowered; if the size of the bore is increased at the position of a velocity antinode (pressure node), the frequency of that mode of oscillation will be increased. This effect is involved in fixing the amount of flare of the clarinet bell, which actually extends about halfway up the instrument, to provide for the closest approximation of the overtones to exact harmonicity. The bore of the clarinet near the mouthpiece end must also be adjusted from the cylindrical shape, an effect usually accomplished in the mouthpiece and adjacent barrel sections.

Such modifications are particularly significant in conical instruments: Small changes in the volume of the small end of the cone (near the reed) can create very large intonation deviations. As a result, the shape and size of the saxophone mouthpiece can have a very dramatic effect on its intonation, and a mouthpiece must be chosen very carefully. This problem is more acute in the oboe and particularly in the bassoon because of their small bores. It is necessary to control very carefully the volume of air that is contained within the reed, and to minimize any changes in the reed shape as breath and lip pressure are changed in the normal course of playing over the entire range of the instrument. Some notes on the bassoon can be varied up and down by over a semitone by changing breath pressure and lip pressure on the reed.

A long-standing conflict involves the effect of choice of material on tone quality. While there is apparently a difference between the wooden flute of the early 1900s and the modern cylindrical metal flute, this can be largely explained by differences in the goemetry of the bore and mouthpiece. On the other hand, it is more difficult to explain the difference between otherwise identical flutes that are plated with, say gold, silver, platinum, or aluminum. Calculations have been made which indicate that there should be no noticeable difference between these metals with respect to their damping of the oscillations of the air column or their vibratory properties. Yet flutists claim substantial differences in tone among these metals.

Another controversy involves the choice of wood, hard rubber, or other "artificial" materials in constructing clarinets. Physically, there should be negligible differences between instruments made from these materials, yet strong claims of differences are made, and expert clarinetists often express strong preferences. A partial explanation of this certainly lies with tradition. As long as clarinetists do not accept instruments of materials other than wood, makers will not apply the highest quality craftsmanship to other materials. As long as the substitute materials are finished with less adequate workmanship, they will continue to receive less than total acceptance. Thus the circle continues. Perhaps we shall some day see "artificial" materials with excellent acoustical properties and superior physical properties, like resistance to temperature and moisture and superior strength, finished with exquisite workmanship. If this occurs, acceptance may come with time.

10.3 *Recorders*

We shall look in some detail at the acoustical properties and problems of the recorder, since the recorder is a very basic instrument and is extremely popular. Most of the acoustical problems discussed can be applied in a more sophisticated way to modern woodwind instruments.

The recorder derived its English name from its use in the Renaissance to teach songs to birds. An early form of the sopranino recorder called the *bird flageolet* was used to play short songs repeatedly for birds, thus teaching them the songs by rote. This process, called *recording*, was extremely popular for several centuries, and hundreds of short songs were composed and taught to dozens of different types of birds.

The recorder, being a fipple flute, achieves its sound by resonance of the noise created by the airstream from the windway striking the lip and feedback of the vibrating air column to drive the oscillating air jet. Since both the fipple end and the bell end are acoustically open ends, the recorder behaves like an open tube, with one end near the fipple opening and the other end located either near the bell or near the open hole closest to the fipple opening.

The earliest Renaissance recorders were simple cylindrical tubes with eight finger holes; in the baroque the bore developed a slight conical taper with the larger diameter at the fipple end. In addition, the two lowest holes became double holes, used to produce the chromatic notes one-half step above each of the two lowest diatonic notes. The range of the Renaissance recorder was about one octave plus a sixth; the improved bore and finger hole geometry of the baroque recorder extended that range to about two octaves plus one note. Figure 10–8 shows the typical fingering layout of a Baroque recorder with English fingering, and Fig. 10–9 is a standard fingering chart for such an instrument.

To achieve the most uniform tone down to the lowest note, the holes should become larger toward the bell end of the tube, and the spacing between holes should become greater. To adapt these requirements to the shape of the human hand, a compromise is necessary: The holes get smaller and slightly closer together near the bell end. By making the last few holes gradually smaller while concomitantly tapering the bore to a smaller radius at the bell end, the most uniform voicing of the notes (quality and loudness) is obtained, although the notes near the bottom of the range are somewhat weaker. This effect is not so pronounced in large recorders in which keys are used for the last few notes, so the holes can remain relatively large. By use of keys this weakness of the low tones has been largely eliminated from contemporary woodwinds.

Ideally, it is necessary to have one hole for each note on the instrument, where the size and position of each hole is chosen to give the correct intonation and uniform tone quality or voicing over the entire range. The most obvious limitation of the recorder, since it has no keys, is that one finger can only cover one hole, limiting the instrument to about eight useful holes (two of which are often made as double holes). Obviously, the first acoustical compromise that must be made is that each hole must have a substantial effect on more than one note, and fork fingerings must be employed for all chromatic notes as well as some diatonic notes.

Another necessary compromise is to use the thumb as an octave key. Opening the thumb hole allows air to flow in and out of the tube, creating an antinode at or near that point. This single thumb hole must serve as an octave hole for a range of notes, as shown in the fingering chart. This is an additional compromise, since the position of the vent hole can be the exact position of an antinode for only one of the notes in the upper octave. In fact, it is used to produce the octave for an interval of over a fourth, from A^b to D on the alto recorder. As can be seen on the fingering chart, there is a great deal of use in the recorder of fork fingerings, particularly for the higher notes. In actual practice this fingering chart will serve only as a guide for many of these fork-fingered notes, since they are extremely sensitive to fingering changes, sometimes requiring half-holes at the bottom of the instrument. Many of these fork fingerings will differ in detail among similar recorders, even those by the same maker, owing to extremely small variations in dimensions when they were manufactured or changes in dimensions with the climate. A significant improvement in modern instruments is greater resistance to the effects of humidity and different kinds of weather so that modern instruments are not so critically affected by these factors.

Some recorders are made with small variations in the bore diameter to provide small final adjustments in the pitch of certain notes. One of the major problems with inexpensive recorders and replicas scaled to high pitch is that the thumb hole must be very carefully half-covered for the upper notes; small changes in the thumb opening will drastically affect both the pitch of the upper notes and how well they speak. A slight constriction of the bore between the left thumb hole and the fipple opening has been suggested for some recorders as a way to improve the ease with which the last few notes in the upper range of the instrument will speak. This might allow these notes to be played in tune more easily and permit greater leeway in the thumb hole opening to obtain good high notes.

The correct hole positions for the recorder were obtained historically in a sequence of "choices," as indicated in Sec. 10.3. First, the sizes and positions of the holes are adjusted to provide a diatonic scale that is in tune using the fingering shown on the chart for the low octave. Then, using the thumb hole half-opened as an "octave" hole, the hole sizes and positions are adjusted to tune the upper octave. Chromatic notes are played by finding the most accurately tuned fork fingerings, as illustrated in the fingering chart. The use of fork fingerings to "force" the positions of the antinodes to the proper location results in a tone that is different in quality from the diatonic notes. This results in considerable variation in voicing over the entire range of the instrument and increases the technical complexity of performance. It is necessary to modify further the hole sizes and positions to obtain the best compromise between making the upper octave in tune, making the chromatic notes in tune with reasonable fork fingerings, and obtaining the best overall uniform voicing.

Several problems remain in the recorder because of the acoustical compromises in its design. Some of the most important compromises deal with intonation, simply because one hole must be a strong determinant for more than one note. One such tuning compromise is between the notes D and F$^\sharp$ in the middle range of the alto recorder. Referring to the fingering chart, we see that the note F$^\sharp$ is obtained by fingering D and opening the left thumb hole. However, the thumb hole position has already been (over) prescribed by the dual requirements that it make the note G in tune and that it properly act as an octave vent over a considerable frequency range. As a result, if D is in tune, F$^\sharp$ will be flat, or if F$^\sharp$ is in tune, then D will be sharp. On most baroque replica recorders, D is made exactly in tune, since that note is played much more often than F$^\sharp$. F$^\sharp$ must be either fingered differently or blown harder than normal in order to raise its pitch. In a recorder designed to play more chromatic modern music, a compromise is reached in which D is slightly sharp and F$^\sharp$ slightly flat. It is then necessary to underblow D and to overblow F$^\sharp$ so that they will be in tune. All recorders are built with this intonation compromise, and an accomplished player learns how to correct the pitch by adjusting breath pressure.

Possibly the most limiting feature of the recorder is that the player has very little flexibility in control of the airstream. Changes in air pressure strongly affect the intonation, leaving very little possible adjustment in volume, and thus limited possibility for "dynamic expression" in music.

Another complication in recorder design results from the overall rise in pitch standard over the centuries since the baroque. Unfortunately, despite the great detail involved in specifying the proper dimensions for recorder bore, holes, fipple, and bell geometry, these dimensions do not scale well with pitch. Slight adjustments in hole positions can correct for minor intonation problems resulting from shortening the design length of the recorder so that it plays at A = 440 Hz rather than A = 415 Hz, the "standard" baroque pitch, which is about one half-step below A = 440 Hz. However, while the intonation can be readily adjusted, the voicing, and the ease with which the notes can be played over the entire range are very strongly affected by scaling. It is extremely difficult and time consuming to correct these problems. In this era of high labor costs, very few craftsmen are willing to take the time required to modify instrument design at high (A = 440) pitch to achieve response similar to that obtained in the baroque. On the other hand, by reproducing instruments according to baroque specifications, at low pitch, the problem of redesign is avoided. This is a strong motivation for performance of baroque music at low pitch on low-pitch instruments, as is now done by several groups in Europe and the United States. An additional factor is the more authentic tone quality produced by instruments played at low pitch. While the problems arising from the pitch change since the baroque appear in many instruments and vocal music as well, it is possible that the effect is greatest in recorders, because they were probably the most highly developed wind instruments in that period, despite their severe limitations. Many makers of baroque instruments now produce recorders, and other instruments as well, at low pitch, generally A = 415 Hz, and performance at low pitch is becoming more commonplace.

The recorder comes in a family of sizes, typical of Renaissance instruments. The soprano in C, alto in F, tenor in C, and bass in F form the basic set, as with most other families. This is supplemented by the great bass in C and the double great bass in F (very rare) on the low end and the sopranino in F on the upper end. An even smaller recorder also appears irregularly in a variety of sizes. The ranges are given in Table 10-1 for various sizes of baroque recorders.

The recorder is not a transposing instrument; all present-day recorders sound at the written pitch. Two sets of fingerings are used, one set for the F instruments, and one set for the C instruments. The F fingering is similar to the fingering of the low register of the clarinet; the C fingering is similar to the fingering in the clarinet middle register. For all the C instruments, some octave of C is the note produced with all fingers down; on the F instruments, some octave of the F is produced with all the holes closed.

This description of recorder problems is of considerable relevance. Because modern instruments are complicated by the addition of more holes and machinery and have greatly extended ranges, similar problems exist in all woodwind instruments, only at a somewhat more sophisticated level.

10.4   *Flutes*

The basic acoustical principles of the flute are very similar to those of the recorder. The single most important acoustical difference is that the player is in direct control of the angle at which the air from the lips strikes the embouchure hole. This flexibility, along with control of the rate of air flow, provides the flute player with a greater range of volume and expression. As the recorder, the flute was a straight cylindrical wooden tube during the Renaissance. The range was extended by making the baroque flute in the shape of a slightly tapered cone, with the small end at the bell. This modification, along with some improvements in the tone hole design

TABLE 10–1 **Ranges of selected woodwind instruments.**

Instrument	Lowest note	Approximate highest note
Soprano recorder in C	C_5	D_7
Alto recorder in F	F_4	G_6
Tenor recorder in C	C_4	D_6
Bass recorder in F	F_3	G_5
Piccolo in C	D_5	C_8
Flute	C_4	D_7
Bass flute	G_3	G_6
Sopranino clarinet in E$\flat$	G_3	$B^\flat_6$
Soprano clarinet in B$\flat$	D_3	$B^\flat_6$
Soprano clarinet in A	$C^\sharp_3$	A_6
Alto clarinet in E$\flat$	G_2	$B^\flat_5$
Bassett horn in F	F_2	C_6
Bass clarinet in B$\flat$	D_2	F_5
Soprano saxophone in B$\flat$	$A^\flat_3$	$D^\flat_6$
Alto saxophone in E$\flat$	$D^\flat_3$	$A^\flat_5$
Tenor saxophone in B$\flat$	$A^\flat_2$	$E^\flat_5$
Baritone saxophone in E$\flat$	$D^\flat_2$	$A^\flat_4$
Bass saxophone in B$\flat$	$A^\flat_1$	$D^\flat_4$
Oboe	$B^\flat_3$	G_6
English horn	E_3	A_5
Bassoon	$B^\flat_1$	$E^\flat_5$
Contrabassoon	$B^\flat_0$	$E^\flat_3$

and cork joint, also allowed somewhat greater expression and improved voicing in the baroque flute.

During the nineteenth century several additional modifications were made. In the early nineteenth century more holes and keys were added to the baroque two-key and five-key flutes, improving the intonation of the flute and vastly simplifying the fingering of the chromatic notes. This also made the voicing and tone quality more uniform by eliminating many fork fingerings for chromatic notes. One result of these additional keys was a rather extensive array of complicated machinery. The flute bore again became cylindrical in the middle of the nineteenth century, and metal began to replace the traditional wood around the beginning of the twentieth century. Modifications of the head joint and the movable cork improved the intonation in the third octave, thus increasing the range by almost one octave.

One of the most important improvements in the flute was the Böhm fingering system, developed by Theobald Böhm over the first half of the nineteenth century. The basic design of the contemporary flute was carried out by Böhm, including the hole layout, design of the head joint and cork position, and details of which keys and rings are attached to each other and controlled with each finger.

The Böhm fingering system organized the machinery to provide a relatively straightforward and systematic set of fingerings by specifying how the keys and rings should be interconnected to provide the simplest basic fingering plan. By providing the appropriate normally open keys, large tone holes could be employed, thus allowing greater volume to be produced by the flute.

Figure 10-10 shows the cross section of a modern Böhm flute. The primary reason for the extension of the usable range of the flute to its modern limits lies in the details of the construction and placement of the cork. Although a movable cork was used in the baroque flute, its normal position was changed dramatically when the flute bore was returned to its original cylindrical shape by Böhm. Böhm added a slight parabolic taper to the section of pipe above the embouchure hole that holds the cork, with the length of the head joint from the embouchure hole to cork equal to the bore diameter at the embouchure hole. The existence of the additional tone holes and the machinery with which they could be closed allowed the entire chromatic scale to be played with uniform voicing and without use of complicated fork fingerings. Böhm used large holes to obtain greater volume from the flute and covered these holes with keys when they became too big for fingers to cover.

The contributions of Böhm to the musical world are important to more than just the flute, as the Böhm fingering system has been extended to the clarinet and saxophone families as well. It also appears that some of the problems of the oboe and bassoon might be helped by introduction of a modified Böhm fingering system. One desirable feature of the Böhm system is that after learning one Böhm instrument any other is relatively easy to learn, because of the similarity in fingering and key placement.

Very few acoustical changes in the flute have been made since Böhm designed

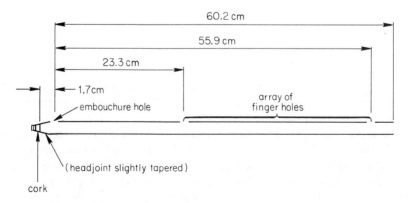

Figure 10-10 Cross section and basic dimensions of a Böhm flute.

his original model in the middle of the nineteenth century. In fact, some makers advertise that they use a design as close as possible to the Böhm original. However, during the intervening century the standard of pitch has risen from about A = 435 Hz, the standard at which Böhm designed his flute, to A = 440 Hz. As a result, certain intonation defects have developed. Measurements on contemporary flutes by Coltman show that relative to the equal-tempered scale, the notes at the bottom of the flute, with most fingers down, are low in pitch, while those with most fingers up are high in pitch, even for high-quality flutes. This error repeats in the second octave, as shown in Fig. 10–11 for several modern Böhm flutes tuned to A = 440 Hz. Performers normally compensate for this error by rolling the instrument about the lip to adjust the angle at which the air strikes the embouchure hole, and perhaps blowing louder or softer. Such performer compensation was not included in Coltman's measurements.

If a longer tuning slide is used and the flute is extended between the embouchure hole and the first finger hole to play at a pitch level about A = 430 to 435 Hz, the tuning error disappears and the intonation across the octave break is more continuous. This requires an additional length of some 8 to 13 millimeters (up to about ½ inch). It is hypothesized that as the standard pitch level rose the flute was "corrected" by simply shortening its length by removing some tubing between the embouchure hole and the finger holes, while the spacings between the finger holes remained unchanged. If this simple shortening were the only adjustment made, it would overcorrect the high notes in each octave, leaving them high in pitch, and undercorrect the low notes in each octave, leaving them low in pitch, as seen in Fig. 10–11. The effect of extending the tuning slide on one such contemporary flute is shown in Fig. 10–12. The notes become in tune for all octaves and the break between octaves is dramatically reduced.

The tone hole spacing can be adjusted to compensate for the tuning errors obtained when the flute is shortened by simply removing tubing from the tuning slide. The upper holes must be moved slightly farther down the tube, while the lower holes must be moved slightly up the tube, both corrections being a few millimeters. This has been done by the contemporary English flute maker Albert Cooper. Figure 10–13 shows the deviations in pitch of a flute manufactured to the Cooper scale standard; it is obviously much better over its entire range.

It appears that the Cooper scale flute is becoming more popular, though it is with hesitancy that it is being adopted by flute players and manufacturers. In this case, the design change appears to be a dramatic improvement. However, tradition seems to dictate that the Böhm design, having been in use for over 100 years, is the only acceptable design, and any deviation therefrom must be bad. Flute players have apparently learned to compensate for this defect automatically and tend to play out of tune on "properly" constructed flutes, or at least the flute has a different feel to the player.

It should be pointed out that any further rise in the standard of pitch will only increase the magnitude of this problem, and make the flute harder to play in tune, unless it is accompanied by an appropriate additional design modification.

Two additional members of the flute family are used in the contemporary

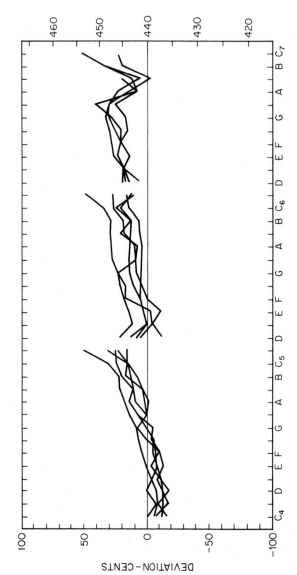

Figure 10-11 Tuning errors obtained with several modern Böhm flutes. A deviation of 100 cents corresponds to one half-step. (Reprinted from *The Instrumentalist* (March 1975, pp. 78, 80). © *The Instrumentalist Company* 1975. Used by permission of *The Instrumentalist Company*.)

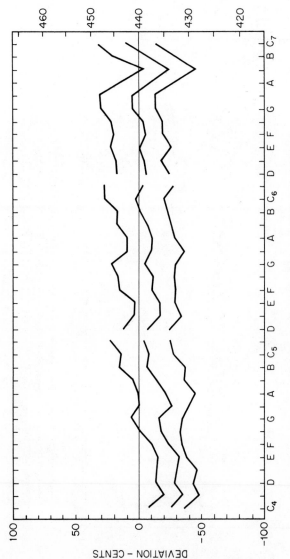

Figure 10–12 Tuning errors in a Haynes Böhm (top), and the errors on the same flute extended between the embouchure hole and the first finger hole by 8 mm (middle) and 13 mm (bottom). The flute is better in tune when it is extended to play at a lower pitch. (Reprinted from *The Instrumentalist* (March 1975, pp. 78, 80) © *The Instrumentalist Company 1975.* Used by permission of *The Instrumentalist Company.*)

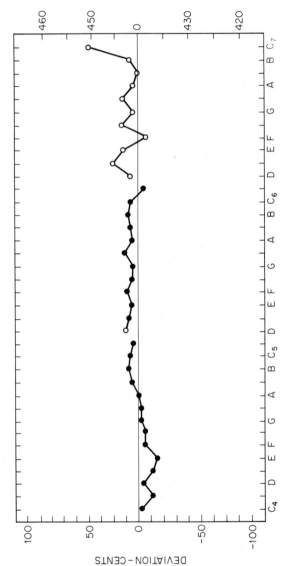

Figure 10-13 Tuning errors obtained on a Cooper scale flute, tuned to $A_4 = 440$ Hz, but with the Böhm finger hole spacing modified for the increase in pitch level. (Reprinted from *The Instrumentalist* (March 1975, pp. 78, 80). © *The Instrumentalist Company* 1975. Used by permission of *The Instrumentalist Company*.)

band and orchestra. The piccolo plays one octave higher than the flute and is in common use. Less common is the flute a fourth lower than the orchestral flute, called either the alto or bass flute in various scores. Their ranges are given in Table 10–1.

10.5   *Clarinets*

J. C. Denner (1655–1707) made significant modifications and improvements in the narrow-bore single-reed woodwind instrument, the chalumeau, thus developing the precursor to the modern clarinet. The tone of the chalumeau was very soft and possessed the typical "woody" character of the lower register of the clarinet, called the chalumeau register. The middle register of the clarinet is often called the clarion register, in recognition of the clear, more bell-like characteristic of its tone quality, which in its early stages was likened to that of the trumpet. The clarinet emerged as a distinct instrument in the late seventeenth century and took its place in the symphony orchestra in the late eighteenth century.

The clarinet is unique among the contemporary woodwinds in that it is the only instrument that behaves acoustically like a closed tube, the reed end being the closed end. A reed instrument can behave like a closed tube only if the bore is predominantly cylindrical. The clarinet bore is very nearly cylindrical from the barrel, just below the mouthpiece, to just below the middle joint, about halfway down the instrument.

There are several important implications of the clarinet's acoustical behavior as a closed tube. As in the closed tubes studied in Chap. 4, the resonance curve has its largest peaks at the odd harmonics of the fundamental frequency. The Fourier spectrum of typical clarinet tones is dominated by the odd harmonics, especially in the first several overtones. This is not true for higher harmonics, say above the fifth or sixth, when the even harmonics become as large as the odd harmonics. The clarinet shares this emphasis on odd harmonics with the square wave, which has only odd harmonics; this is the reason for the similarity between the tone of the clarinet and that of the square wave.

Because the clarinet is effectively a closed tube, its fundamental wavelength will be four times the distance from the reed or node end to the open end, either the bell or the highest open tone hole. The wavelength of the lowest note on the instrument is about four times the length of the clarinet. The flute, on the other hand, is an open tube, so its lowest note has a wavelength of twice the length of the flute. Although the flute and clarinet are approximately the same length, this means that the lowest note on the clarinet is about one octave below that of the flute.

The clarinet overblows at the third harmonic, a musical interval of a twelfth, rather than the octave, as in all other woodwind instruments. This can be understood by referring to Fig. 10–14. The fundamental mode in a closed tube has a velocity (or displacement) node at the closed end and an antinode at the open end,

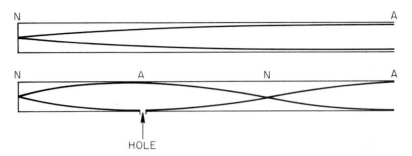

Figure 10-14 First two modes of a closed tube. A small hole is put in the tube at the point of the antinode "A." When the hole is open, an antinode is forced there and the "overblown" resonance results.

as shown. By venting the tube at one-third of its length from the closed end, an antinode is forced at that position. The overblown standing wave is just that of the third harmonic, or a musical interval of a twelfth, the next possible resonant mode of the closed tube. This hole is called the register hole on a clarinet, since the clarion (middle) register is up one-twelfth, not an octave from the chalumeau (lower) register. The altissimo (upper) register consists entirely of forked fingerings.

Figure 10-15 shows, roughly to scale, the positions of the tone holes and register hole of a clarinet. Notice that, as in the recorder, the register key provides the venting to force an antinode over a wide range of notes, where the effective tube length varies by over a factor of 2. The total length of the clarinet from the reed end to the bell is about 66 cm, so for the lowest note the register key should be about 22 cm from the tip of the reed. The highest note on the clarion register has the hole open 25 cm from the reed end, so the register key should be about 8 cm from the tip of the reed. The total change of position of the antinode is almost 17 cm (over $6\frac{1}{2}$ inches) in this simplified approximation.

Another result of the fact that a clarinet overblows at a twelfth is that 11 notes

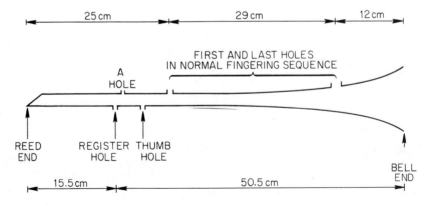

Figure 10-15 Cross section of a clarinet.

in the normal fingering sequence must be provided before the register is changed by opening the register key. This is more complicated than in other woodwinds, such as the recorder or flute, which overblow at the octave and thus require only seven notes. It is necessary, therefore, to use the left thumb hole and a special key (A hole) operated by the left index finger to extend upward the range of the lower register. The range of both registers is also extended downward one half-step by addition of a key controlled by the left little finger to close the lowest hole on the instrument, providing the note B. The B^b between these two notes is obtained by opening the register hole while the left thumb hole and the A key are simultaneously open. The notes A and B^b on the clarinet are called the *throat tones* and have a tone quality somewhat different from either of the other registers. This is due, in part, to the fact that for most notes on the clarinet there are closed finger holes along the tube that affect the tone, but there are fewer or none remaining above these holes to affect the tone of the throat tones. In addition, the register key is really at the wrong position and is the wrong size to produce B^b with the same voicing as the other notes, and it comes out very muffled. The clarinet has an alternate key to produce a B^b with better voicing, but it is inconvenient to use in even moderately fast passages and is generally used only as a trill key. A clarinet would be more adequately voiced and easier to play if two register keys were used to provide for the throat tone B^b and the venting of the entire upper register. Such a double key could be made to open automatically the proper hole, depending upon which note was being played in its clarion register, or if throat-tone notes were being played.

Despite these problems, the clarinet is in some respects the easiest reed instrument to learn to play acceptably, even with an inexpensive instrument. The fingering has been standardized using the Böhm system, making the clarinet easier to play than the oboe or bassoon. In contrast to the conical instruments, as the player changes pressure on the clarinet reed there is relatively little change in the effective length of the instrument, so it is easier for the beginner to play in tune. On the other hand, the fact that the clarinet overblows at the musical interval of a twelfth means that the notes one octave apart are fingered differently.

The two lower registers of the clarinet encompass the two sets of fingerings used in most woodwind instruments. Virtually all woodwinds, including all transposing instruments, are fingered the same as the clarion register, while the F recorders and some other Renaissance woodwinds are fingered like the chalumeau register of the clarinet. As in the recorders and flutes, there is a unique set of fork fingerings required to obtain the notes in the highest register, above the highest note in the clarion register.

Although the clarinet is acoustically a cylindrical instrument, it begins to flare toward the bell just below its midpoint. This flare is necessary to keep the overtones of all notes harmonic and to keep the frequencies of the overblown third harmonics of the clarion register in the correct ratio to the frequencies of the fundamentals in the chalumeau register. These problems arise in part from tuning errors introduced by the single register key, because it cannot be properly positioned for all the notes in the clarion register. Double register keys would significantly reduce this intonation problem, but such a dramatic change would deviate from the existing clarinet tradition and change the sound of some notes, which might be deemed undesirable.

The clarinet comes in a family of transposing instruments, the $E^{\flat}$ sopranino, the $B^{\flat}$ and A sopranos, the $E^{\flat}$ alto, and the $B^{\flat}$ bass being the most common. The $B^{\flat}$ soprano, $E^{\flat}$ alto, and $B^{\flat}$ bass clarinets are used as standard members of the concert band; the $B^{\flat}$ and A clarinets are the standard members of the symphony orchestra. Ranges of the members of the clarinet family are given in Table 10–1.

10.6   *Saxophones*

The saxophone, invented by the French musician Antoine Joseph (known as Adolphe) Sax (1814–1894) in the middle of the nineteenth century, is a relative newcomer to the musical scene and is sometimes scorned by musical purists as a "hybrid" instrument. Because of its recent development, it has not achieved general acceptance in the symphony orchestra, but it is used in the orchestra on occasion and has achieved popularity in the concert band and especially in dance and jazz bands.

The saxophone is a conical single-reed instrument with large bore diameter and very large tone holes, all covered by pads. The large bore diameter allows it to produce loud sounds appropriate for jazz and dance bands. The conical bore shape results in resonances at all harmonics, and the saxophone tone has all harmonics present, different from the cylindrical bore of the clarinet, which emphasizes the odd harmonics. The saxophone overblows at the octave. It has also been standardized to the Böhm system, making its fingering somewhat simpler than that of the clarinet.

One important feature of the saxophone, an improvement over the clarinet, is that the saxophone is equipped with two octave holes. Figure 10–16 shows the approximate dimensions of the saxophone, along with the rough positions of some of the tone holes and the octave holes. The effective length of the saxophone to produce its lowest note, $D^{\flat}_3$ (about 140 Hz) is slightly longer than the instrument. Ideally, the distance of the octave key from the effective end should be one-half the distance from the effective end to the first open hole. Octave hole H2, about 30 cm from the end, serves for a range of lengths from 52 to about 88 cm. Octave hole H1, about 14 cm from the end, serves for a range of lengths from about 28 to about 48 cm. However, the effective end extends the instrument several centimeters beyond the mouthpiece back toward the apex of the cone, placing the effective position of the hole H1 at the halfway point for a hole lying somewhere between those tone holes 28 and 48 cm from the actual reed tip. There is only one octave key; the appropriate octave hole is opened by rings and levers attached to the tone holes whenever the octave key is pressed. Thus, when the instrument is fingered up its scale, opening of the lowest finger hole requiring use of octave hole H1 automatic-

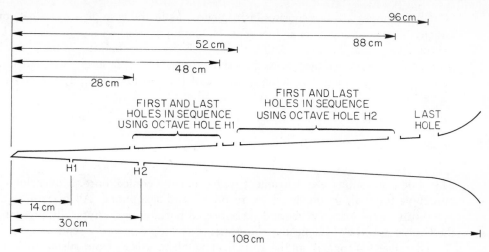

Figure 10-16   Cross-sectional view of a straightened out E^b alto saxophone.

ally opens that octave hole and closes H2. This automatic double octave hole system is a decided improvement over the system used in the clarinet and improves the saxophone intonation significantly.

The saxophone is probably the easiest reed instrument on which to obtain a tone, but some acoustical design problems arise for the beginner that are not present in the flute or clarinet. One characteristic of conical instruments is that small changes in the bore of the instrument near its narrow end can result in unexpectedly serious variations in the effective length of the instrument, changing its pitch considerably. Because the saxophone has a rather large mouthpiece, changes in embouchure, that is, the position and pressure of the mouth and lips, can cause rather large changes in the volume of the air space near the apex of the cone, which can detrimentally affect the pitch. Although this sensitivity of pitch to embouchure may hinder the beginner, accomplished jazz saxophonists use it for expressive purposes by "bending" notes. Tuning the instrument by sliding the mouthpiece in or out changes the effective length by a greater amount than normally would be expected, and the intonation could easily vary throughout the range of the instrument more than in the case of either the flute or the clarinet. These problems are substantially exacerbated by any dimensional errors in the mouthpiece or narrow part of the bore of the saxophone, which occur in poorly made instruments.

Because there is little saxophone music composed by the masters, the saxophone is relatively easy to play by beginners, and the sounds they are likely to get are often very loud and out of tune, the saxophone has perhaps unfairly achieved a bad reputation among serious musicians. An expert player with a good saxophone and properly matched mouthpiece can achieve a remarkable variety of sound, volume, and expression.

Saxophones are transposing instruments. As in the case of the clarinet, one basic set of fingering covers all instruments. Common sizes of saxophones found in the full concert band are the E^b alto, B^b tenor, E^b baritone, and the B^b bass. The B^b soprano and several larger sizes are used primarily in jazz ensembles and large dance bands. The ranges of some members of the saxophone family are given in Table 10–1.

10.7   *Oboes*

The oboe is a double-reed instrument with a narrow conical bore. It overblows at an octave interval, as do the flute, recorder, and saxophone. All harmonics are present in its resonant curve, and therefore all harmonics are typically present in the Fourier spectra of its steady-state tone.

Because it is conical, all the acoustical problems arising from mismatching the reed cavity to the instrument are present, as in the saxophone. However, because the bore is narrow, intonation problems arising from changes in the effective length are severe. Furthermore, although the double reed is acoustically very similar in its operation to the single reed, it is inherently more susceptible to variations in its volume as the embouchure of the player is changed. It is probably these inherent acoustical problems that make the oboe a very difficult instrument to play in tune with good tone, giving origin to its description as an "ill wind that never blows good."

The performance problems of the oboe are significantly exacerbated by the fact that the fingering system has never been converted to the Böhm system. The fingering system remains more difficult to learn and use compared with that of Böhm instruments.

An interesting experiment with the oboe is to substitute a small clarinet mouthpiece for the traditional double reed. The sound of the oboe remains very nearly the same, while it becomes easier to blow due to the simpler embouchure technique for single reed instruments. This verifies the hypothesis that the physical behavior of single and double reeds is very similar, and that the tone quality of a reed instrument is determined primarily by the bore shape and size. Clarinet-type mouthpieces have been used on oboes to simplify the process of learning to finger the oboe while a student is converting from clarinet to oboe.

The alto member of the oboe family, the English horn, is unique among the woodwinds in that it has a region near the bell end where the diameter of the instrument is significantly enlarged. This roughly spherical enlargement acts as a type of resonator and is probably responsible for the formants in the English horn tone, which are believed by some physicists to be the origin of its typical plaintive sound. Actually, the English horn is neither English nor a horn. Its name probably derives from the old French word for "angle," since the instrument has a sharp bend near the mouthpiece, and for horn, because it originally looked somewhat like an animal's horn.

The oboe is not a transposing instrument, but the English horn sounds a fifth below its written notes. The ranges of the oboe and English horn are given in Table 10-1.

10.8 *Bassoons*

If the oboe is a bit antiquated, the bassoon is a true relic of the past, having been the subject of very few significant acoustical improvements since it emerged as a member of the orchestra in the eighteenth century. The bassoon, like the oboe, is a double-reed conical bore instrument. The ratio of its length to its bore diameter is greater than that of any other woodwind instrument. Its very narrow bore makes the bassoon an acoustical nightmare, its intonation problems dwarfing even those of the oboe. Many notes on the bassoon can be easily lipped up and down by more than one half-tone, some as much as a minor third.

The bassoon is over 8 feet long, and its tube must be bent twice in order to be playable. It is fitted with a metal tube called the *bocal* on which the reed is mounted, which increases the length of the instrument and makes it more comfortable to play. The bocal is part of the narrow end of the resonant conical tube, as opposed to the bocals of some large recorders, which simply transmit the air stream from the player's mouth to the windway, and which do not have the standing waves within them. The small diameter of the bocal at the reed end, typically a few millimeters, only serves to increase the sensitivity of the instrument to variations in the effective volume of the reed cavity and can result in considerable pitch variation for a single note.

Due to the extreme length of the bassoon, it must be equipped with three different octave keys, the uppermost of which, called the *whisper key*, is mounted on the bocal near its entry to the wooden tube. Because the bassoon has not been modernized to the Böhm fingering system, the three octave keys must be independently operated by the left thumb. A significant improvement would be to interconnect the octave keys by means of levers and rings, as in the saxophone.

Several of the holes in the bassoon are not covered by rings and pads, but closed directly by the player's fingers. In order that the bassoon be playable, some holes must be located at compromise positions, modified from the ideal diameter, and drilled into the bore at large angles. Owing to the lack of a modern system of interconnected rings and keys, fork fingerings are regularly required on the bassoon. As a result of these tone-hole limitations, the bassoon is very cumbersome to play and suffers from considerable intonation and voicing defects.

The contrabassoon, which is twice the length of the normal bassoon and therefore sounds one octave lower, is the only other member of the bassoon family in use in the modern symphony orchestra. An interesting experiment in tone projection involves use of the contrabassoon doubling the part of the bass viol section of an orchestra. Even though the contrabassoon cannot be heard as a distinct entity when it plays with the basses, it gives their tone a substantial solidity, and a great difference may be noticed when the contrabassoon drops out. Perhaps this effect would be an appropriate subject for further investigation.

The bassoon is not a transposing instrument, but the contrabassoon sounds one octave below its written pitch. Their ranges are given in Table 10–1.

10.9   *Pipe Organs*

Although several large volumes have been written about the organ, we shall confine ourselves to a brief summary of some of its basic features, especially the sound-producing parts, as they relate to the woodwind instruments.

A typical pipe organ will have two or more keyboards, called *manuals*, and a single set of pedals played by the feet. A pipe organ is usually equipped with several complete sets, or *ranks*, of pipes; some large organs have 50 or 60 ranks. A single rank includes one pipe for each note on one of the manuals and, in general, each rank will be a different type of pipe. Which ranks are actuated by each manual is controlled by a set of levers called *stops*. Various stops will usually allow most of the ranks of pipes to be played by any of the manuals, and two or more manuals can be set to control the same rank of pipes.

Organ pipes operate acoustically as either fipple flutes or Renaissance capped-reed instruments, and because the air pressure is kept constant to keep the pitch level constant, there is no possibility for change in volume of any single rank of pipes. Changes of volume are obtained by two techniques. One of the manuals, called the *great* manual, adds additional ranks of pipes as a volume pedal is pushed by the foot. The *swell* manual has its pipes in a large enclosure, and as the swell foot pedal is pushed a set of baffles or dampers in front of the box is opened, allowing more of the sound to escape, and thus increasing the volume.

Two types of action, or techniques for controlling the pipes by the keys, are in common use. In *tracker action*, the key mechanism extends directly to a valve that opens the air to the pipe to be sounded, allowing direct control of the attack transients by the player. In an *electrical action*, pressing the key closes an electrical switch that operates valves to allow air flow to the pipes. Tracker action was used exclusively until the twentieth century, and many organists even today prefer the closer control and more rapid response that can be obtained in a good tracker action organ.

There are two basic types of pipes, flue pipes and reed pipes. A *flue pipe* operates by the same basic technique as a fipple flute and may be either open or stopped (closed) at the end opposite the fipple. The basic parts of the typical flue pipe are shown in Fig. 10–17. Air enters the pipe through the foot, usually positioned at the bottom of the vertical pipe. The air stream is then directed between the lower lip and languid in the case of a metal pipe, or between the cap and block in the case of a wooden pipe, onto the top lip. The top lip is sharp, as is the lip on a recorder; the noise created by the air stream striking the upper lip and the feedback

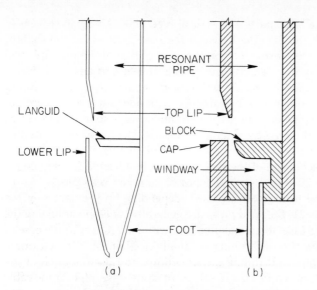

RESONANT PIPE

LANGUID

TOP LIP

BLOCK

LOWER LIP

CAP

WINDWAY

FOOT

(a)   (b)

Figure 10-17 Basic sound-producing mechanism for (a) metal and (b) wooden flue pipes.

of the oscillating air in the standing wave within the pipe then create the sound of that pipe. Air is stored under pressure in the wind chest, a large reservoir that is required to keep the pressure steady while the various combinations of notes being played are changed.

Figure 10-18 shows several types of common flue pipes. Ranks of pipes are often labeled by a length, such as 4 feet, 8 feet, 32 feet, and so on. This length refers to the approximate length of the open pipe in the rank that sounds when the key C_2 is depressed. When the 8-foot rank is engaged and the C_2 key depressed, an 8-foot long open pipe sounds. The wavelength of the note emitted is twice the length of the open tube, about 16 feet, and has a frequency equal to that of C_2 on the piano. Thus the 8-foot rank, or unison rank, plays at the written pitch. If, instead, the 16-foot rank is engaged and C_2 depressed, a 16-foot open pipe sounds. This is an octave lower than the note written. The same principle applies to all ranks: The name of the rank corresponds to the length of the open tube sounded when the C_2 key is depressed.

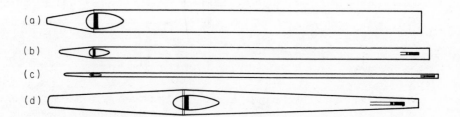

(a)

(b)

(c)

(d)

Figure 10-18 Four examples of flue pipes: (a) diapason; (b) dulciana; (c) viole d'orchestre; and (d) spitz flute. (From Audsley, *The Art of Organ Building*, 1965)

Figure 10–19 shows the basic parts of a typical reed pipe. Air again collects in the foot and passes through the opening between the reed tongue and the shallot. This causes vibration of the reed through a combination of Bernoulli's principle and feedback of pressure pulses in the standing wave, as in the case of any reed instrument. The enclosure of the reed in the air box is very similar to the situation in Renaissance capped-reed and other reed instruments, with one important exception. Since the reed is fixed to one pipe of a specific length and always sounds at the same pitch, it is common to tune the *reed* by sliding the tuning wire up or down its surface. The reed is tuned so that it will sound the proper pitch even if the pipe is not connected. The "tuning" mechanism of the pipe is generally adjusted to provide for uniform voicing. Figure 10–20 shows several types of common reed pipes.

Several types of tuning techniques are used, depending on the nature of the pipe, as illustrated in Fig. 10–21. Stopped pipes are generally tuned by adjusting the position of a metal cap in the case of a metal pipe or a wooden plunger in the case of a wooden pipe. For open metal pipes, tuning is usually accomplished by cutting a strip out of the side of the pipe and rolling the strip down to raise the pitch of the pipe to the proper level, as shown in Fig. 10–21(c). In some open pipes a movable section, added to the top of the pipe, can be slid in or out to vary the pipe length, just as the cap does for a closed pipe. Another tuning device used on open pipes is a small flap that partially covers the end of the pipe, as seen in Fig. 10–22(e) and (f). Such a flap controls the position of the pressure node at the open end.

Many metal organ pipes were tuned during the baroque by a technique known as *bell tuning*. The entire end of the pipe was tapered in to lower the pitch or tapered out to raise the pitch. This type of tuning provides the most stability over a

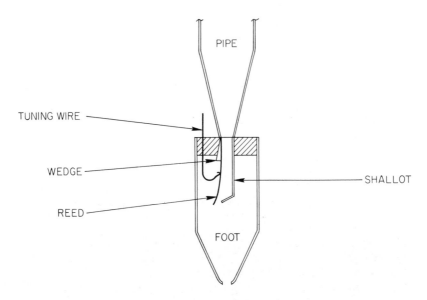

Figure 10-19 Sound-producing mechanism of a reed organ pipe.

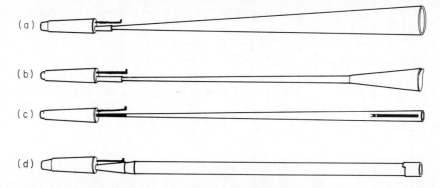

Figure 10-20 Four examples of reed pipes: (a) one type of chorus reed; (b) organ oboe; (c) orchestral oboe; and (d) clarinet. (From Audsley, *The Art of Organ Building,* 1965)

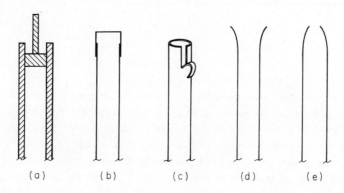

Figure 10-21 Tuning techniques for flue pipes: (a) wooden and (b) metal stopped pipes; (c) standard modern tuning for metal open pipes; and bell tuning to (d) raise and (e) lower pitch of pipe.

long time, but requires exceptional care in order that the end of the pipe not be damaged. Significant information about early pitch level comes from analysis of bell-tuned organ pipes, because in the absence of physical abuse they can remain relatively well in tune for centuries.

Because the reed in a reed pipe, and not the pipe itself, is tuned, reed pipes react somewhat differently from flue pipes to changes in temperature. A reed pipe is more able to remain in tune despite small variations in temperature than is a flue pipe, so that reed pipes will appear to be out of tune with flue pipes when the temperature changes. A change of about 5°C (9°F) will change the pitch of flue pipes by about one-sixth of a semitone, sufficient to sound badly out of tune to even most musical amateurs. In this age of energy conservation, churches and concert halls are being heated somewhat differently than they might have been otherwise.

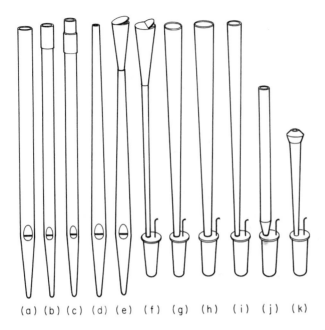

Figure 10-22 Examples of different types of pipes which sound at the same pitch: (a) open diapason; (b) salicional diapason; (c) dulciana; (d) gemshorn; (e) bell gamba; (f) oboe; (g) trumpet; (h) tuba; (i) cornopean; (j) clarinet; and (k) vox humana. The covers on (e) and (f) are used to tune the pipes. (From Audsley, *The Art of Organ Building*, 1965)

Since it can take a long period to uniformly heat or cool an entire set of organ pipes, they must sometimes be tuned at some temperature different from "normal." Many good organ tuners will adjust the pitch level at which they tune an organ on any given day so that the pitch will be exactly A = 440 at, say, 70°F.

The variety of sound obtained with different ranks of organ pipes depends on several physical features of the pipes. Flue pipes can be open, to produce all harmonics, or stopped, to emphasize the odd harmonics. In addition, the shape of the languid and lips, the pressure and volume of air, and the material, size, and shape of the pipe are important. For reed pipes, the details of reed material and shape, the tuning technique, and type of shallot all affect pitch and tone quality. Some of the different types of organ pipes that sound at the same pitch are shown in Fig. 10–22.

EXERCISES

1. Examine a flute, a clarinet, and a saxophone. Notice the similarities in their actions and fingerings, which are due to the Böhm fingering system they all use.
2. Design and construct a simple transverse flute using a piece of plastic tube with a cork or rubber stopper in one end. Design and construct a simple "clarinet," using a piece of plastic tubing and a standard clarinet mouthpiece. Be sure to consider the register key. What problems do you observe when

playing your instruments, and how do standard flutes and clarinets avoid them?

3. Play a chromatic scale on a Renaissance instrument like a krummhorn, shawm, or recorder. Notice and describe the problems in fingering and voicing. What other problems do you notice?

4. List the basic acoustical features of the various woodwind instruments discussed in the text. Which of these features do you think are most important in determining the sound quality of each instrument?

5. Examine the pipes from a pipe organ. Listen to the sounds of various ranks of pipes and relate them to the different sound-production mechanisms of the different pipes. Listen to and compare electric and tracker organs.

6. Attach a clarinet mouthpiece to a flute; play it. Describe the tone obtained and explain how this tone quality arises. Discuss intonation problems that may appear.

7. Attach a flute embouchure joint to a clarinet; play it. Describe the tone obtained and explain how this tone quality arises. Discuss intonation problems that may arise.

REFERENCES

ANDERSON, PAUL-GEHARD. *Organ Building and Design.* Translated from the Danish by Joanne Curnutt. New York: Oxford University Press, Inc. 1969.

One of the more recent books on the organ; it contains some good information on recent innovations.

AUDSLEY, GEORGE A. *The Art of Organ-Building, Volumes I and II.* New York: Dover Publications, Inc., 1965.

Contains a wealth of detailed information on organ construction and operation.

BACKUS, JOHN. *The Acoustical Foundations of Music.* 2nd ed. New York: W. W. Norton & Co., Inc., 1977.

See his reference list.

BAINES, ANTHONY. *Woodwind Instruments and Their History.* New York: W. W. Norton & Company, Inc., 1963.

A classic book on woodwind instruments written for musicians.

BARNES, WILLIAM. *The Contemporary American Organ—Its Evolution, Design and Construction.* 3rd ed. New York: J. Fischer and Bro., 1937.

The organ classic, containing a vast wealth of excellent information; easy to read and contains many excellent drawings and photographs.

BATE, PHILIP. *The Flute: A Study of Its History, Development and Construction.* New York: W. W. Norton & Company, Inc., 1969.

One in a set on the instruments of the orchestra, written for the musician.

BENADE, ARTHUR H. *Fundamentals of Musical Acoustics.* New York: Oxford University Press, Inc., 1976.

See his reference list.

————. "The Physics of Woodwinds." *Scientific American*, October 1960.

An excellent article dealing with the basic acoustical features of woodwind instruments, written by one of the foremost authors in the field.

BÖHM, THEOBALD. *The Flute and Flute Playing*. New York: Dover Publications, Inc., 1964.

The magnum opus for flutes in which the design of the flute dimensions and the Böhm ring and key system is explained.

COLTMAN, JOHN W. "Acoustics of the Flute." *Physics Today*, November 1968, p. 25.

A good general article on flute acoustics.

————. "Effect of Material on Tone Quality." *Journal of the Acoustical Society of America* 49, no. 2, part 2 (1971): 520–523.

This excellent article deals with the physics of the effect of various types of materials on the sound of the flute, including some experimental data.

————. "Flute Scales—Pitch and Intonation (Part I)." *Instrumentalist*, May 1976, p. 39.

Discusses the early history of the flute and gives background for three additional articles in the set discussing flute acoustics.

————. "The Intonation of Antique and Modern Flutes (Part IV)." *Instrumentalist*, March 1975, p. 77.

A discussion of tuning problems in the flute leading up to the proposal by Murray on how to improve the intonation of modern flutes by modifying the Böhm finger-hole geometry.

————. "Sounding Mechanism of the Flute and Organ Pipe." *Journal of the Acoustical Society of America* 44, no. 4 (October 1968): 983–992.

Deals with the edge tone mechanism, the source of sound in the flute, recorder, and organ flue pipe.

LANGWILL, LYNDESAY G. *The Bassoon and Contrabassoon*. New York: W. W. Norton & Company, Inc., 1965.

The basic volume on the bassoon, written for the musician as part of the Benn/Norton series on musical instruments.

RENDALL, F. GEOFFREY. *The Clarinet, Some Notes upon Its History and Construction*. 2nd ed. London: Ernest Benn Ltd. New York: W. W. Norton & Company, Inc., 1957.

Contains the basic historical and contemporary information on the clarinet, written for musicians.

TOFF, NANCY. *The Development of the Modern Flute*. New York: Taplinger Publishing Company, Inc., 1979.

Discusses recent developments in the flute, including the Cooper scale modification to the Böhm flute.

11

Brass Instruments

The instruments of the contemporary brass family are very similar in their acoustical properties and problems. We shall look briefly at the history of brass instruments, which contains considerably more variety than remains in the contemporary brasses, and then discuss some of the basic acoustics of brasses. Finally, we shall consider some specific modern brass instruments and survey some of the features of the remaining members of the family.

11.1 *History of Brass Instruments*

The earliest brass instruments were probably created by primitive people buzzing their lips into the end of animal horns. Many centuries ago this simple "horn" was improved by adding a specially shaped mouthpiece to one end of a tube and a flared "bell" to the other end. These ancient horns have been pictured and discussed throughout antiquity, and are, along with the harp, the instruments most often pictured with angels. Figure 11–1 shows some ancestors of the contemporary brass instruments, which we shall discuss in this section.

In the Middle Ages and early Renaissance the vast growth in the varieties of brass instruments began. Two developments of that time were particularly important, the use of finger holes and the invention of the slide.

Instruments of the *cornetto* family consist of conical tubes, usually curved, blown at the narrow end with a very small mouthpiece. These instruments had

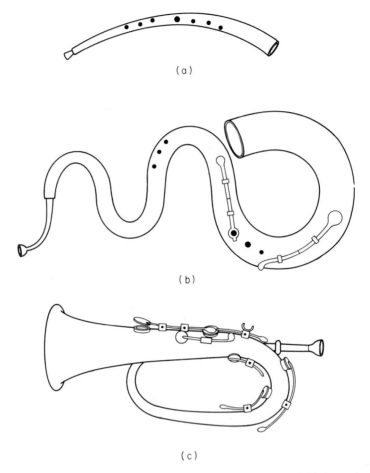

(a)

(b)

(c)

Figure 11-1 Some early "brass" instruments: (a) the cornetto, and (b) the serpent, two members of the fingered Renaissance "brasses," and (c) the keyed bugle. (Part (c) reproduced from *The History of Musical Instruments* by Curt Sachs, by permission, Inc. Copyright 1950 by W. W. Norton & Company, Inc. Copyright renewed 1968 by Irene Sachs.)

finger holes and were fingered similarly to the recorder and other woodwinds, so they are sometimes classified with the Renaissance woodwinds. Paintings of the period often show the cornetto played out of the side of the mouth, not the center as with contemporary brasses. Although this type of instrument was known as early as the eighth century, it reached its pinnacle in the late Renaissance and achieved a reputation for incredible technical capacity and flourish in the performance of typical Renaissance divisions and ornamentation. The cornetto came in a family of sizes, as did the Renaissance woodwinds. The bass instrument in the cornetto family is the serpent, which has a large bore with several curves and a broad bell. The serpent is particularly difficult to play, more so even than the small cornetto, owing to the variation in pitch that can occur when any note is fingered. These difficulties are also present to a lesser extent in the cornetto and are probably why there are very few excellent cornetto players at present.

The slide, first introduced in the fifteenth century, was used on both the trumpet-size and trombone-size instruments. The early slide trombone was called the *sackbut*, derived from the French "pull-push," and had a narrow bore, a small bell, and came in several sizes. The sackbut provided a more reliable bass than any of the woodwinds or the serpent, and consorts of cornettos (sometimes called *cornets*, but not the same as the modern band cornet) and sackbuts were very common in the late Renaissance. The slide trumpet was less popular than the sackbut, probably due to the immense popularity of the cornettos. It was used into the eighteenth century and even appeared in some of Bach's scores. The trombone as we know it today developed directly from the sackbut, with an increase in bore diameter and a larger bell, to provide a somewhat louder and more brilliant sound.

The modern trumpet, however, went through several additional steps in its evolution. Much of the trumpet music in the baroque involved a highly developed technique of playing by which an almost complete diatonic scale could be obtained without use of valves or slide. The baroque trumpet was roughly twice the length of the contemporary trumpet, so its overtones have about half the spacing of its modern counterpart when played in their normal ranges. Futhermore, the typical baroque trumpet technique called for the instrument to be played primarily in its upper octave, where the overtones are spaced as closely as a whole step over much of the range. This very high tessitura trumpet playing became a fine art in the high baroque, but was almost immediately lost with the advent of the classical orchestra, which either called for no trumpets or had the trumpets simply playing the notes of the overtone series in their lower range. The orchestral literature for the brass instruments remained about the same in style and technique until early in the nineteenth century, when the trumpet and the horn were fitted with valves. The trumpet used three valves, which lowered the pitch by two, one, and three semitones, respectively, and which could be played together to obtain the cumulative effect of lowering the pitch by six semitones. An entire chromatic scale could be obtained in this way, even in the lower register. This development dramatically expanded the capability of the trumpet and its role in the orchestra, beginning in the early romantic era.

The trumpet and trombone both use basically cylindrical tubes. As in the woodwinds, conical tubes are also used in brasses, although the acoustical difference between conical and cylindrical tubes is not as dramatic for brass instruments as it is for reed instruments. Both conical and cylindrical brass instruments support all harmonics of a fundamental. Conical reed instruments support all harmonics, whereas cylindrical reed instruments (the clarinet) support predominantly odd harmonics.

The most important early brass instrument using a conical tube was the French horn, which was given its name by the British when it was introduced to

England during the seventeenth century. The French horn is actually a modified conical hunting horn, which is, like the baroque trumpet, played in its upper range where the harmonics are closely spaced. Because the French horn is over twice the length of the baroque trumpet, it plays more than one octave lower in its normal range. French horn technique developed during the eighteenth century, and involved inserting the hand into the bell to change the frequencies of the notes; this is the source of the characteristic muffled tone quality of the French horn. This hand-in-bell stopping technique greatly extended the useful diatonic range of the French horn, and it could even be played chromatically over much of its range. The hand-stopping technique was not entirely satisfactory, however, and when the valve was invented it was immediately applied to the French horn as well as the trumpet.

An additional family of conical brass instruments arose in the nineteenth century, consisting of the modern cornet (not the Renaissance cornetto), the flügelhorn, the euphonium, and the tuba. These instruments, played normally in the lower part of their ranges, use valves to achieve complete scales and differ primarily in their length. The flügelhorn is used sometimes today in marching bands, and the euphonium is used occasionally in the symphony orchestra. The baritone, a conical horn that plays at the same pitch as the euphonium, but is slightly smaller in diameter and therefore has a slightly softer sound, is used almost solely in the concert band. The tuba is used in the symphony orchestra and the concert band, while the sousaphone, which is acoustically similar to the tuba but easier to carry, is used in concert and marching bands.

Another nineteenth-century innovation, now obsolete, was the use of keys to cover holes, particularly on the bugle. The bugle is a conical instrument, without valves, that is constructed in different sizes to form a family. Whereas the cornet and its relatives used three valves, the keyed bugle used about five keys to obtain the chromatic scale. Like the Renaissance cornetto, the keyed bugle was somewhat more difficult to play than its valved relatives, and the sound was not of the same quality and loudness. The keyed bugle became obsolete by about the end of the nineteenth century, leaving valves as the primary mechanism for obtaining the chromatic scale, along with the slide in the case of the trombone. The modern-day bugle, used in the military and drum-and-bugle marching bands, is an unvalved or a single-valved cone with a mouthpiece and wide-flaring bell. Several different sizes are used in some large drum-and-bugle corps.

There are many types of valves, though they can generally be classified under two categories, *rotary* or *piston*, as illustrated in Fig. 11–2. In a piston valve a straight piston is slid up and down so that at its depressed position an additional section of tubing extends the overall acoustical length of the instrument. When the piston is released, that section of tubing is disconnected. In a rotary valve, the motion of the finger is converted into a rotation of a valve that similarly introduces an additional length of tubing. Three main valves are used on all valved brass instruments. The first valve lowers all the notes of the open (unvalved) instrument by two semitones, the second by one semitone, and the third by three semitones. Today most French horns and some euphoniums and tubas are made with rotary valves, while the other common brass instruments usually use piston valves. While

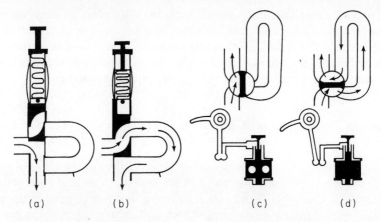

Figure 11-2 Piston valve, (a) open and (b) pressed; and rotary valve, (c) open and (d) pressed. (Reproduced from *The History of Musical Instruments* by Curt Sachs, by permission of W. W. Norton & Company, Inc. Copyright 1950 by W. W. Norton & Company, Inc. Copyright renewed 1968 by Irene Sachs.)

the valve apparently has won the war with keys as the most acceptable method for obtaining the chromatic scale on brass instruments, valves have a basic intonation defect that will be discussed in the next section.

By the middle of the twentieth century, families of brass instruments had largely disappeared, while woodwind families remained. The trumpet is used in the modern symphony orchestra, along with the French horn, trombone, and tuba, while the concert band also uses the cornet, baritone or euphonium, and sousaphone. The ranges of some of these instruments are shown in Table 11–1.

TABLE 11-1 **Range of selected brass instruments. Lower limits of pedal tones are given in parentheses.**

Instrument	Lowest note	Approximate highest note
D trumpet	$G_3^{\sharp}$	D_6
C trumpet	$F_3^{\sharp}$	C_6
$B^{\flat}$ trumpet (cornet)	E_3	$B_6^{\flat}$
Flügelhorn in $E^{\flat}$	A_2	$E_6^{\flat}$
French horn in F	B_1	F_5
French horn in $E^{\flat}$	A_1	$E_5^{\flat}$
Trombone in $B^{\flat}$	$E_2(E_1)$	D_5
Bass trombone in F	$B_1(B_0)$	F_4
Euphonium (baritone)	$B_1^{\flat}$	$B_4^{\flat}$
Tuba in $BB^{\flat}$ (sousaphone)	$B_0^{\flat}$	$B_3^{\flat}$

Many of the brass instruments are, like some woodwinds, transposing instruments. The brasses, also like the woodwinds, can be classified by their shape: cylindrical (the trumpet and trombone) or conical (the cornet, flügelhorn, baritone, euphonium, and tuba). The French horn is theoretically conical, but the modern French horn includes a long section of cylindrical tubing.

11.2   Sound Production in Brass Instruments

The woodwinds have a wide variety of acoustical properties and sounds; the brass instruments, on the other hand, have similar tones and tone production mechanisms. The only modern brass instrument that has somewhat unusual acoustics is the French horn. We shall begin with a discussion of how the buzzing of the lips into the end of a cylindrical tube produces sound and then summarize the effects of the mouthpiece and bell. Finally, we shall consider the acoustical properties of the valves and problems that occur when the valves are used in combination.

The simplest "brass" instrument, a length of straight tube, can be played by buzzing the lips into its end. The standing-wave oscillation in the tube drives the lips like the oscillating air column drives the reed in the woodwinds. When a burst of air is emitted from the lips into the tube, there is a pressure decrease in the space between the lips where the air flow is most rapid, and the lips are drawn together owing to Bernoulli's principle and lip tension. Eventually, the burst reflects back to the lips and forces them open again, allowing another air burst to be emitted from the lips. Then, the process repeats. The lip end behaves acoustically like a closed end of an air column, as a reed end does, so a velocity node (or a pressure antinode) appears there, while the bell end behaves acoustically like an open end and has a velocity antinode (or a pressure node). The tube behaves acoustically like a closed tube, with the fundamental mode having a wavelength approximately four times the length of the tube.

We can perform a series of experiments in which we first blow into a length of plastic tube alone, and then repeat the experiment after inserting a trumpet mouthpiece onto one end of the tube and a metal funnel to the other end, as shown in Fig. 11–3. Here we shall use a tube about 130 cm long, producing a fundamental

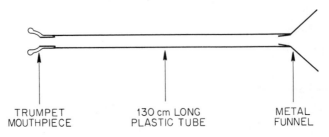

Figure 11–3   Simple experimental "trumpet."

frequency of about 66 Hz. The possible resonant modes in the unmodified tube are the odd harmonics, as shown in Fig. 11-4. Addition of the mouthpiece to one end and the funnel to the other end adjusts the frequencies of the harmonics as shown in Fig. 11-5. The result is to increase slightly the effective length of the tube and to cause the instrument to behave acoustically like an open tube of the new length. It then produces a set of resonances including *all* the notes of the overtone series. The fundamental mode of this new overtone series is slightly lower in frequency than twice the fundamental frequency of the original tube.

The size and shape of the mouthpiece affects predominantly the frequencies of the upper harmonics, causing the harmonics to decrease in frequency, and thus reducing their spacing. The bell has two important effects: (1) it provides the proper match between the instrument and the outside air to transfer the sound most efficiently, making the sound considerably louder, and (2) it modifies the frequencies and stability of the harmonics, especially the lower harmonics. The fundamental is extremely difficult to play, and its frequency is too low relative to the other notes of the overtone series produced by the tube plus mouthpiece plus bell system. The effect of the bell on the lowest playable note, the second harmonic, is to raise its pitch from that of the third harmonic of the original closed tube (G) to that of the second harmonic of the final overtone series (B$^\flat$) in Fig. 11-5. The third harmonic note (F) is also moved slightly upward by the bell. By varying the size and shape of the mouthpiece and the flare and length of the bell, the frequencies of the resonances can be adjusted to most nearly match the harmonics of the overtone series, except the fundamental. The player can adjust the frequencies of the notes by varying the lip tension to tune the notes.

The most obvious effect of the bell is to increase the volume of sound from the instrument. The bell provides, in technical terms, a good "impedance match" or coupling for conducting the sound vibrations in the horn to the surrounding space

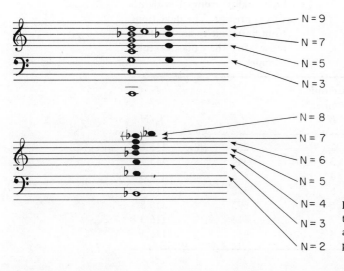

Figure 11-4 Black notes are those obtained by buzzing lips into one end of a 130-cm tube. All the notes of the complete overtone series of the fundamental of a 130-cm long closed tube are shown to the left.

Figure 11-5 Notes obtained when a trumpet mouthpiece and a metal funnel are inserted into the ends of the 130-cm plastic tube of Fig. 11-3.

most efficiently. This is very clear in the case of our plastic tube described previously. In addition, the bell helps to project the sound in a single direction.

The frequencies of the notes produced by a high-quality trumpet are given in Table 11–2. Although they are neither perfect harmonics nor notes of the equal-tempered scale, they are considerably closer to exact harmonicity than the resonant frequencies of our simple plastic tube with mouthpiece and funnel. This is because the size and shape of the mouthpiece and bell of the trumpet have been carefully chosen to provide the closest approximation to the notes of the harmonic series. It is clear that to play a brass instrument in tune (even without using the valves) it is necessary to listen carefully and adjust the pitch of the notes during performance.

We now have developed a brass instrument that can play only the notes of the overtone series. To play the complete chromatic scale, valves must be added. Pressing a valve increases the effective length of the instrument by introducing an extra section of tube. Each valve introduces a different length of tubing and changes the pitch accordingly. First, consider the effect of each valve individually, assuming that the open notes (no valves pressed) have perfect intonation in the equal-tempered scale. (They do not, as can be seen by reference to Table 11–2.) Table 11–3 shows the frequency ratios of the notes of a valved instrument if the open notes are perfectly in tune in the equal-tempered scale. Also shown in the table are the relative frequencies of the notes produced by pressing each valve individually. Table 11–4 gives the number of half-steps the pitch is lowered for various valve configurations. To lower the pitch by one half-step, a length of tube equal to 5.95 percent of the length of the open tube must be added; this is done with the second valve. The first valve lowers the pitch by one full step or two half-steps by adding 12.2 percent to the length of the open tube. The third valve adds 18.9 percent to the

TABLE 11–2 Frequencies obtained with a high-quality trumpet with no valves pressed. The deviations from harmonicity are less than those in the plastic tube trumpet.

Harmonic number N	f_n	f_n/N
2	230	115.0
3	344	114.7
4	458	114.5
5	578	115.6
6	695	115.8
7	814	116.3
8	931	116.4

TABLE 11-3   Frequency ratios obtained on valved brass instruments in the ideal case where (a) the three valves are individually made to the correct length, (b) valve 1 + 3 combination is made correct, and (c) valve 2 + 3 combination is correct. The frequency ratios in the equal-tempered scale are shown in the second column. Boxes indicate notes that are out of tune.

Note	Ratio equal-tempered scale	Valve configuration	Valve configurations tuned exactly		
			(a) 1, 2, 3	(b) 1, 2, and 1 + 3	(c) 1, 2, and 2 + 3
C	2.0000	0	2.0000	2.0000	2.0000
B	1.8877	1	1.8877	1.8877	1.8877
B♭/A♯	1.7818	2	1.7818	1.7818	1.7818
A	1.6818	1 + 2	1.6922	1.6922	1.6922
		3	1.6818	1.6496	1.6660
A♭/G♯	1.5874	2 + 3	1.6017	1.5725	1.5874
G	1.4983	0	1.4983	1.4983	1.4983
		1 + 3	1.5248	1.4983	1.5118
G♭/F♯	1.4142	2	1.4142	1.4142	1.4142
F	1.3348	1	1.3348	1.3348	1.3348
E	1.2599	1 + 2	1.2677	1.2677	1.2677
		3	1.2599	1.2358	1.2481
E♭/D♯	1.1892	2 + 3	1.1999	1.1780	1.1892
D	1.1225	1 + 3	1.1423	1.1225	1.1326
D♭/C♯	1.0595	1 + 2 + 3	1.0927	1.0746	1.0839
C	1.0000	0	1.0000	1.0000	1.0000

length of the tube to obtain a pitch reduction of three half-steps or a minor third. If it were only necessary to use one valve at a time, each could be made the proper length to produce its notes exactly in tune with respect to the open notes by choosing the proper length of tube. However, there is a problem when the valves are used together.

If the first valve is pressed, lowering the pitch of an open note by two half-steps, the second valve is used to obtain the note an additional half-step lower. To lower the pitch by this additional half-step, however, the valve must add 5.95 percent to the length of the open horn plus the length of the first valve. Because the second valve is already made equal in length to exactly 5.95 percent of the open horn, it is therefore short by 5.95 percent of the length of the first valve, or about 0.7 percent too short. Thus, with valves one and two pressed, the instrument would produce notes about 0.7 percent high in frequency, about one-eighth of a half-step; this

TABLE 11–4 **Effect of trombone slide position and trumpet valve configuration on pitch of overtone series produced by the instrument.**

Trombone slide position	Trumpet valve combination	Pitch reduction in half-steps	Musical interval
1	0	0	Unison
2	2	1	Half step
3	1	2	Whole step
4	1 + 2	3	Minor third
5	2 + 3	4	Major third
6	1 + 3	5	Perfect fourth
7	1 + 2 + 3	6	Augmented fourth

can easily be detected by a good musician. The notes obtained by valves 1 and 2 in combination, which are high in pitch, could be produced exactly by using valve 3 alone if it were made to the exact length. If valves 2 and 3 were used together, the notes would be 5.95 percent of 18.9 percent high, or about 1.1 percent high, over one-sixth of a semitone. If valves 1 and 3 were used together, the notes would be over one-third of a semitone high; if all three valves were used simultaneously, the notes would be over one-half a semitone high in pitch.

Table 11–3 shows the frequency ratios obtained for different valve tuning compromises. For instance, the column for valves 1, 2, 3 all tuned exactly shows the ratios of frequencies for an ideal instrument with valves that individually lower the pitch from the unvalved or open configuration by exactly a whole step, a half-step, and a minor third. As we have seen, any valve combination will then deviate from the equal-tempered scale. The column for 1, 2, and 1 + 3 shows the ratios of frequencies obtained when the first two valves are tuned individually, but the third valve is tuned so that, in combination with valve 1, the interval of a tempered perfect fourth is obtained. The various tuning schemes produce different notes in tune.

This problem could be approached in several other ways. First, the brass player must adjust the pitch of the notes by changing lip tension. This will be necessary, anyway, because even the open notes on the instrument are not exactly in tune. However, the errors introduced by the valve length discrepancies become so large that lipping the note in tune is very difficult or impossible, and cannot be used. The tube sections inserted by the individual valves could be made longer so that each is low in pitch when used alone, but more nearly in tune when used in combination, but this is usually not done. A third alternative involves moving an additional slide out when the valves are used in certain combinations to lower the pitch of those notes, particularly when the third valve is used. Another possibility is to provide a fourth valve, which is actually an elongated third valve, to be used in lieu of the third valve whenever it is to be used in combinations, or to provide an additional valve that lowers the pitch of the instrument by a perfect fourth. The

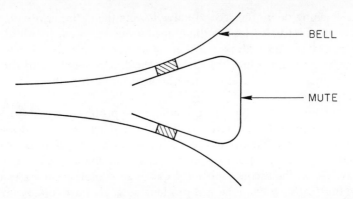

Figure 11–6 A mute for a brass instrument, in position in the bell. The cross hatches indicate the positions of small cork retainers, used to hold the mute in place.

corrections most often used will be discussed in the sections dealing with the various instruments.

Brass instruments, like woodwinds, are made in two basic bore shapes. The trumpet and trombone are cylindrical throughout much of their length, with the end section gradually flaring to the bell. The other common brass instruments are flared throughout, and can be considered approximately conical. One effect of the flare, in the bell region for all brasses, and throughout the instrument for the conical brasses, is that higher-frequency harmonics reflect preferentially within the instrument. Thus, for conical instruments with considerable flare, the higher harmonics are more confined within the instrument and never reach the bell. Therefore, the tone of the instruments with greater flare is lacking in higher harmonics, and they sound somewhat more mellow than their cylindrical counterparts.

A mute, shown in Fig. 11–6, is used to muffle the sound of many brass instruments in a way similar to a French horn player inserting his hand into the bell. Because the mute leaves only a small opening at the end of the instrument, the higher harmonics pass out of the instrument more readily than do the lower harmonics. Furthermore, the mute has a certain volume and acts as a Helmholtz-type resonator emphasizing harmonics in some high-frequency region. Different types of mutes can produce somewhat different sounds.

11.3 *Trumpets*

The trumpet is the treble instrument in the contemporary brasses and is usually built as a transposing instrument. While the trumpets in C and D were popular in the baroque, the B♭ trumpet is normally used in bands and orchestras today, with the C trumpet used less often. In a B♭ trumpet the notes sound one whole step lower than written; that is, a C played on a B♭ trumpet sounds at B♭ on the piano, and so on. Pushing down the second valve lowers the entire harmonic series by one half-step; the first valve lowers the notes by one full step; the third valve alone or the first and second valves together, a minor third; the second and third valves together, a major third; the first plus third, a perfect fourth; and all three together, an augmented fourth. The notes obtained for each of these valve configurations are given in Fig. 11-7.

If the three valves were made the appropriate length to be in tune individually, the combination notes would be very sharp, as we have seen. Therefore, only the first and second valves are made the correct length to be exactly in tune when used individually, and the third valve is often made of such a length that the notes played with the second and third valves in combination are exactly in tune. Thus the third valve used alone will produce notes that are too low in pitch, so the third valve is not used as an alternative for the first and second valves in combination. The notes produced by the first and second valves in combination are still sharp, however, and must be adjusted in performance. The intonation of this configuration of valve lengths is shown in the right column of Table 11-3.

The notes produced when the first and third valves are used together, and especially the notes produced when all three valves are used together, are still quite sharp and require special treatment. Most trumpets are equipped with a tuning slide for each valve, which can be used to adjust the length of tube added to the horn when that valve is pressed. In general, these tuning slides are not adjusted after they are set to their proper positions. However, many trumpet players will have their third valve tuning slide fixed so that it moves very easily with one of the fingers on the left hand, which is normally used only to hold the instrument. This slide can then be pushed out for notes using either the combination of first and third valves or all three valves, when these notes are to be held for a long time and the intonation defect is likely to be noticed. Most trumpets of less than professional quality will not have a third valve tuning slide that can be moved easily, so it is usually necessary to align the two tubes of the slide so that it can be moved easily during performance. Repair shops are usually familiar with this standard trumpet modification. The frequency ratios of notes played in the lower octave of a trumpet with the third valve length chosen in the preceding manner are shown in Table 11-3.

The assumption in Table 11-3 that the open notes (the overtone series) are in tune is not true. The third is low compared to the note of the equal-tempered scale,

VALVES PRESSED ⟶ 0   2   1   1+2   2+3   1+3   1+2+3
                              or 3   (all)

Figure 11-7 Notes which can be played using each of the seven trumpet valve configurations.

while the fifth is slightly high. Each of the series of notes for a given valve configuration deviates from the equal-tempered scale in this way. A trumpet player must compensate during performance for these intonation defects using a combination of lip pressure and alternate fingerings.

The range of the orchestral B$^\flat$ trumpet is given in Table 11–1.

11.4   *Trombones*

The trombone is one of the simplest brass instruments, acoustically and mechanically, and the only brass or woodwind instrument that can be played perfectly in tune without requiring adjustment of the player's embouchure. The slide of the trombone replaces the valves of the trumpet, giving infinitely adjustable pitch control. Continuous sliding notes or glissandos can be played on the trombone. There are seven approximate slide "positions" on the trombone, shown in Fig. 11–8. The correspondence between slide position and valves pressed on a valved instrument is shown in Table 11–4. The trombone, like the trumpet, is a cylindrical instrument, and because the trombone has an effective length twice that of the trumpet, it sounds one octave lower in first position than the trumpet with no valves pressed.

Whereas the trumpet has a usable range down to the second harmonic for each valve configuration, the larger size of the trombone mouthpiece permits the

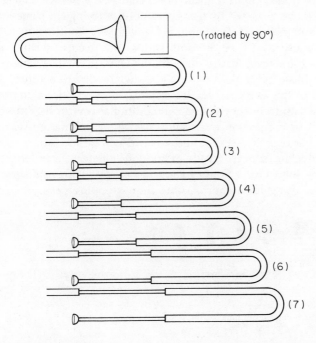

Figure 11–8 The seven slide positions on the trombone. The increase in slide length between positions 6 and 7 is approximately double that between positions 1 and 2.

lips to vibrate at the fundamental frequency of the trombone. The very low notes, known as *pedal tones*, are quite stable and usable. The notes playable by a trombone in each of the seven positions, including the pedal tones, are shown in Fig. 11–9.

Some modern professional-quality trombones can be shortened slightly above first position during performance by pulling the slide in against the force of a spring. This allows the player to obtain a fifth harmonic (the D above middle C), which is sharp with respect to the exact harmonic of the overtone series but in tune in the equal-tempered scale.

The trombone is not a transposing instrument. The standard orchestral and band instrument is called the tenor or B♭ trombone, since the fundamental note of the instrument in first position is a B♭. The bass trombone, playing a fourth lower, is either built larger by a factor of one-third or equipped with a valve, played with the left hand, that introduces a section of tube to lower the pitch by a musical fourth. In this case the slide positions must all be lengthened to account for the additional length of the tube in first position.

An interesting effect, called *horn chords* or *multiphonics*, can be obtained by simultaneously playing and singing into a low brass instrument; while it can be done on several of the brasses, it has been most popular on the trombone. By playing one note and simultaneously singing another note that is a member of the same overtone series, a third note is produced. The extra note is the note whose frequency is the sum of the two original frequencies. The third note is nearly as loud as the two original notes, forming a chord whose quality is determined by the tone quality of the two original notes. Some typical horn chords are shown in Fig. 11–10. Additional chords can be obtained by using two notes forming the same intervals, so the harmonics obtained by this technique are virtually limitless.

Horn chords are produced as a type of combination tone created when the two original notes and their harmonics mix in the standing wave in the instrument. They are not due to a nonlinearity of the ear, as were the combination tones of Chap. 6. The combination of the two complex waves creates an additional set of harmonics belonging to the sum frequency of the two additional waves. If the two original notes are members of a common overtone series, the combination wave will also be a member of that overtone series, and a chord will be formed.

Some well-known traveling concert bands of the late nineteenth and early twentieth centuries, like the John Philip Sousa band, had players who could perform remarkable harmonies using horn chords. One show-stopping performance

Figure 11–9 Notes which can be obtained in the seven trombone slide positions.

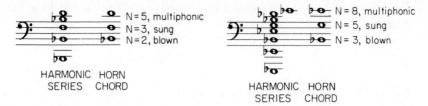

Figure 11–10 Examples of horn chords, or multiphonics. The notes of the relevant harmonic series are shown to the left of each chord. Blowing the lowest note of the chord, while singing the middle note, produces the multiphonic shown for each case.

involved a trumpet player doing sets of very complex variations, accompanied by a single trombone player playing horn chords. Horn chords or multiphonics have been used occasionally in serious contemporary music for the trombone.

The ranges of the tenor and bass trombones are given in Table 11–1.

11.5 *French Horns*

The French horn is the member of the brass family whose physics is somewhat different from that of the other members. The French horn bore is cylindrical over much of its length, but the flare of the bell is relatively large. Relative to the other brass instruments, the normal range of the French horn is higher in its overtone series. During the eighteenth century, before the invention of the valve, a player could change the pitch of the French horn by over a half-step by properly inserting a hand into the bell. This created new notes and allowed accurate tuning of all the notes in the normal range of the instrument. In addition, this hand-stopping technique increased the stability of some of the notes. Horn players of today still follow this tradition of hand-stopping.

Although nearly a complete diatonic scale can be obtained on the French horn without use of valves, its rotary valves improve intonation and permit easier changes from one note to another. Compared with other brass instruments, the small mouthpiece and closely spaced harmonics of the French horn make it difficult to play any desired note. Large leaps to high notes are particularly treacherous. Because of this, as well as the problem of exactly how far into the bell the hand should penetrate, it is very important for the horn player to listen carefully to pitch and tone quality and adjust appropriately.

The modern double horn is equipped with an additional valve, to insert additional tubing and put the horn in either the key of F or the key of $B^\flat$, a fifth lower. This flexibility would introduce additional intonation problems, since the valves cannot be the correct length for both horns. Therefore, two separate sets of valve tubing are provided, one for each horn (F or $B^\flat$). The set of three valve tubes to be used is automatically determined by whether the $B^\flat$ horn valve has been pressed.

Performers still must compensate by adjusting embouchure and the hand-stopping position.

The range of the French horn is given in Table 11–1.

11.6 *Other Contemporary Brass Instruments*

Several other members of the modern brass family merit additional discussion; there is great variety in the details of how brass instruments are constructed and used. The trumpet is the principal high voice for the brass choir. However, the modern cornet, a conical instrument similar in almost every other respect to the trumpet, is often used in bands. Because of its flare, the cornet has relatively less amplitude in its upper harmonics and thus a somewhat more mellow tone. The flügelhorn is the alto member of the cornet family.

Additional intermediate low brasses are the baritone and the euphonium. Whereas the trombone is basically cylindrical, the baritone and euphonium are basically conical and must use valves to change their overtone series rather than slide positions. All three of these instruments have virtually the same range, but the trombone has a more brilliant tone owing to its cylindrical shape. The euphonium has a somewhat greater flare and larger bore than the baritone. It therefore produces a bigger, fuller tone. Because a large bore exacerbates the valve tubing length problem, the euphonium suffers from this problem more than the baritone does. To compensate, most modern euphoniums are equipped with a fourth valve, which lowers the pitch of the instruments by a perfect fourth and provides some alternate fingerings to improve the intonation of notes requiring valve combinations including the third valve.

The tuba is generally pitched one octave below the trombone and euphonium; it has a very large bore and flare. The intonation problems of tubas are much greater than for the higher brasses, and some tubas are therefore equipped with two additional valves, the fourth valve being similar to the fourth valve on a euphonium. To play in tune, most tuba players find it necessary to not only use all valves in exotic combinations, but also to be continually adjusting several of the individual valve tuning slides.

The ranges of some of these instruments are given in Table 11–1.

EXERCISES

1. Classify the brass instruments commonly used in the band and orchestra by range and by bore shape (cylindrical or conical). Discuss the differences in tone between instruments of the same range that can be attributed to bore shape and size.

2. Attach a trumpet mouthpiece to a section of rigid plastic tube about 40 cm long. Drill finger holes, as if making a flute, and play it like a Renaissance cornetto. Describe how the instrument plays, and discuss some of its acoustical problems.
3. Blow a length of rigid plastic tubing by buzzing your lips into one end without a mouthpiece; measure the resulting frequencies with a frequency counter or by comparing them to the notes on the piano (read the frequencies from Appendix A). Calculate the frequencies expected for a closed tube of this length. Why does this tube behave acoustically like a closed tube? Do the calculated and experimental frequencies agree? What frequencies would be expected if a trumpet mouthpiece were added to one end and blown in the same manner?
4. Make a "trumpet" by attaching a trumpet mouthpiece to a long section of tubing or hose. Add a bell made from a metal funnel. Measure the frequencies of the harmonics, and discuss the intonation problems that would be met when playing this instrument with an organ tuned in equal temperament. Play the trumpet with and without the funnel, and discuss the effects of the bell. Calculate the expected frequencies of your trumpet, based on its measured length. How well do the calculated and experimental frequencies agree?
5. Discuss the valve tubing length problem. What compromises can be made with regard to it?
6. Examine the horn chords given in Fig. 11–10. Using your knowledge of the overtone series, determine the notes that you must sing and play to produce other horn chords. Try producing some horn chords on a trombone or a baritone.

REFERENCES

BACKUS, JOHN. *The Acoustical Foundations of Music.* 2nd ed. New York: W. W. Norton & Co., Inc., 1977.

See his reference list.

———, and T. C. HUNDLEY. "Harmonic Generation in the Trumpet." *Journal of the Acoustical Society of America* 49, no. 509 (1971): p. 309.

A somewhat detailed description of sound production in a trumpet.

BAINES, ANTHONY. *Brass Instruments: Their History and Development.* London: Faber & Faber, Ltd., 1976.

An important reference on brass instruments, written for musicians.

BATE, PHILIP. *The Trumpet and the Trombone.* New York: W. W. Norton & Company, Inc., 1966.

Describes the basic brass instruments in terms easily readable by the musician.

BENADE, ARTHUR H. *Fundamentals of Musical Acoustics.* New York: Oxford University Press, Inc., 1976.

See his reference list.

————. *Horns, Strings, and Harmony.* Garden City, N.Y.: Doubleday & Company, Inc., 1960.

Contains some elementary but excellent qualitative concepts on how the trumpet works.

————. "The Physics of Brasses." *Scientific American,* July 1973.

Contains considerable detailed information on the nature of the trumpet and how its sounds are produced, as well as some of their characteristics.

MORLEY-PEGGE, R. *The French Horn, Some Notes on the Evolution of the Instrument and of Its Techniques.* New York: Philosophical Library, Inc., 1960.

Describes the basic features of the French horn and its history, written for the horn player.

12

String Instruments

The history of string instruments is as diverse as the history of music itself. Surprisingly, out of the incredible variety of historical string instruments, only one family of four sizes of violins and few varieties of plucked strings are all that can be called "contemporary."

The violin family is the sole surviving family of bowed strings, while the guitar (acoustical and electronic) and harp are the main plucked strings in modern use, although the banjo, ukelele, and dulcimer are often used for folk and country music. In addition to developing in their own right, the historical string families laid the foundation for the clavichord, harpsichord, and their modern descendant, the piano.

In this chapter we shall first summarize the history of some string instruments. We shall then discuss the basic theory of bowed strings and apply this to the contemporary violin family. Finally, we shall briefly look at the acoustics of some plucked string instruments.

12.1   *History of String Instruments*

Plucking a stretched fiber to obtain a musical tone is probably as old as music. As early as several centuries B.C. artists made drawings of string instruments, many similar to the lyre and the harp. Pythagoras and other Greek philosophers discussed the sounds of a stretched string, as in the monochord, and the effects of stopping the string to obtain musical intervals. These intervals could be consonant or dissonant, depending on the ratio of lengths of the two strings; the very concept of consonance was largely seen as a matter of philosophy and mathematical relationships.

By the medieval era, the harp and lyre had become associated with the music of heaven and were regularly depicted as being held or played by angels. Additional plucked string instruments had evolved, using primarily animal shells, skulls, or other naturally occurring objects as the resonators, over which the wires or gut strings were stretched. A modern-day descendant of such an instrument is the charango, a Peruvian folk instrument that uses an armadillo shell as its resonator.

During the medieval era two new techniques evolved for creating the sound in a string instrument: hammering and bowing. In the hammered dulcimer, small hammers with heads of leather or a similar material were used to strike the strings. This technique was later used in the clavichord and finally adapted to the piano. The technique of bowing or scraping the strings took two directions. The hurdy-gurdy, or organistrum, used a disc that was turned by a hand crank so that its edge scraped the strings. The hurdy-gurdy often had three strings, two of which might be drones, while the third could be stopped, using a set of keys, to play a melody; its sound was similar to that of the bagpipe. Early hurdy-gurdy playing required two people, since the instrument was some 5 feet long, with the crank at one end and keys along one side.

A group of bowed string instruments similar to the later violin family also evolved around the latter part of the twelfth century. They generally had from one to six strings, with four being a popular compromise; most were unfretted. Among the more popular of these were the rebec and the medieval fiddle or vielle.

An important technical development during the medieval period was that of improved woodworking capabilities. The new woodworking facilities allowed construction of a wide-ranging variety of new string instruments, like the rebec and fiedel, with superior physical properties. The lute was probably the most important of the new families of string instruments. The lute was a loose descendant of the Arabic ud, a plucked string instrument with many strings. The word ud, and its relative, lute, come from the Arabic word for wood, and the instruments are so named because they were among the earliest with wooden resonating boxes, in contrast to those made with natural resonators like animal shells. The lute had between 6 and 24 strings, which, with the exception of the highest string, were tuned with adjacent pairs in unison. Despite its limitations, the lute was destined to become by far the most important string instrument of the European Renaissance, retaining its vast popularity for almost 500 years.

No actual medieval string instruments are extant; the primary source of information on them is works of art in which they are depicted. One type of medieval fiddle is known largely from a single painting done by Hans Memling (ca. 1480) on the case of a Spanish organ. One of about ten figures in the painting depicts the playing of a fiedel in some detail, including the bow and bowing technique, and the way in which the instrument was held, as well as some construction details of the instrument and the bow. The same paintings show the psaltery, tromba marina or nun's fiddle, lute, and harp, as well as several reed and brass instruments and the portative organ.

During the Renaissance several new string instruments were developed, usually appearing in families of three of four sizes, and often in many varieties. Most of the new varieties of bowed strings were fretted, as is the modern-day guitar. The viola da gamba, perhaps the most popular family of Renaissance bowed string instruments, was held between the knees and bowed in the style of contemporary French bass violin. Gambas had six strings, came in three sizes, and were often played in groups of up to six voices. The viola da braccio was played in a manner similar to that of our modern violin. The viola d'amore was unique in possessing a set of strings that were not bowed, but vibrated sympathetically as the bowed strings sounded.

In addition to the development of the lute, several other varieties of plucked strings arose. The guitar took roughly its contemporary form. The cittern, which had an oval belly and back, and pandora (or bandora), which had a three-lobe, scallop-shaped body, were prominent throughout Europe, and the vihuela became popular in Spain for a lengthy period.

During the Renaissance the keyboard was developed to its present form and applied to two types of string instruments. In the clavichord, the key controls a tangent, which strikes the string. In the harpsichord the key controls a plectrum, which plucks the string. Use of the harp was expanded, while many other string instruments fell into oblivion when they could not meet the increasing technical demands or requirements for expression.

During the late Renaissance and early baroque, many changes occurred in the popularity of various string instruments. For centuries, the lute had been popular as both a melody and accompaniment instrument and could be called upon to play or double any instrumental or vocal line. Despite its immense popularity, it had overwhelming technical problems, particularly in tuning. It used gut strings, which are very sensitive to temperature and humidity variations, and which tend to stretch easily and therefore to become flat in pitch rather quickly after they are tuned. There is a saying of the time to the effect that "if the lutenist shall attain the age of ninety, he shall have spent seventy of those years tuning his lute." With the advent of reliable harpsichord tuning and action, the harpsichord often replaced the lute by playing chords to accompany solo singers, instruments, or small ensembles. The greater technical capacity of keyboard string instruments also led them to supplant other plucked strings in popularity.

A similar change occurred in bowed strings. The desire for greater expression in early baroque music required a rather more versatile instrument than the viol families. In response, the violin family was developed. The new instruments were played under the chin, and, being unfretted, could be played somewhat more expressively than the viols. They also had generally greater technical capacity; for instance, they could use vibrato, glissando, and harmonics. An attempt was made to prolong the life of the viols by removing the frets and modifying their bowing technique, but they nevertheless were rapidly supplanted by the violin family. One remnant of the viol family is the string bass, which is in many respects more like a member of the viol family than the violin family. It is tuned in fourths, as are the viols, rather than fifths, as are the members of the violin family, so that scales can be played with minimal shifts of hand position. It has sloping shoulders and is generally shaped like a viol, and its bowing technique is more like that of the gamba than that of the cello. In fact, the string bass is sometimes called the *bass viol.*

During the baroque most plucked instruments decreased in popularity, as did the lute. The guitar and, later, several other plucked instruments, such as the banjo, ukelele, and dulcimer, increased in popularity, and to this day are prominent in folk, dance, and country music. The only plucked instrument to be used in the contemporary orchestra is the harp, which has grown to massive proportions in comparison with its medieval ancestor. The guitar is growing in use in serious music, especially in chamber and solo compositions.

Two additional developments affected the acoustics of the violin family. First, the orchestra grew substantially from the baroque to the present, increasing the sound of each of the string sections. Second, the instruments required significant modifications as a result of the rise in pitch over the past three centuries, which gave them a more brilliant sound. These changes will be discussed in the section on the violin family.

A significant recent development is the use of electronic pickup and amplification in instruments like the guitar. The *acoustic* guitar simply uses the vibration of the instrument as its source of sound, while the *electric* guitar makes use of special pickup devices called *transducers* that convert the sound into electronic signals, which are amplified to drive a loudspeaker.

Figure 12–1 shows some of the early string instruments that acted as predecessors to our contemporary bowed and plucked string families. Table 12–1

TABLE 12–1 **Ranges of some bowed and plucked string instruments.**

Instrument	Open strings	Lowest note	Approximate highest note
Violin	G_3 D_4 A_4 E_5	G_3	E_7 $(A_7)^b$
Viola	C_3 G_3 D_4 A_4	C_3	A_6 $(D_7)^b$
Cello	C_2 G_2 D_3 A_3	C_2	E_5 $(G_6)^b$
Double bass	E_1 A_1 D_2 G_2	E_1 $(C_1)^a$	D $(F_5)^b$
Treble viol	D_3 G_3 C_4 E_4 A_4 D_5	D_3	D_6
Tenor viol	G_2 C_3 F_3 A_3 D_4 G_4	G_2	G_5
Viola da gamba	D_2 G_2 C_3 E_3 A_3 D_4	D_2	D_5
Harp	(All)	C_1^b	$G_7^{\sharp}$
Guitar	E_2 A_2 D_3 G_3 B_3 E_4	E_2	A_5

[a] Denotes with use of extender.
[b] ( ) Denotes upper ranges using harmonics.

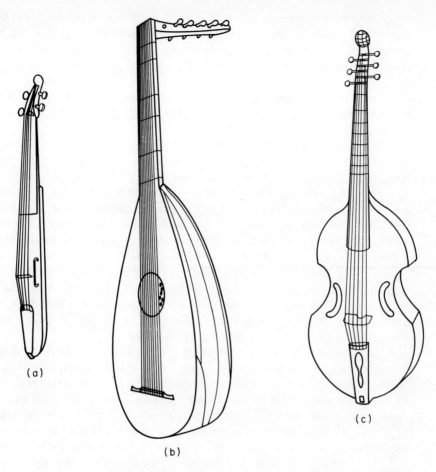

Figure 12–1 Some early string instruments: (a) the rebec, an unfretted medieval bowed instrument with four strings; (b) the lute, a fretted Renaissance plucked instrument with a large number of strings; and (c) the viola da gamba, a six-stringed, fretted, bowed Renaissance instrument which came in three sizes. (Drawings are not to scale.)

gives the normal ranges for the instruments in the violin family and the notes of their open strings, along with the ranges of selected string instruments, both bowed and plucked.

12.2   *Theory of Bowed Instruments*

The basic sound-producing mechanism in all string instruments is virtually the same; the only significant difference is that some are bowed and others are plucked to obtain the original source of the vibration. After the vibrations in a string are produced, these vibrations are transmitted by a bridge to the belly and back of the instrument, causing these large plates, generally made of wood, to oscillate with the frequencies present in the vibrating string. The bridge must act as an efficient coupler between the string and the body, and if the body is to resonate effectively there must be actual resonant frequencies in the plate similar to the resonances in Chladni plates discussed in Chap. 3. Furthermore, it is desirable to have low-frequency resonances in the wooden plates to help produce a full tone in the low range.

The object of the instrument is to transfer the sound of the vibrating string and wood to the air, so there must be an effective means for this coupling. The front and back plates of the instrument are set in vibration so they alternately force air out of and into the hole or holes cut in the front plate. Each instrument has its own geometry for its hole or holes. The hole is circular in the case of the guitar, while in the violin family there are two holes, called f holes because of their shape. The sizes and shapes of the holes and body determine a resonance similar to that of a Helmholtz resonator. In the case of most string instruments this resonance is rather broad, although it is isolated as is the primary resonance of a Helmholtz resonator. The frequency of this Helmholtz-type air resonance is usually in the lower range of the instrument, so it will act to improve the tone and intensity of the instrument in its lower range. This effect can be viewed as similar to the low-frequency air resonance in a tuned-port loudspeaker.

In general, the string instruments of today have developed with these characteristics. Those instruments lacking any of these characteristics produced a less than adequate sound for the music of the time and were rendered obsolete. For example, the medieval rebec has a very small air volume and small wooden plates. As a result, the air and wood resonances are too high relative to the normal range of pitches at which the instruments play, and the tone is too nasal and squeaky for modern taste. Thus, the violin family, which has survived as the bowed string family of the orchestra, is the family that has the best set of air and wood resonances to produce good tone over their entire range.

Let us look in more detail at the mechanism by which the drawn bow causes vibrations in a string. The strands of the bow are generally made of hair from the tail of a horse, which is scaly and grips the string during bow strokes. The bow hair is coated with resin, further increasing the friction between bow and string. As the bow is drawn across the string it tends to grab the string, pulling the string along at the speed of the bow. At some point the tension in the string due to its displacement will be too much and will overcome the frictional force between the bow and the string. At this point the string will snap rapidly back to a point on the other side of its equilibrium position. The bow then begins to pull the string back in the original direction. As time passes, the displacement of the point on the string at which the bowing occurs is thus approximately the shape of a sawtooth wave. The frequency with which this alternating motion of the string occurs is controlled by the length, tension, and size of the string, according to Mersenne's laws, as discussed in

Chap. 3; it is exactly the fundamental frequency of the string, for normal bowing pressure and bow velocity. If too light a bowing pressure is used, the tension in the displaced string will overcome the frictional force more quickly, and the string will snap back sooner. Its frequency, again controlled by the resonances of the string, becomes that of the second harmonic. This is why inexperienced string players, playing timidly and without sufficient bow pressure, often produce squeaky notes an octave too high.

The string actually vibrates in a rather complex manner, more complex than our simple idea of the sum of simultaneous additive motion at the frequencies of the harmonics, as described in Chaps. 3 and 4. Instead, the shape of the string remains roughly triangular, with the vertex of the triangle tracing out an envelope that looks like the string vibrating in its fundamental mode. This is illustrated in Fig. 12-2. This is a type of standing wave, which, in the case of the violin string bowed

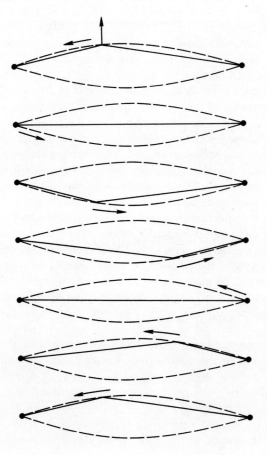

Figure 12-2　A string bowed at the point shown at top vibrates in the unusual "standing wave" shown here. The bow creates a triangular displacement of the string. During one full period, the vertex of the triangle traces out a figure which has the shape of the envelope of the normal fundamental mode of a freely vibrating string. (After Schelling, 1973)

with normal pressure at the normal bowing point, contains all the harmonics with amplitudes that decrease with harmonic number.

The position along the string at which the bow is drawn affects the harmonics produced. If the bow were very thin, and if it produced the exact pull-and-release motion of the string as described, the number of harmonics present in the vibration of the string would be limited as follows. If the point at which the string was bowed were $1/N$ of the length of string, it would produce an antinode in the vibration at that point. Therefore, the Nth harmonic would be missing, since it must have a node at that point. For example, bowing a string at its midpoint would not produce the second harmonic, or bowing a string at one-third of the distance from the end would fail to produce the third harmonic. Bowing one-tenth of the distance from one end would produce no tenth harmonic. Furthermore, the bowing point would be close to a node for the ninth harmonic, so the amplitude of the ninth harmonic would be small. The eighth harmonic would be slightly greater in amplitude, the seventh greater yet, and so forth, down to the fundamental.

In reality, the situation is not so simple. Through an additional interaction between the bow and the string, the Nth harmonic is actually produced with a moderate amplitude even when the bow is at $1/N$ of the distance from the end. However, when a string is bowed normally, the amplitudes of the harmonics generally decrease until the Nth harmonic.

If the bow is moved closer to the bridge (nearer the end of the string) than normal, the N value will be greater and the tone will have more harmonics. This bowing technique is called *sul ponticello* (on the bridge). If, on the other hand, the string is bowed farther from the bridge (or closer to the middle of the string), the N value will be less, and fewer harmonics will be produced. This bowing technique is called *sul tasto* (on the fingerboard), because it usually involves moving the bow above the fingerboard. Thus, by moving the bowing point along the string, the player can produce either a rich or a more pure tone.

One additional effect is important in determining the amplitudes of harmonics in the vibrating string: the damping of the string by the fingers when stopped notes (as opposed to open strings) are played. Finger damping decreases the amplitudes of the higher harmonics and causes the tone to be less brilliant. By avoiding open string notes, the violin tone can be made more uniform note to note; where a violin player has the option of playing a note either on an open string or stopped with a finger on a lower string, the stopped note is generally considered preferable. Stopped notes also permit vibrato, while open notes do not. On fretted bowed instruments like the Renaissance viol family, the frets were designed so as to produce a sound more nearly like that of the open strings for *all* notes, giving the viols a uniquely nasal quality.

When the string is bowed with great pressure by the bow, the very large amplitude oscillations elongate the string and cause an increase in the tension of the string. This increased tension leads to a rise in the frequency of the string, according to Mersenne's laws. Thus, bowing a string with great pressure causes a rise in the frequency of the string oscillation.

12.3 Violin Family

Figure 12–3 shows some views of the interior and exterior of the violin. The bow is held perpendicular to the strings and drawn across the string at a point between the bridge and the fingerboard. This causes the string to vibrate primarily in the plane defined by the string and the bow. These vibrations are transmitted to the bridge, causing it to rock rapidly back and forth, as indicated in Fig. 12–3. These vibrations must now be transmitted to the wooden back and belly.

The sound post is a rigid wooden dowel wedged between the two plates of the violin. As the bridge rocks back and forth, it flexes the front surface of the violin, causing small changes in the volume of the air space within the instrument. The bass bar has two functions. First, it reinforces the belly of the instrument so that the force due to the bridge will not smash the belly. Its most important acoustical function is to extend the vibration of the front surface over a wider area, and thus produce greater oscillatory change in the volume of the air cavity. As the front surface oscillates in and out at the frequency of the fundamental being sounded, it causes a change in the volume of the air space and thereby alternately forces air out of the

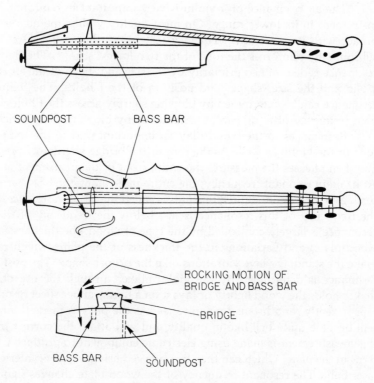

SOUNDPOST   BASS BAR

ROCKING MOTION OF
BRIDGE AND BASS BAR

BRIDGE

BASS BAR   SOUNDPOST

Figure 12-3 Sectional views of the violin.

instrument and draws the air back in. This motion of the string-bridge-front plate system creates motion of the air at the sounding frequency of the violin.

Perhaps the bridge (its size, shape, mass, elasticity, and placement) has more of an effect on the violin tone quality than any other single component. The motion of the bridge as outlined is oversimplified; it also includes some longitudinal vibration at twice the string frequency. This vibration is prevented from coupling to the belly of the violin by exact matching of the bridge feet to the plate. If the bridge were not held down by the other strings, it would rattle, as in the case of the tromba marina, a bowed medieval instrument with a single string mounted asymetrically on the bridge.

Bridge height is clearly a factor in coupling the string motion to the belly; the height of the bridge was increased when the baroque violin was modified to its modern form, to help increase the volume and brilliance of the violin tone. The bridge can affect the wolf tone in the cello (see discussion later in this section), and certain changes of the bridge have been proposed to help in eliminating the wolf tone. The position of the bridge with respect to the sound post is crucial in minimizing the wolf tone, as it is in obtaining the best tone from a violin. The exotic shape of the violin bridge probably helps to suppress undesired modes of oscillation, while maximizing coupling of the desired modes.

The air resonance of a violin is very important in reinforcing the tone of the instrument in its lower range. On most quality violins the air resonance is rather broad, with its peak at a frequency just below that of the D string. It therefore substantially reinforces the tone of the two lowest strings. The frequency of the air resonance is determined primarily by the size and shape of the air cavity within the violin and the size, shape, and position of the f holes. The frequency of the air resonance can be determined by blowing sharply across the f holes, just as the resonant frequency of a jug can be determined by blowing sharply across its mouth.

Resonances in the wood play an important part in the reinforcement of the tone of the violin as well. As the foot of the bridge alternately presses on the belly and then releases the pressure, the entire front of the instrument moves in and out. At its main resonant frequency, the entire plate along the bass bar moves back and forth as a long antinodal region. This is called the *wood resonance*; for most violins the frequency of this resonance is just below the frequency of the A string. This resonance is largely controlled by the type of wood, the thickness and shape of the wood plate, and variations in the thickness of the plate, which can be chosen to make the standing-wave vibrations take the proper shape. The position of the wood resonance is determined by the violin maker through the use of tap tones. The maker holds the wood lightly or lays it on a foam rubber sheet or other soft surface, taps it gently, and listens for tap tones. As the plate is made thinner, the tap tone will become quite bell-like in quality and will attain the correct frequency. Often this measurement is made using electronic equipment to produce Chladni-type patterns in the plate, which can then be observed using a type of interference pattern of laser light. The resonant frequency of the wood plate changes significantly when it is glued to the sides of the instrument, as the nodal lines move toward the edges where the plate is glued. The wood resonance just below the frequency of the A string is with the instrument assembled. The main wood resonance also reinforces the notes within a few half-steps of the low note of the instrument; the second harmonics of these notes are resonated by the main wood resonance. This resonance effect has been called the *wood prime* resonance. Additional resonances of the wood exist in the form of Chladni-type resonances that are closely spaced throughout the audible frequency range. This type of resonance is illustrated in Fig. 12–4 using a metal plate in the shape of a violin.

The sum effect of all these resonances can be seen in the *loudness curve.* If the violin is bowed up the scale chromatically from its lowest note with equal bowing intensity, the notes will vary in volume owing to the differing intensities of the resonances. In a good violin this loudness should not vary much; nonetheless, the air and wood resonances will be clearly observable. A violin with a good tone will have a relatively level loudness curve with a gradual drop in level to about one octave above the E string; this drop in the loudness curve apparently gives a more mellow and less squeaky tone quality to the notes around the open E. Figure 12–5 shows the loudness curves for a good violin and a poor one. Good violins differ from poor ones in that the loudness curve is more uniform for a good violin, and more sound is generally available on a better instrument.

When string instruments are simply scaled down (half-size, quarter-size) for use by small children, their tone usually suffers drastically. Because the instruments are smaller, their wood and air resonances are higher in frequency; as a result, the low notes are weak. Furthermore, the higher harmonics will be over emphasized, and the tone will therefore be too brittle or squeaky. Current work in this area is apparently paving the way for significant improvements in the tone quality of small-sized violins.

The choice of wood type and the way in which the wood is cut and trimmed are clearly important to violin tone. Norway spruce has been very popular owing to its density and elastic properties. While there has been some recent experimentation with substitute materials like plastics and resins, the usual choice for the back, front, and sides of the violin is limited to the types of wood that have been traditional for centuries. The strings of the violin are usually wire-wound gut for the lower three strings and thin solid steel for the highest (E) string. Wound Perlon, a type of plastic fiber, has recently emerged as a quality substitute for gut owing to its

Figure 12–4 Chladni figures in a violin-shaped $\frac{1}{4}$-inch thick aluminum plate driven at the central point. Nodal lines are shown. The figure on the left resulted when a relatively low driving frequency was used; its nodal lines are separated by large distances. The figure on the right resulted when a high driving frequency was used; its nodal lines are more closely spaced.

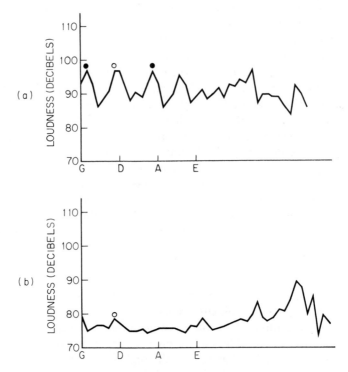

Figure 12-5 Loudness curves for a good violin (a) and a poor violin (b). The black dots mark the wood (upper) and wood prime (lower) resonances; the circle marks the air resonance. (From "The Physics of Violins" by Carleen Maley Hutchins. Copyright © 1962 by Scientific American, Inc. All rights reserved.)

superior ability to remain in tune. The use of steel E strings causes the sound to be extremely brilliant in the upper range of the violin, and is one reason for the requirement that the loudness curve decrease in amplitude for notes going up the scale above the frequency of the open E string.

Several other factors have been suggested as being important to the tone of the modern violin. For example, the type of varnish, the type of glue, and certain techniques for finishing the surface are believed to be important by many experts. Whether the finishing oil and varnish penetrate the surface of the wood is believed to be important; oils used in the eighteenth century were significantly less penetrating than some oils used today.

Regular playing throughout its life improves the tone quality of a violin. It is of course necessary to construct the instrument of high-quality wood of the appropriate type that has been properly aged. The vibrations of the wood produced when it is regularly played apparently increase its resiliency, and therefore improve the response of the resonances in the wood. Experiments have been performed where one of two identical instruments is played while the other is allowed to sit idle. The one that is played develops a tone quality and responsiveness that are better than in the unplayed instrument.

The changes in the construction and sound of the violin over the last three centuries have important acoustical implications. The early violins by Stradivarius and his contemporaries were pitched about a half-step lower (approximately $A_4 = 415$ Hz, the baroque average), had much shorter fingerboards that were set straight into the body, and used gut strings exclusively. The tension in the strings was considerably less than that of modern violins, with important consequences; the violin tone was much more mellow and less brilliant. The bass bar was also shorter and less massive.

As the orchestra grew in size, more wind instruments were added, and the tone of the winds became louder and more brilliant. In response, more brilliance was required from the strings. This was achieved by lengthening the fingerboard and the violin strings and considerably increasing the string tension. The fingerboard was also angled more sharply to the body. These modifications result in a substantial increase in the force that the bridge exerts on the belly. To keep the leg of the bridge from smashing through the belly and to increase the amplitude of the vibration of the wood, it was necessary to increase the size and strength of the bass bar. Further increases in string tension were used to obtain additional rises in pitch level over the past century or so. This also introduced increasing brilliance into the tone quality of the violin. Because gut strings, particularly the E (highest) string could not withstand this additional tension, it was necessary to use a thin steel string in place of gut. To match the increased brilliance of the steel E string, the other strings are now generally made of wire-wound gut or a synthetic substitute.

The air and wood resonances in old violins remained at about the same frequency over the centuries, while the pitch level at which the instrument is played rose slightly over one half-step. Therefore, the frequencies of the resonances with respect to the frequencies of the open strings have changed by about one half-step from their positions in the seventeenth and eighteenth centuries. This seems to indicate that the positions of these resonances is not critical, as long as they are close to the open strings and have all the desirable physical characteristics such as the correct width and intensity. This is a point of considerable interest, though, and many modern violin makers attempt to duplicate the resonant properties of Stradivarius violins because they sound so good.

The other members of the violin family are the viola, the violoncello or cello, and the string bass. While the viola can generally be made as a larger scaled version of the violin, the cello and bass would be too large to play if they were properly scaled. Because the cello is smaller than a scaled-up violin of equal pitch, the air and wood resonances appear in different frequencies relative to the frequencies of the open strings. For the cello, the main wood resonance generally appears very close to the note $F_2^{\sharp}$, often with serious consequences. When the cellist plays the note $F_2^{\sharp}$, the main wood resonance is driven (at its frequency) as the cello sounds the frequency of the note $F_2^{\sharp}$. A loud beating sound results between these nearby frequencies; this is known as the *wolf note*, because it is a growling tone. In fact, the mechanism by which the cello wolf note is produced is rather complicated. The wood resonance appears to be split into two frequencies by the driving force of the sounding string. These two periodic resonances then beat with each other. This wolf note must be eliminated or significantly reduced in amplitude in order for the cello to play the nearby notes without fear of driving it. This can often be accomplished by modifying the cello front plate or moving the sound post. Some experimentation is being done with other techniques of reducing the wolf note or moving its frequency farther from one of the chromatic notes so that it will not normally be strongly driven.

The string bass is in many respects more like a member of the Renaissance viol family than the violin family (see Sec. 12.1). The string bass is sometimes equipped with an extender on its lowest (E) string to allow a low C to be played. In the modern orchestra, the basses often double the cello parts, sounding one octave lower. By using this extender, it is possible for the basses to double the cellos down to the lowest note on the cello. If a bass were merely a scaled-up version of a violin, it would be enormous. Instead, the bass has had to be made somewhat smaller, and as a result the bass tone can be rather weak in its lower range.

An interesting effect used in music for the modern violin family is harmonics. There are two types, natural (or open-string) harmonics and artificial (or stopped-string) harmonics. To produce a natural harmonic, the string is lightly touched at a distance of about $1/N$ from the bridge, and the string is bowed very lightly between the bridge and the point being touched. This produces the Nth harmonic of that string. The tone obtained is very pure and soft, and generally well tuned, since it only depends on the string having been tuned properly; vibrato is generally not used on these harmonics. It is often possible to play a rapid arpeggio ending with a natural harmonic more easily than a regular finger-stopped note, because to produce these notes the finger need only barely touch the string in an extremely high position, perhaps beyond the fingerboard. This increases the effective range of the instrument slightly and gives some variation to the sound. Artificial harmonics are produced by playing a normal finger-stopped note and simultaneously touching the string with another finger at $1/N$ of the length of the stopped string; one advantage of artificial harmonics is that they can be played with the hand in a relatively low position. An interesting effect, called the "sea gull," can be obtained by using artificial harmonics on a string bass or cello. An artificial harmonic is produced in a high position. Keeping the thumb (stopping finger) and little finger (lightly touching the string) with the same relative spacing, the hand is slid up along the string. This results in a sequence of short downward glissandos, each lower in pitch, characteristic of the sound of the sea gull.

12.4   Plucked String Instruments

After the onset of the plucked tone, the tone of a plucked string instrument consists entirely of transients. Usually the string must be plucked rather strongly to retain an audible sound for a reasonable time. Therefore, as in the case of the violin bowed with great pressure, the string tension is increased substantially above that of the string at rest or vibrating with small amplitude. Thus the fundamental of the initial tone will be high in frequency, as is the fundamental of a violin string when bowed with above-normal pressure. Because the higher harmonics are relatively unaffected by this apparent increase in tension, they will remain nearly constant in frequency as the tone dies out. In most plucked instruments, therefore, the decay transient consists of a general damping out of the tone accompanied by continuous changes in the wave shape as the overtones decay at different rates and as the overtones shift in phase with respect to the fundamental. The tone of a guitar is an excellent example of this. Although the guitar tone lasts a relatively long time, the wave form continually changes, because the overtones are continually changing in phase and amplitude with respect to the fundamental.

The details of how the string motion couples through the bridge to the wooden plate are somewhat different for plucked strings. The bridge on plucked instruments is lower, and the instruments usually do not have sound posts. The coupling mechanism is more like that of a piano than that of the violin family. The bridge of plucked instruments couples oscillations in the plane perpendicular to the plate more efficiently than those parallel to the plate, giving prompt and sustained components to the decaying sound. This can be heard by listening carefully to the tone of the guitar.

The pizzicato technique for a violin family instrument must be somewhat different from that for the plucked strings because of the mechanism by which the bridge couples the oscillations to the body. A violin player is taught to pluck with a finger motion parallel to the front plate, not outward, away from the plate. This creates a string vibration parallel to the plate, which can then couple to the front plate by the same rocking motion of the bridge. Plucking the violin perpendicular to the front plate will not produce this rocking motion of the bridge, and will therefore yield a weaker plucked sound.

The harp is the only true plucked string instrument regularly found in the modern orchestra. The big sound obtained by a modern harp is due to vibrations in the frame, rather than a sounding board or plates, as in the other string instruments. Whereas the medieval harp was limited to a diatonic scale with a range of about two octaves, the modern orchestral harp can play chromatically over a range of several octaves. To obtain sharp and flat notes, seven double-action foot pedals are used. Each pedal controls all the strings with a given letter name and has three fixed positions (sharp, natural, and flat), which control stops that move against the string, as shown in Fig. 12–6.

Virtually all plucked instruments except the harp are fretted; the positions of the frets are set to produce the correct intonation of the fretted notes. Two effects are present in determining the frequency of stopped notes: the position of the frets and the slight increase in tension of the strings when they are pushed down against the frets.

The "rule of 18" has been used since medieval times in determining fret positions. According to this procedure, each successive half-step is obtained by positioning a fret at one-eighteenth of the remaining length of the string. This produces

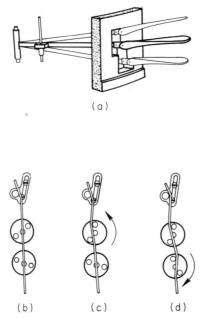

(a)

(b)   (c)   (d)

Figure 12–6 Tuning mechanism used in the double-pedal harp. When the pedal is in its central position, as shown in (a), the tuning disks are in the positions shown in (c), and the strings are tuned to a C major scale. Pushing a pedal up and down rotates the disks on all strings having the same letter name. A lowered string is shown in (b), and a raised string is shown in (d). (Reproduced from *The History of Musical Instruments* by Curt Sachs, by permission of W. W. Norton & Company, Inc. Copyright 1950 by W. W. Norton & Company, Inc. Copyright renewed 1968 by Irene Sachs.)

a ratio of frequencies of approximately 1.0588 for notes a half-step apart, a reasonable approximation to the equal-tempered half-step, which is 1.0595. However, due to the difference between strings the frequency ratios do not scale identically for all strings, and playing using higher frets will often introduce successively greater intonation errors between the strings. Attempts were made to obtain proper intonation, including some irregular temperaments, on some lutes and other early plucked strings by slanting the frets and other techniques. The additional tension created in the string when it is held down against a fret also affects the intonation. Intonation problems seem to appear to some extent in all fretted instruments, the bowed viols as well as plucked instruments. The best that can be done is to situate the frets in their best compromise positions.

The most significant recent development in plucked string instruments is the use of electronic pickups, particularly with the guitar. The acoustic guitar simply uses the normal wood and air resonances of the instrument to amplify the sound of the string. In an electric guitar, a transducer, or an electronic pickup, or perhaps several such pickups mounted on the wood plates of the guitar, convert the sound vibration into an electronic signal. This signal is then fed into a preamplifier and power amplifier, and then to a speaker system. The electric guitar has become immensely popular in recent years, especially in show music, jazz, and rock and roll. A great variety in sizes, shapes, and basic designs for electric guitars has appeared recently, some of which bear little resemblance to the traditional acoustic guitar.

EXERCISES

1. List and describe the resonances that play an important role in the violin. What happens when these resonances occur at the wrong frequencies?
2. Calculate the sizes of "full-size" cellos and basses by assuming the size of the instrument should scale with the wavelength of the lowest note. Compare this with a normal cello and bass. Roughly, how much would a hand have to move to play an octave on one string of these "full-size" instruments?
3. Compare the sound produced by a guitar when it is plucked with motion parallel to the front and with motion perpendicular to the front, that is, away from the plate. Compare the pizzicato sound obtained on a violin using these two types of motion.
4. Remove the sound post from an inexpensive violin; then play the violin. Compare the sounds before and after the sound post is removed. Explain the difference you hear.
5. Listen to a recording of a baroque-style violin. Compare this sound to that of a contemporary violin.
6. Play some natural and artificial harmonics on a violin. Describe the sound quality and explain its origin.
7. Examine some folk instruments, such as the psaltery, banjo, ukelele, and dulcimer. Describe the mechanisms by which string vibrations are transformed into sound waves in the air.
8. Use a pair of cords to suspend a bow above a horizontal violin so that the bow is perpendicular to one of the strings and gently touches one of them at a point. Hold or fasten the violin in its horizontal position and bow the string with a second bow. Observe any motion of the suspended bow for various positions along the string. What conclusions can be drawn from this experiment concerning bowed violin strings? Where is the motion the greatest? How does the position of the bow used to bow the string affect the motion of the string?
9. Put cotton in the f holes of a violin or cover the f holes with masking tape (completely or partially) to investigate the role of air resonance in the production of violin tone. How do your modifications affect volume and tone quality at different frequencies (different notes)? Note particularly changes near the frequency of the air resonance.

REFERENCES

BACKUS, JOHN. *The Acoustical Foundations of Music.* 2nd ed. New York: W. W. Norton & Co., Inc., 1977.

See his reference list.

BENADE, ARTHUR H. *Fundamentals of Musical Acoustics.* New York: Oxford University Press, Inc., 1976.

See his reference list.

FIRTH, IAN M., AND J. MICHAEL BUCHANAN. "The Wolf in the Cello." *Journal of the Acoustical Society of America*, 53, no. 2 (1973).

A somewhat detailed description of the physical mechanism by which the cello wolf tone is created.

FORD, CHARLES. Ed. *Making Musical Instruments—Strings and Keyboard.* New York: Pantheon Books, Inc., 1979.

Contains much of the detailed information necessary to appreciate the construction and workings of string instruments. It contains excellent and complete information on how to build many instruments.

HUTCHINS, CARLEEN, M. "The Physics of Violins." *Scientific American*, November, 1962.

An early survey article by one of the important researchers in the area of acoustics of string instruments.

SCHELLING, JOHN C. "The Bowed String and the Player." *Journal of the Acoustical Society of America*, 53, no. 3 (1973).

Deals with several aspects of the physics of bowed string instruments.

———. "The Physics of the Bowed String." *Scientific American*, January, 1974.

Deals primarily with the mechanism by which the bow imparts energy to the string of a bowed string instrument.

WHEELER, THOMAS H. Rev. ed. *The Guitar Book—A Handbook for Electric and Acoustic Guitarists.* New York: Harper & Row, Publishers, Inc., 1978.

Contains a wealth of information on all kinds of acoustical and electric guitars, including some history.

13

The Piano

In this chapter, we shall cover the history of the development of the modern piano, which involves other stringed keyboard instruments, such as the harpsichord and clavichord. We shall discuss the construction and operation of the modern piano and relate this to basic physical principles, some of which have been treated earlier, such as Mersenne's laws, while others may be new to the reader. Performance technique will be discussed when appropriate.

13.1   *History of the Piano*

The modern piano, or piano-forte, has historical roots stretching back to antiquity, with the discovery that a stretched string emits a tone when plucked. Although the origins of the classical lyre are unknown, several inscriptions suggest the ancient Assyrians had some form of this primitive string instrument. The Greeks also had simple lyres for which the sound box was a hollow tortoise shell; they also had citharas, which consisted of a wooden soundbox, two curved arms, and a crosspiece that held taut the several strings connected to the sound box. The number of strings on citharas was as high as 11 or more by the fifth century B.C. Both lyres and citharas were plucked with a hand-held implement called a plectrum.

Around 582 B.C., Pythagoras constructed a monochord, an instrument that consisted of a single string stretched across a resonator box. It had a movable wooden bridge that could divide the string into two sections each of definite length, and thus definite pitch. Pythagoras used his monochord to investigate the relationship between pitch of the fundamental and the length of the string, and discovered that sections of string whose lengths were related by integers emitted tones that sounded pleasant together.

The psaltery, mentioned in the Bible, consisted of a number of strings stretched across a resonator box. The performer could pluck several strings at a time, when desired. Keyboards, essential in further development toward the modern piano, were first used with pipe organs in about the second century B.C.

Around the beginning of the eleventh century a three-stringed instrument, the hurdy-gurdy or organistrum arose, in which one of the strings could be pressed by metal tangents at different positions, producing different pitches while the other strings were drones. Church musicians of the period used monochords to help teach singers chants. These monochords were then grouped together on the same sound box to form polychords. Near the beginning of the fifteenth century, polychords were fitted with keys that controlled tangents similar to those of the organistrum. This was the first clavichord.

In the clavichord there were several tangents per string, each tangent striking a different point along the string. One end of the string was damped with a cloth strip. When a key was depressed, its tangent struck the string, set it vibrating, and forced a node at the point of contact. The undamped length vibrated at a frequency determined by its length, while the damped length did not vibrate. When the key was released, the tangent left the string and the tone stopped because of the cloth damping. Figure 13–1 shows the action of a clavichord.

Although the tone of the clavichord is soft, there is a fair dynamic range available. As the key is struck more strongly, more energy is imparted to the string and it sounds louder. An unusual type of vibrato is obtained when the key pressure is changed while its note is sounding; the tension and thus the pitch of the string change slightly. Because the positions of the tangents along the string are fixed, pitch relationships cannot be changed. For that reason, temperament changes required to perform in harmonically distant keys cannot be made. By the end of the

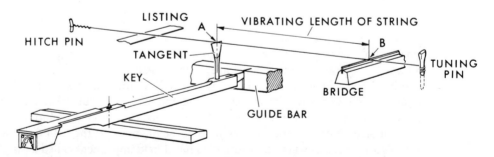

Figure 13–1 Clavichord action. As the key is pressed, the tangent strikes the string and sets the length of string at the right vibrating. (From Apel, *Harvard Dictionary of Music*, Harvard University Press. Reprinted by permission.)

seventeenth century, however, clavichords were made with only one tangent per string, and thus the temperament limitation was avoided.

Parallel to the development of the clavichord was the progression from the psaltery to the harpsichord. The triangular psaltery was fitted with keys, and by the beginning of the sixteenth century an elaborate action had developed. In harpsichords, each string vibrates in its entirety and is excited by one key. As a key is depressed, a vertical jack moves up and a crow quill plectrum plucks the string, as illustrated in Fig. 13–2. As the key is released, the jack descends until the angled, lower side of the plectrum touches the string. The jack then tips sideways, slightly, and the jack can descend farther. Mounted on the jack above the plectrum is a cloth damper, which then damps the string. As with the clavichord, the harpsichord string can vibrate only as long as the key is depressed.

Although generally louder than the clavichord, the harpsichord has one important drawback: Its dynamic range is limited. Striking a key harder does not result in a significantly louder sound because the string moves about the same distance to the side as it is plucked. A partial solution was to add one, two, or more "choirs" of strings, tuned in unison and/or octaves such that one key activated corresponding strings simultaneously, producing a louder sound. A jack slide could be hand-shifted to disengage the plectra of any choir of strings. As a jack slide is moved to the left, in Fig. 13–2, the plectrum is disengaged and the string will not sound when the key is struck. Because only one, two, or three of the choirs can be activated at any one time, the harpsichord dynamics are terraced.

In 1709, Bartolommeo Cristofori, a harpsichord maker in Florence, Italy, produced the first hammer action to replace the quilled jack harpsichord action. This changed the tone quality (his hammers were leather covered), but most importantly it allowed a large, even dynamic range. He called his new instrument *gravicem-*

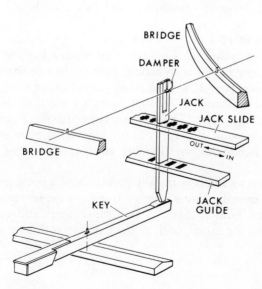

Figure 13–2 Harpsichord action. As the key is pressed, the jack rises and the quill plucks the string. When the key is released, the jack descends and the quill tips to the side so the string is not plucked again. (From Apel, *Harvard Dictionary of Music,* Harvard University Press. Reprinted by permission.)

balo* col piano e forte (soft and loud) expressing the new dynamic spectrum available. This was the origin of the "pianoforte." Cristofori's later, improved action contained many of the essential features found in the actions of today.

As considerable force could be transmitted with the hammer action, strings that were tighter, stronger, and longer could be used to increase the dynamics. Also, multiple strings for each key were used for the same end. These changes required a longer and more solid piano body, and in 1822 Babcock in Boston introduced a cast metal frame, which permitted very high string tension. Modern grand pianos withstand enormous force owing to this tension, a total of some 60,000 pounds, the weight of 20 small cars!

13.2   *Construction and Action of the Piano*

Figure 13-3 shows an exploded view of a modern grand piano. The body or case is made of wood veneer, but recently other materials, such as clear plastic, have been tried. The large flat soundboard is made of spruce planks joined together, is bellied slightly upward, and fits snugly into the case. The pin block is made of very strong, laminated maple and contains about 230 steel tuning pins, or pegs to which the ends of the metal strings are attached. Both the pin block and the soundboard sit under the heavy cast iron frame, with the pin block positioned such that the enormous tension of the strings is transmitted to the frame. The strings are attached to tuning pins that pass through holes in the frame. The strings pass over the bridge, around pegs set at the far end of the frame, then back to the pin block. Some manufacturers use felt strips to damp this unstruck portion of the string between the bridge and pegs. Other manufacturers leave it undamped, and this may help produce a "brighter" sound.

The keyboard consists of 88 keys, 36 black and 52 white, which when depressed initiate a complicated motion of the action that sends a hammer to strike the unison of strings. Each key has its own action, such as the one shown in Fig. 13-4, except the actions of the upper 21 keys of many pianos do not have dampers. Also, the size and weight of the hammers for high notes are less than the size and weight of the hammers for low notes. This permits better *coupling* between the hammer and string, increasing ease of play: A heavy hammer is more effective in exciting a heavy string, a lighter hammer is more effective in exciting a lighter string.

As the performer depresses the key, the key pivots about a point just beyond its midpoint (the section nearest the player is longer), sending the far ends upward, like a seesaw. This upward motion leads to two subsequent motions: (1) the underlever and damper head are raised, and (2) the hammer is thrown or pushed

---

* Gravicembalo is a modification of *clavicembalo,* the Italian word for harpsichord; *gravi* means "lower."

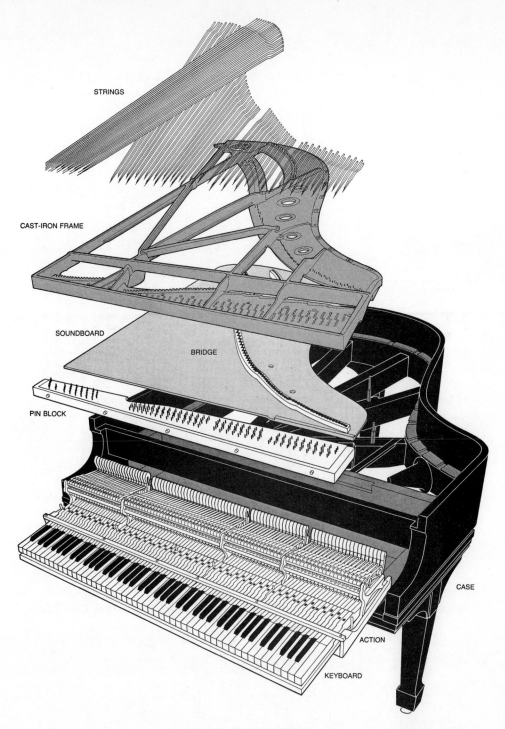

STRINGS

CAST-IRON FRAME

SOUNDBOARD

BRIDGE

PIN BLOCK

ACTION

KEYBOARD

CASE

Figure 13-3 Exploded view of a Steinway Model B piano. (From "The Coupled Motions of Piano Strings" by Gabriel Weinreich. Copyright © 1979 by Scientific American, Inc. All rights reserved.)

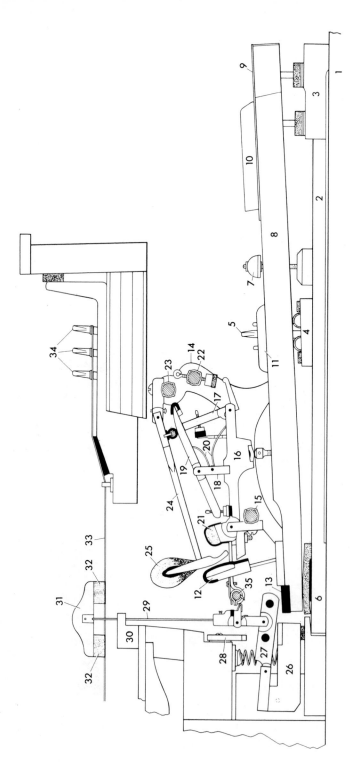

Figure 13-4 Cross section of grand piano action: 1. keybed; 2. keyframe; 3. front rail; 4. balance rail; 5. balance rail stud; 6. back rail; 7. key stop rail; 8. white key; 9. key covering; 10. black key; 11. key button; 12. backcheck; 13. underlever key cushion; 14. action hanger; 15. support rail; 16. support; 17. fly; 18. support top flange; 19. balancer; 20. repetition spring; 21. hammer rest; 22. regulating rail; 23. hammer rail; 24. hammershank; 25. hammer; 26. underlever frame; 27. underlever; 28. damper stop rail; 29. damper wire; 30. damper guide rail; 31. damper head; 32. damper felts; 33. string; 34. tuning pin; 35. sostenuto rod. (From Apel, *Harvard Dictionary of Music*, Harvard University Press. Reprinted by permission.)

upward to strike the strings. As the key is released, the damper returns to the string and the sound stops.

Another important feature of modern piano action is that it allows the player to repeat notes without fully removing the finger from the key. With proper regulation, as the key is released, first the hammer will reset, then the damper drops onto the strings. This feature produces two important results. First, notes can be repeated extremely rapidly on a well-regulated piano, much more so than on the harpsichord, for example. Many composers of the nineteenth century took advantage of this and wrote music requiring this new technical capability. Second, a note can be repeated while it is still sounding, without using the pedal. This can be useful in some sustained passages.

Consider a note played and held for several seconds. No matter what performance technique is used striking the key, the string only "knows" how fast the hammer strikes it, and this speed determines the loudness of the sound. For a given key, many performance techniques (such as playing from the finger, wrist, elbow, or shoulders) can be used to get the same loudness. If different techniques are used but produce the same loudness, the sounds of the string will nonetheless be indistinguishable. In short, there is no way to keep the loudness the same while changing warmth, crispness, or any other attribute of the piano string sound.

The attack technique may have more subtle effects, however, effects that are discernible by a well-trained ear. The noise of the action can differ between notes of equal loudness, depending on performance technique, and the slapping of a finger against the key can sometimes be heard. Although the string motion is only determined by the final speed of the hammer, the noise at the attack need not be strictly determined by this final speed.

This is not to say that the attributes mentioned cannot be obtained, or that the pianist should not learn different performance techniques. Varying the loudness can change the sound quality, as we shall see in the next section, and different techniques are appropriate for different loudnesses. Of course, the duration that the key is depressed can be controlled, as this will result in different effects, and different performance techniques are appropriate for producing different-duration notes. Thus one can have staccato (short) notes and legato (long) notes of the same loudness, but these differences result from differences in the duration of the note and cannot be attributed to the attack per se.

Because the strings in the upright piano are vertical and those in a grand piano horizontal, the corresponding actions and sounds differ. Hammers in grand pianos fall back because of gravity, while a more complicated system is used in uprights, for instance. The sound due to the initial impact of hammer with strings is thrown back, usually toward a wall, for uprights. In the grand piano, however, this sound is projected up and out, reaching the listener before any reflections occur. This leads to a crisper sound.

Because most effort went into perfecting the horizontal piano and the upright was introduced primarily for cost and size considerations, we shall restrict our attention to the grand piano in the following section.

The piano strings are the heart of the piano, and their motions are the most important factor in determining the sound quality of the piano. Their motions are complicated and, in many respects, deviate from the "ideal" as discussed in previous chapters. Some of the most important aspects of piano string motion were only properly understood in the last few years, and some aspects are still not completely understood.

Mersenne's laws can be used to understand the most obvious features of piano strings. The first is that the strings of high notes are shorter than those of lower notes, as described by Mersenne's first law. If length were the only difference between the strings, and the shortest were, say, 6 inches, the longest would have to be over 64 feet long! Instead, the mass per unit length is changed, in accordance with Mersenne's third law, to make the pitch of the lower strings low while keeping the length reasonable. This extra weight is added in the form of multiple wrappings of the strings. Lower strings are wrapped in copper wire, and the very lowest strings have two such wrappings. The tension does not vary significantly from string to string, as this might lead to uneven stress on the frame and perhaps buckling.

Inharmonicities play a very important role in the sound of the piano. Because there is some inherent stiffness in the strings, the overtones produced do not have the simple, integral relationship that we have discussed throughout the book. This inherent stiffness makes the overtones too high, as if the oscillations were in a more tightly stretched string. The higher overtones result from modes in which the string has many loops, and thus a great deal of bending, and these are successively higher in pitch than the simple harmonics of $2f$, $3f$, and so on. In a piano string, the 16th overtone may be as much as a half-tone higher, and the 50th overtone may be as much as seven half-steps higher than in the ideal case. Wrapping the strings in the manner discussed is done to minimize stiffness. A thicker string, like a stiff rod, would raise the overtones unacceptably. Electronically synthesized "piano" sounds that are made of integrally related overtones do not sound as full or rich as when the overtones are inharmonic; the tone does not sound like that of a piano.

This inharmonicity affects piano tuning. If the notes an octave apart were tuned such that their fundamentals had a ratio of frequencies of precisely 2, then the slightly raised second overtone of the lower note would beat with the fundamental of the upper note when played simultaneously. To avoid the beating, notes an octave apart are *stretched*; that is, they are tuned to have a ratio of frequencies slightly greater than 2.

The interaction of the hammer with the strings affects the sound as well. If any string such as a guitar string is plucked or struck near its end, many higher harmonics will be excited, as the string has a sharp bend in it. Guitarists sometimes pluck strings near the bridge to produce this type of sound for variety. To produce uniform tone along the keyboard, then, the hammers must strike their respective strings at the same ratio along their lengths. What should this ratio be? If the hammer were to strike the string at its midpoint, low-frequency overtones would sound;

if the hammer were to strike near the end, higher overtones would predominate. Because the sound is considered more beautiful and rich when many high and low overtones are present, the position is chosen such that many of the overtones sound. Hammers are set to strike strings at a point ranging between one-seventh and one-eighth of their lengths, positions found to yield rich tones when struck. If the hammers were very sharp, a node might be produced at the striking point, as in the clavichord. Because the hammer is not a point, but fairly soft and curved, a node is not produced there and many overtones sound.

The sharpness or hardness of the hammer also affects the sound quality by exciting many high overtones. If the felt on the hammer is worn and hardened, the string will have a sharp cusp in it when struck. It will thus have a large number of higher overtones with appreciable amplitudes. This produces the "tinny" sound of an old, worn, barroom piano. This tinny quality can be reduced by pricking the felt on the hammers with a needle, loosening and softening the felt.

Up to now, we have been discussing single strings, but many important aspects of the piano sound result from the fact that most keys activate several strings tuned very nearly in unison. The first such aspect is loudness. The highest 68 keys have triplets of strings, and the lowest keys have singlets or doublets. This arrangement helps to keep the loudness uniform from note to note, as a small, single upper string is not very loud when sounded alone.

The multiple strings have other, more complicated effects involving the prompt and sustained sounds. The loudness of a single long note drops quickly for the first few seconds. After that, the decay is much more gradual, lasting as long as 30 seconds or more. The existence of the early (prompt) and later (sustained) sounds can be explained using several different mechanisms.

The first of these involves polarization. Recall from Chap. 2 that a transverse wave, such as a wave on a string, can be polarized. For instance, the motion of the string can be up and down, or it can be side to side (or combinations of both). When the hammer strikes a string, most of the energy goes into forcing the string to vibrate in this up-down (vertical) polarization, but because there are slight irregularities in the hammer surface and striking angle, the string vibrates in its side-side (horizontal) polarization as well, although not as much. The sound due to the vertical polarization dies down very quickly as the vertical motion of the string is transmitted downward very effectively by the support to the soundboard, where it is radiated as sound. The horizontal polarization does not radiate its energy as effectively because the string pushes horizontally on the soundboard; thus the sound due to horizontal string motion will last a long time. Even though the sound due to the vertical polarization is initially louder (the prompt sound), there comes a time, about 5 seconds after the string is struck, that the sound due to the horizontal polarization is louder (the sustained sound). This partial explanation of the origin of the prompt and sustained sounds is also applicable to the guitar, as we saw in Sec. 12.4.

The full explanation must take into account the multiple strings of a single key. Consider a key that has a doublet of strings. Just after hammer impact, the strings are moving up and down in phase (we will neglect the horizontal motion for the time being), but due to irregularities in the hammer, as mentioned earlier, one string (call it A) will have a slightly higher amplitude of motion than the other (B). Because they are initially moving in phase, they push and pull hard on the bridge, and sound energy is radiated from the soundboard. Later, as the amplitude of the strings decreases, there comes a point where the amplitude of B is essentially 0, although A is still oscillating. The motion of each string couples to the motion of the other through the bridge, and A begins to drive B; this is an example of driven oscillation. Thus B begins to oscillate, although now it oscillates out of phase with respect to A. A continues to decrease its amplitude and B to increase its amplitude until they are approximately equal. At this time, energy is radiated much more slowly as the bridge does not move much; as A pulls the bridge up, B pulls it down, and vice versa. There is, essentially, no net vertical force on the bridge. This is another cause for the sustained sound. If this were a perfect system, no energy would be lost through the support, only through the strings pushing the air directly. This basic mechanism also occurs with three strings, but it is more complicated and we shall not discuss it in more detail.

For some piano keys these complex interactions may occur more readily with certain overtones than others. The result is that different overtones will decay at different rates. This can be heard by striking and holding the note C_3. The sound of the third overtone, at G_4, decays more slowly than that of the fundamental, as can be easily heard.

We can now understand the function of the left piano pedal, the *una corda*, or soft pedal. During soft passages, the sustained part of a held note may be too soft, softer than the background noise in the concert hall, while the prompt sound is audible. The problem then is to make the sustained part of the note louder without making the prompt sound louder. The una corda pedal solves this problem. When the pedal is pressed, the entire keyboard and action are shifted slightly to the side, so a hammer that would normally strike three strings only strikes two. The two struck strings begin moving in phase and immediately start to drive the unstruck string to move out of phase. Thus, the sustained sound will occur before the amplitudes of the struck strings have dropped very much. Consequently, the sustained sound is louder relative to the prompt sound when the una corda pedal is pressed than when it is not pressed.

On an upright piano or spinet, the *soft pedal* replaces the una corda pedal. Rather than slide the keyboard and action, the soft pedal modifies the action so that the hammer will not hit the strings as hard when the key is pressed with normal finger motion.

The middle pedal, or *sustain pedal*, on a grand piano is missing from most uprights and some grand pianos. The pedal sustains only keys that are being held at the time the pedal is pushed. For a few grand pianos and most uprights that have a middle pedal, it simply acts as a damper pedal for the notes in the low range, say below about C_3. This is useful for some types of music in which chords must sound while both hands play other notes.

The last mechanism for producing the sustained tone involves several concepts and results from complicated motion of deliberately mistuned strings. First, let us recall some aspects of driven oscillations and resonance as discussed in Chap. 3. When the driver's frequency is slightly different from the natural or resonant frequency of the resonator, only a little energy passes from the driver to the resonator. (Recall the experiments with pendula.) But at the resonant frequency, a great deal of energy is passed from the driver to the resonator, and the driver leads the resonator by 90° in phase.

Consider a vibrating string attached to an idealized frictionless "springy" support on one end, as shown in Fig. 13–5(a). For such a springy support, the motion of the string and support are in phase. No energy is lost by the string, as after one complete oscillation the system is back in its original configuration. The string vibrates as if it had a node out beyond the support, and thus has a lower frequency than if the support were the usual fixed type described previously.

Now consider the idealized, frictionless "massive" support shown in Fig. 13–5(b). This consists of a very heavy object free to slide without friction up and down a rod. Here the motion of the string and the support will be out of phase, and, as before, no energy will be lost by the string to the support. After one complete oscillation, the system is back in its original configuration. Because the node is shifted inward in the massive-support case, the frequency of the string oscillation is higher than if there were a fixed support.

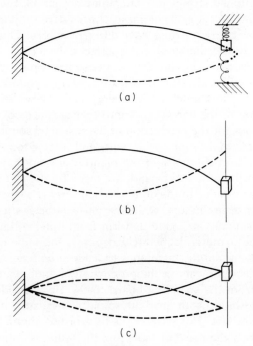

Figure 13–5 String oscillating with one fixed end (at left) and the other end attached to (a) a springy support, (b) a massive support, and (c) a resistive support. (from "The Coupled Motions of Piano Strings" by Gabriel Weinreich. Copyright © 1979 by Scientific American, Inc. All rights reserved.)

A resistive support is different. It does have friction, so after each oscillation the string has lost some energy to the heating of the support by friction. The node, however, is right at the support so that the string's frequency is the same as in the case of a rigid support, as shown in Fig. 13–5(c). The motion of the support is 90° behind that of the force owing to the strings, a point to which we shall soon refer.

These support cases are similar to the driven oscillations mentioned in earlier chapters and will help explain the interaction of the piano string with its bridge. Real bridges have friction and are, for the most part, resistive supports.

Imagine two piano strings, slightly mistuned, that are oscillating out of phase with each other. As we saw earlier, the bridge will not move appreciably as the forces due to the strings cancel. As the string with the slightly higher frequency (call it *A*) oscillates, it gains in phase over the other string (*B*). The strings are no longer out of phase, and a net force results on the bridge. In fact, this force differs in phase by 90° compared to either string, ahead of one, behind the other. The resistive bridge responds by oscillating 90° out of phase with respect to the force, as we saw previously; it is moving in phase with the higher-frequency string (*A*) and out of phase with *B*. Thus *A* "feels" a springy support and its frequency is lowered; *B* "feels" a massive support and its frequency is raised. The two strings reach the same intermediate frequency and sound in perfect unison. We saw how two ideal identically tuned strings oscillating out of phase would not lose energy to their support. Here, with the slightly mistuned strings, energy can be slowly transmitted to the support because the bridge moves.

Consider now the mistuned strings just after they are struck by a hammer; they begin oscillating in phase. As *A* gains in phase, the resistive support cannot be 90° behind both *A* and *B*, as the two strings have different phases. Thus the initial energy loss is reduced in this case compared to the perfect unison case. Also, as one string "feels" springy support and the other string a massive support, the frequencies of the strings are made identical. Thus the mistuned strings sound at the same pitch, and because they never lost much initial energy (as they would have were they perfectly in tune), more of the energy goes into the sustained sound.

Of the three mechanisms for the production of the sustained sound—horizontal polarization, out-of-phase motion of two strings, and interaction of mistuned strings—the first two depend on subtleties of the piano itself and may vary from note to note. The mistunings, on the other hand, are made by the piano tuner and can compensate for irregularities due to the other mechanisms. That the mistunings set by an expert piano tuner seem random, while the tone characteristics, including the prompt and sustained sounds, are quite uniform from note to note, substantiates this view. If the same piano is deliberately detuned, the same expert tuner will, in general, make the same subtle mistunings to achieve uniform tone.

One final mechanism is significant in the production of good piano tone, the coupling of the vibration from the sounding board to the air. Consider a tuning fork, which you strike, and hold up to hear. Its sound is very soft, often hardly audible at a distance of a few feet. However, if the base of the tuning fork is held against a piece of wood, like a chair seat or table top, the sound is dramatically increased. This is because the vibrations from the tuning fork are transmitted to the rather larger piece of wood, which can couple the vibrations to the air much more efficiently than can the tuning fork alone. The entire wooden surface oscillates in the manner of a loudspeaker, producing motion of the air adjacent to the surface. The ability of a piece of vibrating wood like a sounding board to couple its vibrations to the air goes up dramatically with size, accounting for the tremendous volume of sound that can be produced by a large grand piano. This is another illustration of impedance matching.

In conclusion, we have seen how the keyboard instruments have evolved throughout history and how many physical principles and processes are involved in the modern piano. It is hoped that the student has a greater appreciation for the piano and recognizes it as truly "grand."

EXERCISES

1. Trace, roughly, the production of a piano tone from the performer's finger motion to the final emergence of the sound.
2. Play and hold a piano note. Listen for the prompt and sustained sounds. Describe three mechanisms by which these sounds are produced. Which of these mechanisms can be controlled by a piano tuner? Which cannot?
3. Define "inharmonicity." What makes piano notes inharmonic? What effects does this have on tuning and tone quality?
4. Listen to harpsichords and clavichords and compare their volume, range, and tone quality. Describe the physical reasons for these differences.
5. Obtain recordings of piano music played on historical pianos, beginning from about 1750. Analyze the sound for each type of music being heard.
6. Listen and compare the effect of the una corda pedal on the tone of a grand piano with the effect of the soft pedal on an upright piano or a spinet. Explain the origin of the differences.

REFERENCES

APEL, WILLI. *Harvard Dictionary of Music.* 2nd ed. Cambridge, Mass.: Harvard University Press, 1973.

A classic dictionary of music and instruments.

BENADE, ARTHUR H. *Fundamentals of Musical Acoustics.* New York: Oxford University Press, Inc., 1976.

Slightly more advanced than the present text.

BLACKHAM, E. DONNELL. "The Physics of the Piano." *Scientific American,* December 1965.

Stresses the role of inharmonicity in piano tone and tone perception.

HART, HARRY C., MELVILLE W. FULLER, AND WALTER S. LUSBY. "A Precision Study of Piano Touch and Tone." *Journal of the Acoustical Society of America*, October 1934, pp. 80–94.

Describes experiments and results showing that the tone of a piano cannot be altered without changing the loudness.

HOLLIS, HELEN R. *The Piano: A Pictorial Account of Its Ancestry and Development.* London: David and Charles Publishers, 1975.

A musician's introduction to the history of the piano.

WEINREICH, GABRIEL. "The Coupled Motions of Piano Strings." *Scientific American*, January 1979, pp. 118–127.

Contains a clear description of the complicated interactions of the strings of a single piano note.

WHITE, WILLIAM B. *Theory and Practice of Piano Construction.* New York: Dover Publications, Inc., 1906, reprinted 1975.

A readable account of how technical innovation and experimentation have affected the piano throughout history.

WOLFENDEN, SAMUEL. *Treatise on the Art of Pianoforte Construction.* Surrey, England: Unwin Brothers Ltd., 1977.

A good starting point for the student who wants to understand the workings, including actions, of a piano.

14

Percussion
Instruments

Of all the instrument families, the percussion family has the greatest variety of sound, both in dynamics and sound quality. Such sounds range from the delicate "ting" of a triangle, through the rich tone of a marimba, the crack of a snare drum, the clear tone of a bell lyra, the majestic peal of church bells, to the thundering roll of timpani and crash of cymbals.

In this chapter, we shall describe the physics of two important classes of percussion instruments, the idiophones (xylophone, vibraphone, chimes, bells, cymbals, and so on) and the membranophones (timpani, drums, tabla, and so on). Chordophones, such as the piano, and aerophones, such as the whistle, are classified with percussion instruments, but will not be treated here. We shall examine the features that characterize the sound of most percussion instruments: significant inharmonicities, a large number of overtones, and the unique attack and decay transients. These characteristics arise by different mechanisms in different instruments.

14.1 Bar Instruments

The bell lyra, shown in Fig. 14–1, is acoustically one of the simplest bar percussion instruments, a subgroup of the idiophones, or struck instruments lacking membranes. In marching bands, a bell lyra is usually held with one hand at the top, the other end being supported at the performer's waist. A hand mallet in the

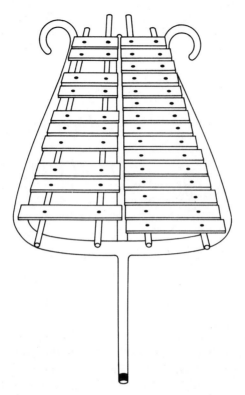

Figure 14-1 The bell lyra.

performer's free hand is used to strike the bars, which are laid out similarly to a piano keyboard. In concert situations, the bell lyra is sometimes laid on a table and played like the more familiar glockenspiel.

When a bell lyra bar is struck, it vibrates transversely at a frequency determined by its dimensions, the properties of the metal, and location of the suspension points. Such a bar is a length of rectangular metal held to supporting felt pads by a loose rivet. As with strings, there are multiple modes of oscillation, although here the bar ends are free to vibrate.

Figure 14-2 shows a side view of a bell lyra and its first four modes of oscillation along with their frequencies. Although the bars have slight arches cut underneath, we shall ignore their effects for the moment. Note particularly that the frequencies are not harmonically related. This is because of the considerable inherent stiffness in the bar, an extreme example of the effect in piano strings. These modes can also exist for oscillations across the width of the bar, as if Fig. 14-2 represented a view of the narrow end of the bar. Because the width of the bar is less than its length, corresponding oscillations across the width will, in general, be higher in frequency than those along the length. In general, none of these modes is harmonically related to the fundamental or to each other.

There are other bar modes. The bar can oscillate torsionally much as the

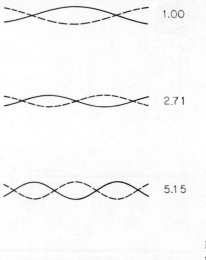

1.00

2.71

5.15

8.43

Figure 14–2 The first four transverse modes in a metal bar. (After Rossing, "Acoustics of Percussion Instruments I," *The Physics Teacher*, December, 1976)

Tacoma Narrows Bridge did before it collapsed. The simplest such mode has diagonally opposite corners moving up while the other diagonal pair moves down, and vice versa, as shown in Fig. 14–3. Other, higher torsional modes have more twists; as a general rule, none is harmonically related to the fundamental.

While the bar is not prevented from oscillating in longitudinal modes, these are not excited significantly by striking the top of the bar with a mallet. Also, the longitudinal modes have frequencies much higher than the transverse modes and do not significantly affect the bar's sound.

When the bar is struck, many transverse modes are excited. The higher modes decay quickly, while the fundamental sounds much longer, as the bar is supported near the nodes of the fundamental mode. Thus, after an initial sharp impact sound, the bar sounds quite clear and pure for several seconds before becoming inaudible.

The marimba, shown in Fig. 14–4, is another bar instrument, but differs from the bell lyra in several important ways. The bars are made of rosewood (or plastic) and are supported by strings that pass through two horizontal holes near the nodes of the fundamental. Each bar has an arch cut out of its bottom, and a tubular resonator sits vertically beneath each bar. These differences influence the sound quality. As wood does not have the high resonance characteristics that metal does, marimba notes die away faster than do bell lyra notes. The vibration of artificial marimba bars does not decay as fast as with wood bars.

Figure 14–3 The simplest torsional mode in a metal bar.

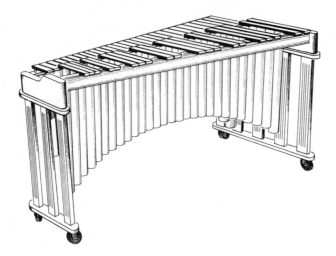

Figure 14-4   The marimba. Note the vertical resonator tubes under the bars. (From Apel, *Harvard Dictionary of Music*, Harvard University Press. Reprinted by permission.)

The arch, shown in Fig. 14-5, is important for several reasons. The deep arch in each low-frequency bar helps keep the size and the cost of the marimba within limits. If the arches were not present, each bar would have to be twice as long as the bar one octave above it, which would result in very long bars for the low notes. Deep arches in the lower bars decrease the wave velocity in the bar. In this way, the arched low bars can be shorter than the unarched bars while still having the proper frequencies. In any bar the arch affects the modes of oscillation. Because the wave speed is lower at the thin center of the bar than at the thick ends, the wavelength of a standing wave of a certain frequency is shorter in the middle of the bar than at the ends. This can be seen by comparing the modes of a marimba bar (see Fig. 14-5) with the modes of a bell lyra bar (see Fig. 14-2).

Another important effect of the arch concerns tuning and tone quality: The first two overtones are approximately two octaves and three octaves plus a minor third above the fundamental.

Finally, each resonator is closed at the bottom and has a length chosen so as to resonate at the fundamental of the corresponding bar. The resonators affect the sound in two ways: First, because energy in the vibrating bar is transferred efficiently to the air, the loudness is increased and the sound decays more rapidly. Second, the resonator reinforces the fundamental, but not other frequencies, as they are not at the resonant frequencies of the tube. A marimba without resonators sounds richer in overtones, almost like a milk bottle being struck by a spoon, and the tone is softer and lasts longer.

Many principles applicable to the marimba are also useful in understanding the xylophone, shown in Fig. 14-6. Xylophone bars are narrower than those of the marimba, and because they are also thicker, they have a higher wave velocity and thus a higher frequency for a bar of the same length. The arch cut in a xylophone bar is shallower than that in a marimba bar, as the first overtone is tuned to three

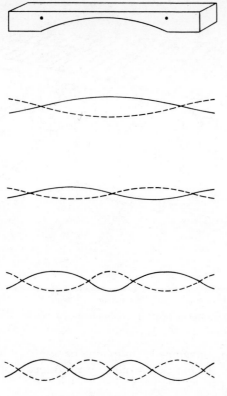

Figure 14–5 The first four transverse modes in a marimba bar. Note that the wavelength of each mode is greater toward the end of the bar than at the center. (After Rossing, "Acoustics of Percussion Instruments I," *The Physics Teacher*, December 1976)

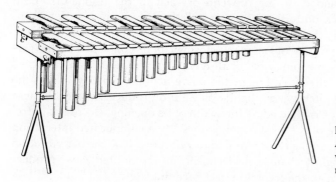

Figure 14–6 The xylophone. (From Apel, *Harvard Dictionary of Music*, Harvard University Press. Reprinted by permission.)

times the fundamental frequency. Both the fundamental and this first overtone will be reinforced by the closed xylophone resonator tubes. Oscillations in the thick bars die away quickly; this and the presence of the reinforced first overtone give the xylophone its sharp, bright sound.

The vibraphone (or vibraharp), shown in Fig. 14–7, has deeply arched aluminum bars in which the first overtone is two octaves above the fundamental.

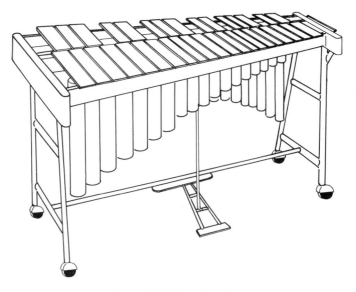

Figure 14-7   The vibraphone.

Because the decay time for undamped vibraphone bars is very long, a horizontal damper rod running under the bars, controlled by a foot pedal, is used when short notes are desired.

A unique effect obtainable with the vibraphone involves motor-driven butterfly valves that open and close the top end of each of the resonators. Each resonator tube has its own butterfly valve, and all the valves operate in unison. When the valves are open, the tubes resonate and the sound is loud; when closed, the sound is soft. The rate of this opening and closing can be controlled by the performer to obtain fast vibe, medium vibe, or slow vibe, or the motor can be turned off to avoid the tremolo effect. Although the valve action primarily affects the volume of the sound, the pitch varies slightly as well to produce vibrato. Because the effective width of a tube is affected by the width of its opening, the resonant frequency changes as the size of the tube opening does. By a complicated interaction between the standing wave in the resonator and the vibrating bar, the bar vibration is affected as well. This slightly changes the frequency of the emitted wave, producing a slight vibrato, although the tremolo effect predominates.

The long decay time and excellent resonance characteristics of the vibraphone permit an effect not attainable with the preceding instruments, that of *bending* a note. If a hard mallet is pressed on a node of an oscillating vibraphone bar, that is, over a support point, there will be little damping. As the mallet is slid toward the center of the bar, however, damping will occur, but, more importantly, the pitch will become lower. The mallet head acts to increase the effective weight of the bar and, by a law analogous to Mersenne's third, the pitch will drop. This technique is sometimes used in contemporary jazz.

Orchestral chimes consist of a set of hollow vertical pipes suspended at the top by thin ropes. The tubes range in length and thickness of wall to obtain, usually, about one and one-half octaves, as shown in Fig. 14–8. Each pipe is struck with hammers on the edge of the top in a motion that is both downward and sideward. The three lowest modes of such a tube have frequencies with ratios of approximately 2 : 3 : 4. The ear then perceives a note an octave below the fundamental—a missing fundamental. Other overtones deviate from this approximate linearity and help give the unique sound quality of the chime. Chime tubes are often fitted with metal plugs in their ends. The plugs damp out many higher overtones, lower the frequencies of the low overtones, and reduce wear on the tube.

    Nonuniformities in the thickness on the different sides of a chime tube can detune the overtones of corresponding transverse modes. If the frequencies of these modes are close, but not equal, beating will result. This problem can be alleviated by squeezing the tube in a vise to make it more symmetric around its perimeter. The frequencies become equal, and the beats are therefore eliminated.

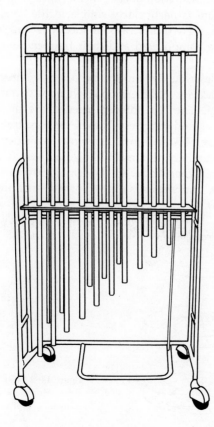

Figure 14–8 Orchestral chimes.

Triangles are steel bars bent into a triangle with one open corner; they are struck with metal beaters. The bends do not affect the tone of the bar significantly, and thus the tone of the triangle is very similar to that of a straight bar of equal length.

When the triangle is struck in the standard way, most of the vibration occurs within the plane of the triangle, although some motion perpendicular to this plane also occurs. As the triangle bar is thick and solid, there is a great deal of stiffness and thus inharmonicity. The initial "click" of metal-on-metal is followed by a tone made up of inharmonic overtones that then die down gradually, as there is little friction with the support. The higher overtones die down more rapidly, leaving the fairly clear tone of the lower overtones.

14.3   *Membranophones*

While strings and air columns are essentially one-dimensional media, surfaces, such as drumheads, are two dimensional. The sound-production mechanism in membranophones begins with the oscillation of such membranes. The overtone structure in a one-dimensional medium is particularly simple in the ideal case; the corresponding overtone structure in two-dimensional media is not so simple. In particular, the modes are more complicated and their frequencies are not related integrally.

In a string we can describe the modes by a single integer, the harmonic number. Because a drumhead is two dimensional, we need two integers to describe each of its modes precisely. The two integers needed are the number of nodal diameters and the number of nodal circles. These numbers are shown along with the modes in Fig. 14–9. Figure 14–9(a) shows the lowest frequency mode. The center of the membrane moves up and down, almost like your chest as you inhale and exhale. Figure 14–9(b) shows a higher mode. As the center moves up, the outer ring of the membrane moves down, and vice versa. Just as there are nodes at the ends of a

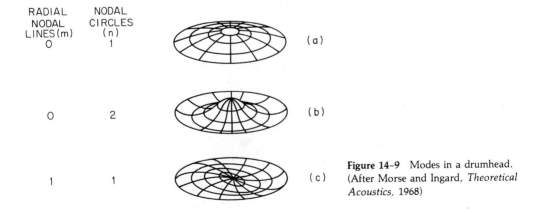

RADIAL NODAL LINES (m) — NODAL CIRCLES (n)

RADIAL NODAL LINES (m)	NODAL CIRCLES (n)	
0	1	(a)
0	2	(b)
1	1	(c)

Figure 14–9   Modes in a drumhead. (After Morse and Ingard, *Theoretical Acoustics*, 1968)

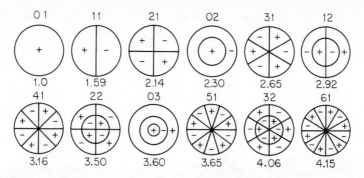

Figure 14-10   Schematic showing modes in a drumhead, the number of radial nodal lines *m*, and the number of circular nodal lines *n*. Also shown are the relative frequencies. (After Rossing, "Acoustics of Percussion Instruments II," *The Physics Teacher*, May 1977)

fixed string, there is a nodal ring around the edge of the membrane where it is held fixed to the edge of the drum. On Fig. 14-9 (b), however, there is another nodal line, a ring around the center. Points along this ring do not move up or down, but remain fixed. Figure 14-9(c) shows a mode that is not symmetric around the center point; as the left side moves up, the right side moves down, and vice versa. Here the nodal line is a straight line running through the center of the membrane. Rather than draw the complicated figures in Fig. 14-9, we can just draw the nodal lines with + and − symbols to represent the relative motion of the different parts of the membrane at any instant, as in Fig. 14-10.

Because the rim of the head is held fixed, all modes will have a nodal ring around the outside. Half a period later, the parts of the membrane will be in the opposite phase, a + becomes a −, and vice versa. Also shown in Fig. 14-10 are the relative frequencies of the modes. The frequency of the lowest mode, the fundamental, has been set equal to 1.

Figure 14-10 depicts an ideal membrane, one that can support tension but does not have any inherent stiffness or complex interaction with the air. The figure is also valid for low-amplitude oscillations in real membranes where these effects are not important.

Just as the tension in a string affects its frequency, the tension in a drumhead affects the pitch produced. The higher the tension, the higher the frequency. A tauter head has a higher wave speed, and for a given wavelength oscillation, a higher frequency.

Two important features of drumhead oscillations should be noticed. The frequencies of the modes are not integrally related, and the modes that have more closely spaced nodal lines have, in general, higher frequencies.

Timpani   Timpani, or kettle drums, shown in Fig. 14-11, are the most important percussion instrument in the standard orchestra; they appear in the music of

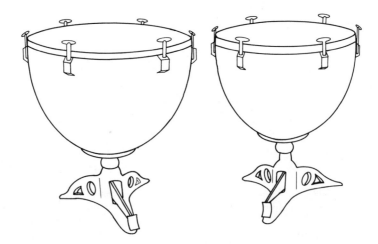

Figure 14-11   The timpani or kettle drums.

composers from before the baroque era to the present. Although they are in the drum tradition that traces back to antiquity, the ancient Persians had bowl-shaped drums that resemble modern timpani. Timpani are played in sets of two to six drums, each sounding at a different pitch. Each drum consists of a skin or plastic head stretched over a copper or fiberglass bowl. Tuning pegs around the rim of the bowl hold the head tight and can be adjusted so that the head is stretched evenly in all directions. The tension on the heads, and thus the pitch of the drum, is controlled by a foot pedal that pulls the head uniformly tighter about the rim.

Timpani are usually struck with wood sticks or sticks whose large ends are covered with a cloth or felt. The head is usually struck about one-quarter of the diameter from the rim. Just as a string can vibrate in many modes simultaneously, the timpani head vibrates in many of the modes shown in Fig. 14-10 simultaneously. We saw that ideal strings had a particularly simple overtone structure, and nonideal strings had a complicated (inharmonic) overtone structure. For timpani heads this situation is reversed; the ideal head supports modes that are inharmonic, but the nonideal head can support modes that are nearly harmonic. An actual timpani differs from the ideal case in two ways: There is inherent stiffness in the drumhead, and there is complicated coupling, or interaction, between the head and the air.

The pitch of the drum is determined by the 11 mode. The lowest frequency mode, the 01 mode, dies down very quickly as the large surface of the head pushes coherently against the air. If the drum is struck in the center (and some contemporary music requires this), the 01, 02, 03, and so on modes are excited, while the 11 is not. The resulting sound is a dull "thud" that dies away very quickly.

The stiffness of the head raises the frequencies of the higher modes, while the complex interaction of the head with the air lowers the frequencies of the low modes. The latter effect dominates and tends to make the overtones nearly harmonic. The pitch heard is determined primarily by the 11 mode, which is lowered slightly because of the complex interaction between the vibrating head and the air. The louder the note, the more this interaction affects the pitch. Careful timpanists take into consideration the loudness of the note when tuning. For instance, in the second movement of Beethoven's 9th Symphony there is an important three-note phrase played by the timpani alone, first loud then very soft. Some timpanists make subtle tuning adjustments so that the phrases sound at the same pitch. The effect is noticeable in a single note, too. A loud timpani note rises slightly in pitch as the amplitude of the head motion decreases and the interaction of the head with the air diminishes.

Omitting the quickly damped 01 mode, the next lowest modes have frequencies with ratios of approximately $2 : 3 : 4 : 5$ (modes 11, 21, 31, and 41), a harmonic series built on a missing fundamental. Some timpanists hear the pitch of the timpani as that which is written; others hearing the same tone claim to hear a note an octave lower. The latter hear a missing fundamental while the former do not. This may be due to the large amplitude of the 11 mode and the low amplitudes and short duration of the higher modes, along with subtle hearing variations between performers. This issue has not been settled.

Tom-toms, snare drums, and bass drums   Tom-toms are drums that consist of a cylindrical shell (like a tin can with the ends missing) with either one head and open bottom, or two heads, one over the top opening and one over the bottom. Complicated interaction of the motion of the struck head and the motion of the head on the bottom tends to give such two-headed drums an indefinite pitch. The depth of the drum (length of the cylinder shell) can affect the tone quality of the sounds produced. In general, a deep tom-tom will have a longer standing wave between its heads and thus a lower tone than a shallow tom-tom.

The snare drum resembles a shallow two-headed tom-tom with one important difference; lengths of gut or wire are stretched along the outside of the bottom head. These snares vibrate against the head when the drum is played and give the crisp sound that is characteristic of the snare drum. The shallow body also aids to this end. A lever is usually present on snare drums to release the snares so that the drum can sound like a tom-tom (when desired).

The bass drum consists of two large heads stretched over a large cylindrical shell. The drum is usually used to add weight to the lower sections of the band or orchestra. Each head has a loosely defined pitch center, but because they are tuned to close but different pitches, the sound of the drum does not have a well-defined pitch center. Thus the bass drum can be used unchanged for music in any key. Large beaters are used to permit good coupling and louder sound. Often, short, loud bass drum notes are desired; these can be obtained by striking on or near the center. This excites the 01 mode, which decays rapidly. Bass drums on drum sets have a beater that is operated by a foot pedal; the beater strikes the head with a direct stroke near the center.

Tabla   The preceding drums all have heads of uniform thickness. Some Asian Indian drums, however, have weighted heads; that is, a circular portion of

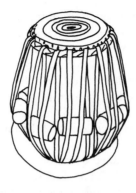

Figure 14-12 Tabla (performer's view). Bayan (left) and dayan (right).

the head is heavier than the rest, as shown in Fig. 14–12. Examples of these Indian drums are called collectively tabla and are often played in duet with the sitar. Figure 14–12 shows these two drums; the right-hand drum is called the dayan (or tabla) and the left-hand drum is called the bayan. The dayan is usually made of wood with a calfskin head held taut by calfskin straps; the larger bayan is bowlshaped and made of metal. The weighting or loading of the membranes is achieved through a mixture of soot, iron rust, and gum that is allowed to dry on the head. The circular area so weighted is called the *ak*. The first four dayan modes are very nearly harmonic because of the loading, which creates the clear ringing tone of this drum. One stroke, called *tin*, is produced by lightly touching the head with the fourth or the little finger, while simultaneously slapping the head a quarter of the way around. This forces the equivalent 11 mode to predominate, yielding a high-frequency tone.

The bayan has a deeper, more "tubby" sound, due to its larger size and lower head tension. The bayan is unusual in that its head tension is often adjusted during the playing of a single note. The heel of the left hand rests on the head, about midway between the center and the edge opposite the ak. To produce the stroke *ga*, the fingertips strike the head on the other side of the ak; then the heel of the hand is quickly pressed into the head, and sometimes slid along the head. This raises the pitch of the struck tone, and a tubby, "shlurping" sound results.

The pitch ranges of the percussion instruments described are shown in Table 14–1.

14.4 *Gongs, Tam-Tams, and Cymbals*

Gongs and tam-tams are circular-shaped metal plates that are struck with large, soft beaters. Gongs (and their Indonesian relative, the gamelan) generally have a domed central area, thick metal, and a general pitch center. Tam-tams, on the other hand, are generally flat, made of thin metal, and have a poorly defined pitch.

When a tam-tam is struck near the middle, low frequencies sound first, and

TABLE 14-1 **Approximate ranges of selected percussion instruments having definite pitch.**

		Written	Sounds
Idiophones	Bell lyra	A_3–A_5	2 octaves higher
	Glockenspiel	G_3–C_6	2 octaves higher
	Xylophone	F_4–C_7	1 octave higher
	Marimba	C_3–C_7	As written
	Vibraphone	F_4–F_6	As written
	Chimes	C_4–F_5	As written
Membranophones	Timpani 30 in.[a]	$D^\flat_2$–A_2	As written
	28 in.	F_2–C_3	See text
	25 in.	$B^\flat_2$–F_3	See text
	23 in.	D_3–A_3	See text
	Tabla: bayan	C_3–C_4	As written
	dayan	$F^\sharp_3$–G_4	As written

[a] Diameter of head.

later the higher modes sound. The attack is thus gradual and consists of a rushing of the initial sound. Because the tam-tam does not speak immediately after it is struck, it is tricky for the percussionist to make it sound at the desired moment. To avoid this problem, percussionists "prime" the tam-tam by lightly tapping it with the beater. Low-frequency, low-amplitude, inaudible vibrations result, which nonetheless allow the tam tam to speak sooner after it is struck than if it were not primed.

Cymbals are circular metal plates, slightly arched with a more strongly arched cup at the center. When supported at the center, cymbals can be struck with soft or hard sticks; soft sticks produce a prolonged "whooshing" sound as many modes are excited. When the cymbal is struck on the edge with a hard wood stick, many higher modes are excited and a brighter sound results.

Crash cymbals are a pair of hand-held cymbals struck sharply against one another. This technique generates many high-frequency overtones, especially during the attack. To project the full sound from the cymbals, the performer holds their faces toward the audience immediately after striking. Because the cymbal plates vibrate perpendicularly to the face, most of the sound projects out. If crash cymbals are played in an anechoic chamber (very absorbent walls), no sound will be heard along directions perpendicular to the face of the cymbals.

Diffraction and interference also affect the tone quality of the cymbal. The out-of-phase waves emitted from the back of the cymbal can diffract around and interfere destructively with the waves emitted by the front; the sound will thus be reduced. Because low frequencies diffract more than the higher frequencies, the low frequencies are not as loud. This effect was described in our speaker-with-baffle experiment in Chap. 2.

14.5 Bells

The vibrations of bells are similar to those in a thick metal plate, with radial nodal lines and circular nodal lines. These modes are similar to those in a beaker or glass flicked with a fingernail. As with plates and drumheads, the lower-frequency modes have fewer nodal lines, while the higher modes have more nodal lines, more closely spaced. Unlike a drumhead, the bell edge is free to vibrate, and thus there can be antinodes there.

The lowest mode that sounds with appreciable amplitude has four meridian nodal lines and produces the "hum" tone. In this mode, opposite sides move inward and outward together while the motion one-fourth of the way around the bell is out of phase with this motion, as can be seen in Fig. 14–13. The strike tone is adjusted to be nearly an octave above the hum tone. Both have four nodal meridians, but the strike tone has a circular nodal line partway up the bell, while the hum tone does not. The pitch of the bell is associated with the striking tone, while the hum tone is usually a vague, low-pitched sound.

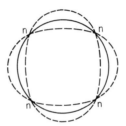

Figure 14–13 Fundamental mode in a bell, viewed from above.

English bells, designed for sequential or *change* ringing, are not as carefully tuned as Flemish bells or carillon bells, which must sound pleasant (and beatless) together. Because the thickness of the bell wall is tailored to adjust the overtones, careful study of the overtones of the bells has suggested different thickness profiles. Bells are often tuned in mean-tone temperament so that notes a major third apart played simultaneously will be beatless.

Metal casting and tuning of bells is slow and expensive, so other techniques have been used to produce bell and carillon tones. The most common employs metal bars or rods that are shaped to give the desired overtones. An electrical pickup converts the motion into electrical signals, which are then amplified and played through a loudspeaker.

EXERCISES

1. Compare and contrast oscillations in bars with those in strings. Be sure to discuss harmonicity, nodes, and modes. What types of modes can exist in a bar that cannot exist in a string?

2. Describe tone production in a triangle.
3. Give two reasons for the arches cut in the bottom of the bars of certain tuned percussion instruments like the xylophone and marimba. How do resonators affect the tone of these instruments?
4. How do the butterfly valves affect the tone of the vibraphone? Using a vibraphone, compare the loudness and duration of a tone produced when the valves are open with a tone produced when the valves are closed.
5. Why do drums have bodies or shells? Why do single-headed tom-toms, for instance, need a cylindrical shell?
6. How is the indefinite pitch achieved in a bass drum?
7. How is the pitch of a bayan changed during a single note? How does the ak affect the tone of the tabla?

REFERENCES

APEL, WILLI. *Harvard Dictionary of Music.* 2nd ed. Cambridge, Mass.: Harvard University Press, 1973.

Classic dictionary of music and instruments.

BLADES, JAMES. *Percussion Instruments and Their History.* New York: Praeger Publishers, Inc., 1970.

A clear presentation of the history of most important percussion instrument groups.

BRINDLE, REGINALD S. *Contemporary Percussion.* New York: Oxford University Press, Inc., 1978.

Gives examples of how composers use different percussion instruments for different effects. It is particularly suited for a composer who does not know very much about percussion instruments.

MORSE, P. M. *Theoretical Acoustics.* New York: McGraw-Hill Book Co., 1968.

A very advanced book that requires considerable advanced mathematics and is not directed toward music or musical instruments per se.

RAMAKRISHNA, B. S., AND M. M. SONDHI. "Vibrations of Indian Drums Regarded as Composite Membranes." *Journal of the Acoustical Society of America* 26 (1954).

This is a very advanced article; students can appreciate the Chladni figures made using tabla heads, though.

ROBERTSON, DONALD. *Tabla: A Rhythmic Introduction to Indian Music.* New York: Peer International, 1968.

A short, clear introduction to the instruments and their parts, as well as a primer for tabla performance.

ROSSING, THOMAS D. "Acoustics of Percussion Instruments I." *Physics Teacher*, December 1976, pp. 546–555.

———. "Acoustics of Percussion Instruments II." *Physics Teacher*, May 1977, pp. 278–288.

An excellent two-part introduction to the physical principles and their consequences when applied to standard and some Eastern percussion instruments.

Appendix A

Elementary Music Theory

This Appendix contains the basic music theory assumed in the discussions of the overtone series. We begin with the correspondence of musical notation with keys on the piano (or organ or music synthesizer) keyboard. This will allow us to identify simple musical intervals and establish the sequence of notes in the overtone series.

Figure A-1 shows the correspondence of the notes of the piano keyboard with notes on the ten horizontal lines of a musical staff system. Notes on the piano keyboard are labeled (for example, C_4) according to the convention suggested by the U.S. Standards Association. The notes are placed on the staff as heavy filled or open ovals either on a line or in a space between two adjacent lines; the staff system most commonly used consists of a bass clef and a treble clef, as shown. The treble clef symbol indicates that the note G_4 is represented by the next to bottom line on that staff, while the bass clef symbol indicates that the note F_3 is the next to top line on that staff. Between the two clefs are three notes, as shown in the figure. The line between the two clefs is C_4, called middle C because it is just about at the middle of the normal piano keyboard. To represent notes above or below the staff, additional lines called ledger lines are used.

The white keys on the piano keyboard are labeled successively by the letters A through G, which then repeat. Keys that occupy the same position relative to the groups of either two or three black keys have the same letter name. For example, any key just below the group of two black keys is labeled by the letter C. The subscripts describe the particular C (of approximately ten within the range of hearing). The black keys can be labeled in two ways. For example, the black key be-

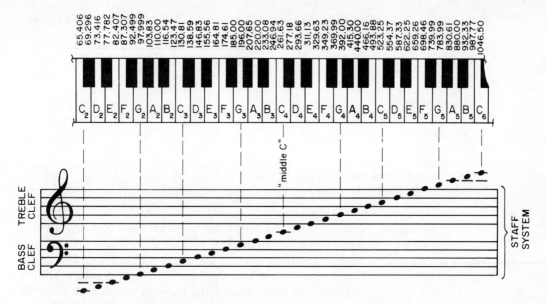

Figure A–1   Correspondence between notes on the musical staff system and the keys on the middle of the keyboard. Shown above each key is its frequency.

tween F and G can be called F♯ (♯ means sharp), meaning it is just above F, or it can be called G♭ (♭ means flat) meaning it is just below G. Because there is no black key between B and C, B♯ is the same as C and C♭ is the same as B; similarly, E♯ is F and F♭ is E.

Two notes that are adjacent on the piano keyboard (either both white or one black and one white) are said to be a musical interval of one half-step apart; for example, the interval between E and F or between G and G♯ is one half-step. A whole step is equal to two half-steps, as in the interval between F and G or between E and F♯.

In this book use of the overtone series and the musical intervals between the notes of the overtone series will be limited to the white keys only. Table A–1 lists the musical intervals useful in constructing the overtone series, along with the number of half-steps in each interval. For example, one octave, which is the musical interval between any two nearest notes with the same letter name, consists of 12 half-steps. There is an interval of a perfect fifth between any two white keys where there are 7 half-steps between the two notes, such as F to C or G to D. B to F is not a perfect fifth, since it contains only 6 half-steps. Similarly, G to C and A to D are perfect fourths, since they contain 5 half-steps, B to D and E to G are minor thirds, containing 3 half-steps, and G to B and F to A are major thirds, separated by 4 half-steps.

Knowledge of scales gives us another way to determine the interval between two notes. The name given to the interval is the number of the upper note in the

TABLE A-1 **Musical intervals, the number of half-steps included in each, and the frequency ratio between the two notes.**

Name of interval	Number of half-steps	Approximate frequency ratios
Octave (8va)	12	2 : 1
Minor seventh	10	7 : 4
Perfect fifth	7	3 : 2
Perfect fourth	5	4 : 3
Major third	4	5 : 4
Minor third	3	6 : 5
Major second (whole step)	2	—

scale beginning on the lower note. An interval of a *major third* contains the first and *third* notes of a *major* scale, for example C to E or G to B. An interval of a *minor third* contains the first and *third* notes of the *minor* scale, for example E to G or B to D. A *perfect fourth* contains the first and fourth notes of the scale, while a *perfect fifth* contains the first and fifth notes of the scale. A perfect fourth can be obtained from either a major or a minor scale, as can a perfect fifth.

For a given musical interval, the frequencies of the two notes involved will always be in a certain ratio; these approximate frequency ratios are also shown in Table A-1. The frequency ratio between two notes one octave apart is *exactly* 2 : 1. That is, if the frequency of A_4 is 440 Hz (vibrations per second), then the frequency of the next highest A, which is A_5, is 880 Hz, a factor of 2 higher. The frequency of A_3 is 220 Hz, a factor of 2 lower. Thus the frequency ratio between any two adjacent A's is 2 : 1. For any musical interval of a perfect fifth, the frequency ratio between the notes is approximately 3 to 2; that is, if the lower note has a frequency of 600 Hz, then the note a perfect fifth higher will have a frequency of approximately 900 Hz. Two notes a major third apart have a frequency ratio of approximately 5 : 4, so if the lower frequency is 600 Hz, then the upper frequency would be about 750 Hz. It is important to remember that the particular ratios hold for the intervals wherever the interval is taken on the keyboard.

In the equal-tempered scale, the octave is the only interval that has an *exact* integral frequency ratio, exactly 2 : 1. Other intervals in this temperament have frequency ratios that are only *approximately* integral. In the equal-tempered scale the frequency ratio between two notes one half-step apart is always 1.0595. This is the twelfth root of 2 (to four decimal places), and use of this factor guarantees that octaves, which contain 12 half-steps, have the ratio of frequencies of exactly 2 : 1. A true perfect fifth above $A_3 = 220$ Hz would be $220 \times \frac{3}{2} = 330$ Hz, but the frequency of E_4 is actually 329.63 Hz, a difference of 0.37 Hz.

The sequence of notes called the overtone series, as illustrated in Fig. A-2, is of particular interest, because the frequency ratios between the notes are the integers 1, 2, . . . as tabulated in Table A-2. Starting with G_2, the note G_3 one octave higher has a frequency two times the frequency of G_2. D_4 has a frequency of 3 : 2

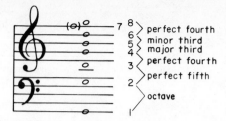

Figure A–2 The first eight notes of the overtone series built on G_2.

times the frequency of G_3 or three times G_2. G_4 has a frequency of $4:3$ times the frequency of D_4 or four times the frequency of G_2. Table A–2 shows these frequencies up through the eighth harmonic for exact frequency multiples and the actual frequencies of piano notes of the equal-tempered scale. For the octaves, the frequencies of the integral multiples are exact; for the third, fifth, and sixth notes, the frequencies of the exact multiples are very close to the frequencies of the notes on the piano. Because the frequency ratio of $7:1$ is only very roughly approximated by $(\sqrt[12]{2})^{34}$ (34 half-steps in the equal-tempered scale), we put the seventh harmonic in parentheses when it is written on the staff as in Fig. A–2. The modern ear, accustomed to the equal-tempered scale and the sound of musical intervals using the note that best approximates the true seventh harmonic, generally finds the unapproximated intervals involving the seventh harmonic foreign and somewhat "out of tune."

The notes of the overtone series are called *harmonics* of the fundamental, or lowest note, and are labeled by their harmonic numbers, which are the integers giving the frequency ratio. In the case discussed here, G_2 is the fundamental or first harmonic and, for example, the fifth harmonic would be B_4.

It is possible to build such an overtone series on any note of the piano, using the intervals shown in Figure A–2. However, for our purposes, we shall only use the overtone series built on the note G_2 as presented here to avoid the use of sharp or flat notes.

TABLE A–2 **Frequencies of exact harmonics of the note G_2 compared with frequencies of the nearest notes of the equal-tempered scale.**

Note	Harmonic number	Exact frequency (Hz)	Piano frequency (Hz)	Difference frequency (Hz)
G_2	1	98.0	98.0	0.0
G_3	2	196.0	196.0	0.0
D_4	3	294.0	293.7	0.3
G_4	4	392.0	392.0	0.0
B_4	5	490.0	493.9	3.9
D_5	6	588.0	587.3	0.7
F_5	7	686.0	698.5	13.5
G_5	8	784.0	784.0	0.0

Appendix B

Terms and Units

Quantity	Symbol	Common units
distance	x, y, z	meter (m); km, cm, mm, mi, ft
time	t	second (sec)
velocity or speed	v	m/sec, cm/sec, mm/ms
speed of sound	S	345 m/sec or 34,500 cm/sec
acceleration	a	m/sec^2, cm/sec^2
acceleration of gravity	g	9.8 m/sec^2 or 980 cm/sec^2
mass	m	gram (g), kg, mg
force	F	Newton (N), pounds (lb)
pressure	p	Pascal (1 Pa = 1 N/m^2), lb/in.^{2}
period	T	sec, ms
frequency	f	Hertz (Hz), kHz, MHz
wavelength	λ (lambda)	m, cm
phase	ϕ (phi)	degree (°)
tension	F	Newton
weight	W	Newton, pound
weight per unit length	W	N/m
harmonic number	N	—
intensity	I	W/m^2
sound intensity level	SIL	decibel (dB)
loudness level	LL	phon
electrical voltage	V	volt (V)
electrical current	I	ampere (A)
electrical resistance	R	ohm (Ω, omega)
electrical power	P	watt (W)
sound absorption	—	sabin

Appendix C

Prefixes with Common Units

Quantity	Symbol	Value	Examples
pico-	p	10^{-12}	1 pA $= 10^{-12}$ A
nano-	n	10^{-9}	1 nm $= 10^{-9}$ m
micro-	μ (mu)	10^{-6}	1 μA $= 10^{-6}$ A
milli-	m	10^{-3}	1 mW $= 10^{-3}$ W
centi-	c	10^{-2}	1 cm $= 10^{-2}$ m
deci-	d	10^{-1}	1 dB $= 10^{-1}$ bel
kilo-	k	10^{3}	1 kΩ $= 10^{3}$ Ω
mega-	M	10^{6}	1 MV $= 10^{6}$ V
giga-	G	10^{9}	1 GHz $= 10^{9}$ Hz

Index

Quarter-comma mean-tone temperament (*see* Temperament, quarter-comma mean-tone)
Quarter-track tape, 194
Quiet region (*see* Shadows)
Quinke's tube, 40–41

Rackett, 246
Radar waves (*see* Wave, electromagnetic)
Radio:
    AM, 193
    FM, 193–194
    FM frequency excursion, 193
    FM noise, 193–194
    FM pilot signal, 193
    FM quadraphonic, 193
    FM stereophonic, 193
    FM subcarrier, 193
Radio waves (*see* Wave, electromagnetic)
Ramp wave:
    synthesis, 94
    timbre of, 100
Range of human hearing (*see* Human ear, Frequency range)
Rank, organ, 276
Rarefaction, 23–25
Rauschpfeife, 246
Ray, 32
Rebec, 302, 305–306
Recorder, 246–247, 259–262
    baroque, fingering, 255–256
    bore diameter, 260
    dynamic expression, 261
    family, 262
    fipple, 81–82
    hole positions, 260–261
    hole size, 260
    octave key, 260
    problems, relevance, 262
    range, 259, 262–263
    roughness experiment, 141
    scaling with pitch, 261–262
    spectrum of, 100, 101
    timbre, 259–260
    tone holes, 257
    tuning compromises, 261
    use of keys, 260
    voicing, 260
        effect of scaling, 261–262
    wave form, 100, 101

Record head (*see* Tape heads, record)
Recording:
    monaural, 134
    quadraphonic, 134
    sound-on-sound, 133–134
    stereophonic, 134
Recording (with birds), 259
Recording Industry Association of America (RIAA), 186
Recording of sound, 168–200
Recording studios, reverberation time, 206
Record player (*see* Disc record player)
Records (*see* Disc records)
Reed:
    clarinet (*see* Clarinet, reed)
    organ, 278
    resonances in, 251
    woodwind instruments, 251
Reed organ pipes (*see* Organ pipes, reed)
Reed tongue (*see* Organ pipes, reed tongue)
Reel-to-reel tape, 194–195
Reference graphs (*see* Lissajous figures, reference graphs)
Reflected sound, diffuse, 216
Reflection, 31–35, 203
    architectural, 216
    boundary, 35
    ellipsoid, 34–35
    geological layers, 35
    irregularly shaped objects, 32–34
    law of, 32
    multiple, 215
    transmission and, 35
Reflection of transverse pulse (*see* Transverse pulse, reflection of)
Reflection of waves, phase changes, 67–68
Reflective walls, 209
Reflector, hyperbolic, 34
Reflectors, and loudspeakers, 34
Refraction, 35–40, 203
    atmospheric, 38–39, 212
    change of direction of wavefront, 35–37
    Huygens's wavelets, description, 37–38

    ocean waves, 37
    thunder, 39–40
    wind, 39
Regulation, of piano action, 325
Release, 129 (*see also* Envelope generator)
Reproduction of sound, 168–200
Reproduction systems, audio, 171–174
Resin, 306
Resistance, 169–171
Resistive support, 330
Resistor, 169–170
Resonance, 12–14, 67–74
    air, in pipes, 86
    architectural, 210–211
    bone, 85–86
    bridge, 85 (*see also* Tacoma Narrows Bridge)
    coupled, 13–14
    curve, 107–114
        closed tube, 110–111
        difference from Fourier spectrum, 110
        open tube, 110–111
    English horn, 274
    forced oscillation, 14
    Helmholtz, 306 (*see also* Helmholtz resonator)
    phase, 13
    piano, 329
    manipulated in vocal tract, 160–161
    room, 220
    seashell, 145
    speaker, 220
    speaker enclosures, 211
    ultrasonic, 86
    wooden plates of string instruments, 306
Resonator:
    marimba, 336
    vibraphone, 338
    xylophone, 337
Reverberation time, 204, 214
    adjustment, 214
    appropriate, 205–206
    Baroque music, 206
    chamber music, 206
    churches, 206
    classical music, 206
    conference rooms, 206

365

baffle (*see* Diffraction, speakers)
  efficiency of, 173
  power transfer to, 192
Spectrum analyzer (*see* Fourier analyzer)
Speed, 2 (*see also* Wave, speed)
Sphere, radius of, and inverse square law, 31
Spiegel der Orgelmacher und Organisten, 227
Spring, extended, 4–5
  force (*see* Hooke's law)
Springy support, in piano, 329
SQ matrix, broadcast, 193
SQ matrix decoder, 189
SQ matrix format, 188–189
Square wave:
  clarinet, 100, 269
  synthesis, 93
Staccato, piano, 325
Stage size, 208
Standard waves:
  Fourier spectra of, 99–100
  synthesis, 91–95
Standing wave, 63–86
  addition to form, 63–66
  antinodes, 65–67
  architectural, 220
  bassoon, 275
  bowed, 307–308
  closed tube, 81–83
    human vocal tract, 159–160
  development by reflection, 68–70
  dome, 84–85
  frequency of, 71–74
  fundamental, 72
  impossible, 70
  longitudinal, 67 (*see also* Longitudinal standing wave)
  nodes, 67
  in slinky, 75–76
  nodes, 65–66
    woodwind, 258
  open tube, 81–83
  organ pipe, 278
  picture, 66
  sound, 82
  stretched rope, 69–70
  transverse, 63–72

wavelength of, 71–72, 81–83
  wave speed, 71
  woodwind instruments, 250, 258
Steinway, 323
Stereo decoder, FM radio, 193
Stereo headphones (*see* Headphones)
Stereophonic records, 188 (*see also* Disc records, stereophonic)
Stereophonic tape recording (*see* Recording, stereophonic)
Stethoscope, 156
  phase sensitivity experiment, 156
  ultrasonic fetal (*see* Ultrasonic fetal stethoscope)
Stiffness, in triangle, 340
Stirrup (*see* Ossicles)
Stops, organ, 276
Stradivarius, 230
  violin, 313
Straightarm, tone arm, 183–184
Stretching octaves, 326
Strike tone, of bell, 346
String:
  guitar, 12
  gut, 303
  ideal, 342
  mistuned, in piano, 329–331
  wire-wound, 313
String bass, 304, 313
  extender, 314
String instruments, 301–316
  banjo, 304
  bass viol, 304
  bowed, 305–308
    sound production, 305–308
  charango, 302
  cittern, 303
  dulcimer, 304
  fiedel, 302
  gamba, 303
  guitar, 303–304
  harp, 302–304
  history, 301–305
  keyboard, 303, 320
    clavichord, 303
    harpsichord, 303
  lute, 302–303, 305
  lyre, 302
  monochord, 301–302, 319–320

organistrum, 302
pandora, 303
plucked, 314–316
  bridge on, 315–316
  decay transients, 315–316
polychord, 320
psaltery, 302–303, 320–321
rebec, 302, 305–306
scaled, 311
standing waves in, 69
string bass, 304
string vibration, 307–308
  harmonics in, 308
timbre, 308
tromba marina, 302–303, 310
ukelele, 304
vihuela, 303
viola da braccio, 303
viola da gamba, 303, 305
viola d'amore, 303
Stylus, 182–183
  elliptical, 189
Subharmonic generator, 200
Subjective fundmental, 149
Subjective listening response, and audio component choice, 191
Subjective loudness, 149
Sul ponticello, 308
Sul tasto, 308
Sum tones, 150–152, 154
Superposition, principle of, 29–30, 46–47
Support:
  massive, 329
  resistive, 330
  springy, 329
Surgery, ear, 158
Sustain, 128–129 (*see also* Envelope generator)
Sustained sound, in piano, 327–331
Sustain pedal, 328
Swell manual, organ, 276
Sympathetic vibrations (*see* Vibrations, sympathetic)
Synchronization of instruments, 208–209
Synthesis, Fourier (*see* Fourier synthesis)
Synthesizer, 105, 118–133
  audio spectrogram of, 165
  block diagram, 126
  computer control, 133